Advances in Experimental Medicine and Biology

Volume 1507

Series Editors

Nima Rezaei (ID), Research Center for Immunodeficiencies,
Children's Medical Center
Tehran University of Medical Sciences
Tehran, Iran
Avia Rosenhouse-Dantsker, Department of Chemistry
University of Illinois at Chicago
Chicago, IL, USA
Robert Gerlai, Department of Psychology
University of Toronto
Mississauga, ON, Canada

Advances in Experimental Medicine and Biology provides a platform for scientific contributions in the main disciplines of the biomedicine and the life sciences. This series publishes thematic volumes on contemporary research in the areas of microbiology, immunology, neurosciences, biochemistry, biomedical engineering, genetics, physiology, and cancer research. Covering emerging topics and techniques in basic and clinical science, it brings together clinicians and researchers from various fields.

Advances in Experimental Medicine and Biology has been publishing exceptional works in the field for over 40 years, and is indexed in SCOPUS, Medline (PubMed), EMBASE, Reaxys, EMBiology, the Chemical Abstracts Service (CAS), and Pathway Studio.

2023 CiteScore: 5.9

Addmore Shonhai
Abidemi Paul Kappo
Jo-Anne de la Mare
Editors

Advances in Biochemistry and Molecular Biology to meet Africa's Needs

Proceedings from the 28th South African Society for Biochemistry and Molecular Biology Congress

 Springer

Editors
Addmore Shonhai (iD)
Biochemistry & Microbiology
Department
University of Venda
Thohoyandou, South Africa

Jo-Anne de la Mare (iD)
Biochemistry, Microbiology &
Bioinformatics
Rhodes University
Makhanda Grahamstown, South Africa

Abidemi Paul Kappo (iD)
Biochemistry Department
University in Johannesburg
Johannesburg, Gauteng, South Africa

South African Society for Biochemistry and Molecular Biology Congress, sasbmb, sasbmb-2024, SASBMB-28, Polokwane, South Africa 2024-7-7-2024-7-10, https://www.sasbmbcongress.co.za/about-the-sasbmb/

ISSN 0065-2598   ISSN 2214-8019   (electronic)
Advances in Experimental Medicine and Biology
ISBN 978-3-032-24253-2   ISBN 978-3-032-24254-9   (eBook)
https://doi.org/10.1007/978-3-032-24254-9

Preface

The 28th Congress of the South African Society for Biochemistry and Molecular Biology (SASBMB) was hosted by the University of Venda and held from 7 to 10 July 2024 at the Protea Ranch Resort in Polokwane. This congress came a year after the 50th anniversary of the society and 83 years after the first instruction of Biochemistry as a formal university subject at the University of Pretoria. This was the largest congress of its kind in the history of the society, with over 400 delegates, and the first time the congress was held in Limpopo province. At the opening ceremony, the gold medal was awarded to Prof. Ed Sturrock of the University of Cape Town in acknowledgment of his outstanding research track record centered on angiotensin-converting enzymes and their inhibitors, as well as his service to the disciplines of biochemistry and molecular biology in the country. The silver medal was awarded to Prof. Thulani Makhalanyane from Stellenbosch University in recognition of his excellent contribution to the field of microbiome ecology in extreme environments. Several distinguished international researchers attended the meeting and delivered plenary lectures, including Profs. Kiaran Kirk from Australian National University (IUBMB-sponsored speaker), Matthias Mayer from Heidelberg University in Germany, Stanley Mukanganyama from the University of Zimbabwe, Greg Blatch from Notre Dame University in Australia, Nikolai Kuhnert from Constructor University in Germany, and Tomohisa Ogawa from Tohoku University in Japan. The thematic areas of the congress included protein biochemistry, computational and structural biology, omics science, enzymology and metabolism, infectious and parasitic diseases, noncommunicable diseases, drug discovery, phytomedicine, and plant, animal, and general biotechnology, under which 12 talks, 87 short talks, and 211 posters were presented.

At the General Meeting of the society, Prof. Francois van der Westhuizen retired as President of the society, and Dr. Jo de la Mare from Rhodes University was inducted into this position, having previously served as Secretary (2018–2022) and Vice-President (2022–2024). Finally, it was announced that the society would be hosting the congress of its parent body, the International Union of Biochemistry and Molecular Biology, in Cape Town from 19 to 23 September 2027. This will be the first time that the congress will be held on African soil.

To celebrate these milestones, we decided to archive some of the papers that were presented at the 28th SASBMB congress in this book. Biochemistry and molecular biology are two branches of science that are critical to solving current and emerging global challenges. Africa is the most genetically diverse continent, and in addition to this, it is a reservoir for several tropical diseases. As such, the biomedical challenges confronting the continent require solutions tailored to these needs. The select chapters presented in this book were part of the presentations made at the 28th SASBMB congress and reflect the great strides that South African scientists are making toward providing solutions to the challenges that the continent faces. The chapters have been published by Springer Nature both as a separate book and as part of the book series, Advances in Experimental Medicine and Biology (AEMB). AEMB is indexed in all the major databases, thus expanding the footprint of the publications. Below, we share a few images from the 28th SASBMB congress.

Prof. Ed Sturrock receiving the SASBMB Gold medal award. From left: Prof. Francois van der Westhuizen (outgoing President), Prof. Ed Sturrock, Prof. Iqbal Parker (Executive Secretary, IUBMB), Dr. Jo de la Mare (incoming President)

Prof. Thulani Makhalanyane receiving the SASBMB Silver Medal. From left: Prof. Francois van der Westhuizen (outgoing President), Prof. Thulani Makhalanyane, Prof. Iqbal Parker (Executive Secretary, IUBMB), Dr. Jo de la Mare (incoming President)

Prof. Addmore Shonhai from the University of Venda, Chair of the local organizing committee

We trust you will enjoy reading this book as much as we enjoyed compiling it.

Thohoyandou, South Africa   Addmore Shonhai
Johannesburg, Gauteng, South Africa   Abidemi Paul Kappo
Makhanda Grahamstown, South Africa   Jo-Anne de la Mare

Contents

Snippets from the 28th Congress of the South African Society for Biochemistry and Molecular Biology

Gregory L. Blatch, Collet Dandara,
Abidemi Paul Kappo, Jo-Anne de la Mare,
Matthias P. Mayer, Stanley Mukanganyama,
Sithandiwe E. Mazibuko-Mbeje,
Addmore Shonhai, Edward Sturrock,
Özlem Tastan Bishop, and Liesl Zühlke

Abstract

The 28th Congress of the South African Society for Biochemistry and Molecular Biology (SASBMB), hosted by the University of Venda, was held at the Protea Hotel Polokwane Ranch Resort from 7–10 July 2024. The theme of the congress was "Biochemistry leading the future." The event featured several plenary presentations and parallel sessions focusing on a broad range of topics, including drug discovery, infectious and parasitic diseases, phytomedicine, non-communicable diseases, protein biochemistry, and computational and structural biology, among others. Here, we highlight key lectures that were presented by the plenary speakers whose presentations set the tone for this meeting. Overall, the congress was a resounding success.

G. L. Blatch
Biomedical Biotechnology Research Unit,
Department of Biochemistry, Microbiology and
Bioinformatics, Rhodes University,
Makhanda, South Africa

Faculty of Health Sciences, Higher Colleges of
Technology, Sharjah, United Arab Emirates

The Vice Chancellery, The University of Notre Dame
Australia, Fremantle, Australia

C. Dandara
Division of Human Genetics, Department of
Pathology & Institute of Infectious Diseases and
Molecular Medicine, University of Cape Town,
Cape Town, South Africa

Platform for Pharmacogenomics Research and
Translation Unit, South African Medical Research
Council, Cape Town, South Africa

A. P. Kappo
Molecular Biophysics and Structural Biology Group,
Department of Biochemistry, University of
Johannesburg, Johannesburg, South Africa

J.-A. de la Mare
Department of Biochemistry, Microbiology and
Bioinformatics, Rhodes University,
Makhanda, South Africa

M. P. Mayer
Center for Molecular Biology of Heidelberg
University (ZMBH), DKFZ-ZMBH-Alliance,
Heidelberg, Germany

S. Mukanganyama
Natural Products Unit, Department of Therapeutics,
The African Institute of Biomedical Science and
Technology, Wilkins Hospital, Harare, Zimbabwe

© The Author(s) 2026

A. Shonhai et al. (eds.), *Advances in Biochemistry and Molecular Biology to meet Africa´s Needs*,
Advances in Experimental Medicine and Biology 1507,
https://doi.org/10.1007/978-3-032-24254-9_1

Keywords

Chaperones · Hsp70 · J domain proteins ·
Malaria · Arabian Peninsula · Missense
mutations · HIV/AIDS · TB ·
Pharmacogenomics · Angiotensin-converting
enzyme · Congenital heart disease · Diabetes
· Natural products

1.1 Hsp70 Chaperones in Protein Folding and Regulation of the Heat Shock Response: Unfolding and Pulling

One of the international guest speakers, Matthias
Mayer, reported on the molecular mechanism of
Hsp70 chaperones and their cooperating co-
chaperones of the J-domain protein (JDP) family. Hsp70s are undeniably the most versatile of
all molecular chaperones being involved in a
wide variety of protein folding processes in the
cell, including de novo folding of nascent poly-
peptides at the ribosome, translocation of proteins across membranes, refolding of stress
denatured proteins, solubilization of protein
aggregates and amyloid fibrils, assembly and
disassembly of protein complexes, and regulation of stability and activity of many native proteins. This versatility is based on the
ATP-regulated interaction of their tweezers-like
substrate binding domain with short degenera-
tive sequence motifs in their substrate polypep-
tides. Hsp70s are targeted to their substrates by
JDPs that interact with the substrates themselves,
transfer them into the substrate binding pocket
of Hsp70, and stimulate the ATPase activity of
Hsp70s, thereby inducing the trapping of the
substrate. The crystal structure of the *E. coli*
Hsp70 in complex with ATP and the J-domain of
its JDP DNAJ revealed how the J-domain interacts with a hydrogen bond network within the
nucleotide-binding domain of Hsp70 to trigger
ATP hydrolysis. Matthias Mayer further showed
how human Hsc70 disassembles the trimeric
human heat shock transcription factor Hsf1 by
unzipping the triple leucine zipper domain,
thereby dissociating Hsf1 from DNA and shut-
ting off the heat shock response. His team identified the binding site of Hsc70 within Hsf1 that is
relevant for this action and demonstrated that the
proximity of this site to the trimerization domain
is essential for monomerization of Hsf1 trimers.
The force required for such an unzipping process
is the focus of the current research endeavor in
Mayer's lab.

1.2 Malaria in the Arabian Peninsula: Public Health Status to Anti-Malarial Drug Discovery

Malaria disproportionately affects Africa, with
the continent accounting for at least 95% of
global deaths due to the disease (World Health
Organization 2024 Malaria Report). While most
countries in the Middle East are considered
malaria-free due to successful elimination efforts,

S. E. Mazibuko-Mbeje
Department of Biochemistry, North-West University,
Mmabatho, South Africa

A. Shonhai (✉)
Department of Biochemistry and Microbiology,
Faculty of Science, Engineering and Agriculture,
University of Venda, Thohoyandou, South Africa
e-mail: addmore.shonhai@univen.ac.za

E. Sturrock
Department of Integrative Biomedical Sciences &
Institute of Infectious Disease and Molecular
Medicine (IDM), Faculty of Health Sciences,
University of Cape Town, Cape Town, South Africa

Ö. T. Bishop
Research Unit in Bioinformatics (RUBi), Department
of Biochemistry, Microbiology and Bioinformatics,
Rhodes University, Makhanda, South Africa

National Institute for Theoretical and Computational
Studies (NITheCS), Stellenbosch, South Africa

L. Zühlke
Office of the Vice-President, South African Medical
Research Council, Faculty of Health Sciences,
University of Cape Town, Cape Town, South Africa

imported cases and the potential for reintroduction remain a concern, especially in countries experiencing conflict or instability. With climate change impacting mosquito habitats and transmission patterns, malaria surveillance, control, and elimination efforts are increasingly becoming important to countries in this region. Gregory Blatch gave a refreshing lecture on malaria with a specific focus on the Arabian Peninsula region. He further merged this subject with a discussion on current anti-malarial drug discovery efforts.

The unicellular protozoan parasite, *Plasmodium falciparum*, invades human cells and transforms them into vehicles of pathology, thereby causing the most virulent form of human malaria. Since 2015, the Eastern Mediterranean Region has experienced an overall increase in annual malaria cases and deaths (Ahmad et al. 2024a). In the Arabian Peninsula, most of the Gulf Cooperation Council countries (GCC) have been declared free of indigenous malaria (Bahrain, Kuwait, Qatar, and the United Arab Emirates); however, there is significant imported malaria in these countries, and malaria is still endemic in two GCC countries (Saudi Arabia and Oman) and their neighbours (Yemen) (Ahmad et al. 2024a). Increasing anti-malarial drug resistance has highlighted the need for a target, or combinations of targets, that can act across the entire parasite life cycle. A few of the major families of heat shock proteins (HSPs) are considered bona fide drug targets (e.g., HSP40, also called J domain protein, JDP; HSP70, and HSP90) (Ahmad et al. 2024b). There is growing evidence that a few *P. falciparum* HSP40s (PfHSP40s) exported into human host cells are important in the trafficking, folding, and functioning of key malaria virulence factors (Ahmad et al. 2024b). Molecular docking-based virtual screening conducted on drug repurposing libraries identified several drug-like compounds that were potentially specific modulators of exported PfHSP40s (Singh et al. 2023). These compounds represent novel candidates for validation and repurposing as anti-malarial drugs.

1.3 Multi-faceted Health and Disease Effects of Missense Mutations: A Protein Structure Perspective

Proteins are dynamic macromolecules that carry multilayered relational information; hence, a change in protein sequence can have drastic structural and functional effects, with either negative or positive consequences. During her presentation, Özlem Tastan Bishop explained how missense mutations and their individual or combined mechanisms of action can affect various biological phenomena, and how incorporating mutation data into the early stages of drug discovery can enhance the overall pipeline (Tastan Bishop et al. 2022). She framed this genomic variation–driven approach around three key aspects of drug discovery. In the first part, she presented examples of deciphering current drug resistance mechanisms, with a specific focus on Human Immunodeficiency Virus (HIV) (Sheik et al. 2018) and *Mycobacterium tuberculosis* (Sheik et al. 2020; Barozi et al. 2022) drug resistance and their underlying molecular basis. In the second part, she explored SARS-CoV-2 as a case study to better understand drug target proteins and their evolutionary capacity at per-residue (atomic) resolution (Barozi et al. 2024; Govender et al. 2025). The final part of her presentation highlighted pharmacogenomics, with emphasis on the importance of population-specific approaches to drug discovery and development (Mwaniki et al. 2025).

1.4 Pharmacogenomics: Integration of "Omics" in Understanding Treatment Response

Pharmacogenomics is relevant worldwide for modern therapeutics and yet needs further uptake in developing countries. There is a paucity of studies with a naturalistic design in real-life clinical practice in patients with co-morbidities and multiple drug treatments. African patients are often burdened with communicable and non-communicable co-morbidities, yet pharmacogenomics in African clinical settings remains limited. Collet Dandara presented a lecture in which he highlighted the underrepresentation of African genomes in global databases and the minuscule inclusion of African populations in clinical trials that result in the development of therapeutic drugs. His work showcased their evaluation of the role and impact of underlying host genetics on treatment responses, and discussed the various approaches currently being used to understand the functional significance of genetic variants either on disease susceptibility or pharmacogenomics. The presentation highlighted how pharmacogenomics cuts across from in vitro assays to in vivo studies and ultimately into the clinical settings where human patients are challenged with relevant therapeutic regimens. His presentation showcased how "omics" has been an integral aspect of their understanding of pharmacogenomics. The work contributed by his group was informed by the disease burden in Africa, the list of commonly used medications and information on observed adverse drug reactions among patients. Herbal medicinal plants also form part of his group's research in order to understand their interaction with conventional medicine and their effects on phenotype conversion. The diseases that have featured in their pharmacogenomics studies and their relevant drugs include HIV (antiretroviral therapy), tuberculosis (anti-tuberculosis therapies), cardiovascular diseases, hypertension and cancer. For example, they have reported on how Efavirenz (EFV) and rifampicin (RMP) alter microRNA expression signatures and expression of drug-metabolizing enzyme genes, such as CYP3A4, CYP3A5, UGT1A1, CYP2B6, and NR1I3, in vitro in HepaRG liver cells. Their work showed that EFV resulted in a significant increase in messenger (mRNA) expression of CYP3A4, CYP3A5, and UGT1A1, whereas NR1I3 expression was significantly decreased (Swart and Dandara 2019). On the other hand, treatment with RMP resulted in a significant increase in mRNA expression for CYP2B6 and CYP3A4, whereas NR1I3 expression decreased. These data point to several important clinical implications through changes in drug/drug interaction risks and achieving optimal therapeutics. These findings show that differential expression of microRNAs after treatment with EFV and RMP adds another layer of complexity that should be incorporated in pharmacogenomic algorithms to render drug response more predictable. He also highlighted the work on the pharmacogenomics of herbal medicine. For example, in one of several studies, two commonly used herbal plants *Hyptis suaveolens* (HS) and *Boerhavia diffusa* (BD), which are used to treat various conditions including boils, dyslipidaemia, eczema, malaria, jaundice and gonorrhoea were evaluated using CYP inhibition assays and the effect of BD was found to be most potent on CYP3A4 (7.36 ± 0.94 μg/mL) compared to CYP2D6 (17.79 ± 1.02 μg/mL). The study revealed that crude aqueous extracts of HS and BD potentially inhibit drug metabolising isozymes CYP1A2, CYP2D6, and CYP3A4 in a reversible and time-dependent manner. Thus, care should be taken when these extracts are co-administered with drugs that are substrates of CYP1A2, CYP2D6, and CYP3A4. He further discussed the comprehensive research his group has done on the pharmacogenomics of drugs such as warfarin and efavirenz. Warfarin dose variability observed in patients is partially attributed to genetic variation, with *CYP2C9* and *VKORC1* being the most characterized. A total of 73 SNPs in 29 pharmacogenes, inclusive of the principal genes involved in pharmacokinetics and pharmacodynamics of warfarin, were evaluated (Ndadza et al. 2021). His group's studies report on novel variants that are important for pharmacogenomics-guided warfarin treatment among Africans, which include CYP2C rs12777823G>A, CYP2C9 c.449G>A (*8), CYP2C9 c.1003C>T (*11), and CYP2C8 c.805A>T (*2), which were significantly associated with warfarin maintenance dose. His presentation concluded with new findings on the pharmacogenomics of hypertension and dyslipidaemia. Overall, he stressed that a thorough understanding of pharmacogenomics as a holistic approach that transcends disciplines, incorporating all the different "omics," is imperative to achieve precision medicine, for effective and safe administration of therapeutics.

1.5 An ACE in the Hole: Challenges and Triumphs of Structure-Based Drug Discovery in Cardiovascular Disease

In Prof Sturrock's SASBMB Gold Medal Award lecture, he discussed some of the challenges and triumphs of structure-based cardiovascular drug discovery. Identification of additional components of the renin-angiotensin system (RAS) and associated vasoactive pathways, as well as new structural and functional insights into established targets, has led to novel therapeutic approaches with the potential to provide improved cardiovascular protection and better blood pressure control. One of Sturrock's research thrusts has been the discovery of dual angiotensin-converting enzyme (ACE) C-domain/NEP inhibitors that are likely to be superior to current ACE inhibitors and thus produce favourable changes in the levels of vasoactive peptides, while at the same time reducing the likelihood of adverse effects. His approach involves iterative rounds of synthesis, kinetic and structural characterization to optimize the druggable properties of early lead compounds.

A breakthrough in understanding the 3D structure, allostery, and dimerization of full-length somatic ACE was achieved when his group published the first cryo-EM structures of the monomeric and dimeric forms of the highly flexible apo enzyme. This work paved the way for the design of novel allosteric ACE inhibitors that act remotely from the active sites and are more likely to succeed in terms of their selectivity and potency. To this end, he has leveraged the ACE structural biology to carry out high-throughput small-molecule screening against allosteric sites. The top-scoring compounds were clustered based on structural similarity, and a

final list of 100 compounds was curated based on predicted selective binding to the ACE C-domain. Four compounds from the virtual screen had IC_{50} values in the low micromolar range and provided good early-lead compounds for the development of allosteric C-domain-selective ACE inhibitors.

1.6 Partnerships for Children's Heart Diseases in Africa (PROTEA): Investigations from Bench to Bedside to Community

Congenital heart Disease (CHD) is the most lethal congenital anomaly, causing over 217,000 deaths globally each year (Kassebaum et al. 2017). It is also the most prevalent anomaly, affecting 3–9 per 1000 live births (Liu et al. 2020). Although significant strides have been made in this field over the past 50 years, with survival rates as much as 97% in high-income and equitable countries such as Denmark, those born with CHD in low and middle-income countries (LMICs) are under-recognized, under-treated, and have a far lower survival rate. CHD is a lifelong disease with significant gaps in care experienced in adolescence and early adulthood, with implications for decreased survival (Alsaied et al. 2017). The Protea-Partnerships for Children with Heart Disease in Africa-Study, based at the University of Cape Town, has focused attention on this disease (Aldersley et al. 2021) with a bespoke integrated database to explore and understand critical research and clinical questions (Aldersley et al. 2023), gather important genomic data (Saacks et al. 2022), and increase capacity development and multidisciplinary and transdisciplinary collaboration (Swanson et al. 2020). Similarly, Rheumatic Heart Disease (RHD) is a disease that continues to be highly prevalent in LMICs compared to high-income countries with substantial morbidity (Zühlke et al. 2016) and mortality (Karthikeyan et al. 2024). Screening for RHD using hand-held echocardiograms (Zühlke and Mayosi 2013), task-shifting and sharing (Rwebembera et al. 2024) have changed the landscape in the past decade, positioning early RHD in the continuum to heart failure for RHD. People with childhood-onset heart disease, including RHD and CHD, living in LMICs need improved data, human resources for health, and targeted policies to improve practices and increase survival (Curry et al. 2018; Zühlke et al. 2019; Hasan et al. 2021).

1.7 Progression of Obesity and Type 2 Diabetes Promotes Brown Adipose Tissue Dysfunction and Impairs the Gene Expression of Batokines

Obesity remains a center of attention in the research community due to its adverse health outcomes, such as type 2 diabetes. Brown adipose tissue (BAT) is increasingly viewed as a potential therapeutic target against obesity-related metabolic dysregulations. Beyond regulating thermogenesis, BAT secretes signalling factors, "batokines," that regulate local and systemic metabolism. Since the BAT activity declines as this tissue accumulates more fat in obesity, it is imperative to elucidate the pathophysiological role of batokines in obesity. To this end, Sithandiwe Mazibuko-Mbeje reported on the research conducted by her group, which involved the determination of BAT morphological changes and the potential pathophysiological role of batokines during obesity progression using both in vivo and in vitro models. Obese-type 2 diabetic *db/db* mice and control littermates *db/+* were monitored from 8, 12, and 18 weeks, and the morphological changes of BAT were assessed using haematoxylin and eosin (H&E) stains. Conversely, T37i brown adipocytes were chronically exposed to various doses (0.25, 0.5, 0.75, and 1 mM) of palmitate for 48 hr. Antidiabetic drug metformin and β3-adrenoreceptor agonist CL-316,243 were used as controls at 1 µM. Cytotoxicity and lipid accumulation were assessed using the lactate dehydrogenase assay and Oil Red O stain, respectively. Gene expression analysis of BAT thermogenic marker, uncoupling protein 1 (UCP-1), and batokines was

determined both in vivo and in vitro using real-time polymerase chain reaction (RT-PCR).

The results showed that BAT from *db/db* mice displays a hypertrophied phenotype that is consistent with reduced gene expression of *Ucp-1* and increased expression of the pro-inflammatory cytokine, tumour necrosis factor-alpha (*TNF-α*). Importantly, gene expression of the batokines that regulate sympathetic neurite outgrowth and vascularization, including bone morphogenic protein 8b (*Bmp8b*), fibroblast growth factor 21 (*Fgf-21*), and neuregulin 4 (Nrg-4), was altered in BAT from *db/db* mice. Likewise, gene expression of meteorin-like (*Metrnl*), growth differentiation factor 15 (*Gdt-15*), and C-X-C motif chemokine-14 (*Cxcl-14*), regulating pro- and anti-inflammation, was altered. In in vitro experiments, T37i brown adipocytes displayed a dose-dependent cytotoxicity, and all the doses increased lipid accumulation that was accompanied by decreased *Ucp-1* expression. These data demonstrated that obesity progression promotes BAT hypertrophy and impairs batokines gene expression, while supporting a potential pathophysiological role of batokines.

1.8 Biochemical Mechanisms of Antimicrobial Action of Natural Products Isolated from Selected Medicinal Plants in Zimbabwe

The World Health Organization (WHO) estimates that approximately 80% of the global population relies almost entirely on traditional medicine for their primary healthcare needs. This reliance stems not only from economic constraints that make conventional pharmaceuticals inaccessible but also from the cultural acceptance and holistic approaches of traditional medical systems, which often address psychological and spiritual aspects inadequately covered by modern medicine. Medicinal plants, therefore, present an untapped reservoir of potential therapeutic agents that could contribute significantly to the sustainable management of infectious diseases, especially in resource-limited settings.

Understanding the mechanisms by which plant-based traditional medicines exert their effects is essential for validating their use, evaluating their efficacy, and assessing potential toxicities, particularly when administered alone or in combination with conventional drugs. Stanley Mukanganyama gave a presentation discussing his group's research on medicinal plants of Zimbabwe. The study involved random selection of the plants based on ethnomedicinal knowledge and reports of biological activity. Extracts from these plants underwent a series of purification steps and biological assays to assess their antimicrobial potency and safety profiles. The biochemical mechanisms of action identified include: (a) disruption of microbial cell membrane integrity, leading to leakage of intracellular contents; (b) inhibition of microbial proteases, impairing essential metabolic pathways; (c) modulation of ATP-dependent drug efflux pumps, enhancing intracellular drug retention; (d) synergistic antimicrobial enhancement, where plant-derived compounds boost the efficacy of other antimicrobial agents; and, (e) inhibition of microbial biofilm formation, a key factor in chronic and resistant infections. These mechanisms reflect only a subset of the diverse biochemical strategies employed by plant-derived natural products to exert antimicrobial effects. The findings provide strong evidence for the biochemical basis of the antimicrobial activity of these compounds. With further development and optimization, these natural products could serve as valuable lead compounds in the development of new therapeutics for treating existing and emerging microbial infections.

1.9 Conclusions

Infectious diseases remain the leading causes of death in Africa. Africa accounts for the following global infectious diseases burden: malaria (94%), HIV/AIDS (68%), and TB (25%) (Niohuru 2023). It is not surprising that the various lectures presented by keynote speakers at the 28th SASBMB congress focused on these important health concerns plaguing the continent. It is worth noting that cases of non-communicable diseases have increased on the continent in the past two decades, with these diseases projected to become the leading causes of disease in sub-Saharan Africa by 2030 (Niohuru 2023). In response to this, several lectures were presented at this congress covering various non-communicable diseases such as congenital heart diseases, diabetes, and cancer. The theme of the 28th Congress of the SASBMB, "Biochemistry leading the future," aligns with current and future efforts aimed at addressing the burden of diseases afflicting Africans through enhancing our basic understanding of the various conditions, as well as developing therapeutic interventions.

Conflict of Interest Author GLB declares that he has no conflict of interest. Author CD declares that he has no conflict of interest. Author APK declares that he has no conflict of interest. Author JdlM declares that she has no conflict of interest. Author MPM declares that he has no conflict of interest. Author SM declares that he has no conflict of interest. Author SEM-M declares that she has no conflict of interest. Author AS declares that he has no conflict of interest. Author ES declares that he has no conflict of interest. Author ÖTB declares that she has no conflict of interest. Author LZ declares that she has no conflict of interest.

Ethical Approval A part of this chapter cites studies conducted with human participants performed by the author LZ and her team. Ethical approval for the respective studies is indicated in the cited literature.

Funding The authors would like to acknowledge the support of the following sponsors towards the 28th SASBMB congress: the University of Venda (UniVen), the International Union of Biochemistry and Molecular Biology (IUBMB), Diplomics (https://www.diplomics.org.za), the National Research Foundation (NRF) of South Africa (Grant No: KIC24043211994), and the South African Medical Research Council (SAMRC).

References

Ahmad T, Alhammadi BA, Almaazmi SY et al (2024a) Malaria public health status: global context and update on Gulf Cooperation Council countries. Preprints. https://doi.org/10.20944/preprints202407.2174.v1

Ahmad T, Alhammadi BA, Almaazmi SY et al (2024b) *Plasmodium falciparum* heat shock proteins as anti-malarial drug targets: an update. Cell Stress Chaperones 29:326–337

Aldersley T, Brooks A, Human P et al (2023) The impact of COVID-19 on a South African pediatric cardiac service: implications and insights into service capacity. Front Public Health 11:1177365. https://doi.org/10.3389/fpubh.2023.1177365

Aldersley T, Lawrenson J, Human P et al (2021) PROTEA, a Southern African multicenter congenital heart disease registry and biorepository: rationale, design, and initial results. Front Pediatr 9:763060. https://doi.org/10.3389/fped.2021.763060

Alsaied T, Bokma JP, Engel ME et al (2017) Predicting long-term mortality after Fontan procedures: a risk score based on 6707 patients from 28 studies. Congenit Heart Dis 12:393–398. https://doi.org/10.1111/chd.12468

Barozi V, Chakraborty S, Govender S et al (2024) Revealing SARS-CoV-2 Mpro mutation cold and hot spots: dynamic residue network analysis meets machine learning. Comput Struct Biotechnol J 23:3800–3816

Barozi V, Musyoka TM, Sheik Amamuddy O et al (2022) Deciphering isoniazid drug resistance mechanisms on dimeric *Mycobacterium tuberculosis* KatG via post-molecular dynamics analyses including combined dynamic residue network metrics. ACS Omega 7:13313–13332

Curry C, Zühlke L, Mocumbi A, Kennedy N (2018) Acquired heart disease in low-income and middle-income countries. Arch Dis Child 103:73–77

Govender S, Morgan E, Ramahala R et al (2025) Transfer learning towards predicting viral missense mutations: a case study on SARS-CoV-2. Comput Struct Biotechnol J 27:1686–1692

Hasan BS, Rasheed MA, Wahid A et al (2021) Generating evidence from contextual clinical research in low- to middle income countries: a roadmap based on theory of change. Front Pediatr 9:764239. https://doi.org/10.3389/fped.2021.764239

Karthikeyan G, Ntsekhe M, Islam S et al (2024) Mortality and morbidity in adults with rheumatic heart disease. JAMA 332:133–140

Kassebaum N, Kyu HH, Zoeckler L, et al (2017) Child and adolescent health from 1990 to 2015: findings from the global burden of diseases, injuries, and risk factors 2015 study. JAMA Pediatr 171:573–592

Liu Y, Chen S, Zühlke L, Babu-Narayan SV et al (2020) Global prevalence of congenital heart disease in school-age children: a meta-analysis and systematic review. BMC Cardiovasc Disord 20:488

Mwaniki RM, Veldman W, Sanyanga A et al (2025) Decoding allosteric effects of missense variations in drug metabolism: Afrocentric CYP3A4 alleles explored. J Mol Biol 437:169160. https://doi.org/10.1016/j.jmb.2025.169160

Ndadza A, Muyambo S, Mntla P et al (2021) Profiling of warfarin pharmacokinetics-associated genetic variants: black Africans portray unique genetic markers important for an African specific warfarin pharmacogenetics-dosing algorithm. J Thromb Haemost 19:2957–2973

Niohuru I (2023) Disease burden and mortality. In: Healthcare and disease burden in Africa. SpringerBriefs in economics. Springer, Cham. https://doi.org/10.1007/978-3-031-19719-2_3

Rwebembera J, Marangou J, Mwita JC et al (2024) World Heart Federation guidelines for the echocardiographic diagnosis of rheumatic heart disease. Nat Rev Cardiol 21:250–263

Saacks NA, Eales J, Spracklen TF et al (2022) Investigation of copy number variation in South African patients with congenital heart defects. Circ Genom Precis Med 15:e003510. https://doi.org/10.1161/CIRCGEN.121.003510

Swanson L, Owen B, Keshmiri A et al (2020) A patient-specific CFD pipeline using Doppler Echocardiography for application in coarctation of the aorta in a limited resource clinical context. Front Bioeng Biotechnol 8:409. https://doi.org/10.3389/fbioe.2020.00409

Swart M, Dandara C (2019) MicroRNA mediated changes in drug metabolism and target gene expression by Efavirenz and rifampicin in vitro: clinical implications. OMICS 23:496–507

Sheik Amamuddy O, Bishop NT, Tastan Bishop Ö (2018) Characterizing early drug resistance-related events using geometric ensembles from HIV protease dynamics. Sci Rep 8:17938

Sheik Amamuddy O, Musyoka TM, Boateng RA et al (2020) Determining the unbinding events and conserved motions associated with the pyrazinamide release due to resistance mutations of *Mycobacterium tuberculosis* pyrazinamidase. Comput Struct Biotechnol J 18:1103–1120

Singh H, Almaazmi SY, Dutta T, Keyzers RA et al (2023) In silico identification of modulators of J domain protein-Hsp70 interactions in *Plasmodium falciparum*: a drug repurposing strategy against malaria. Front Mol Biosci 10:1158912

Tastan Bishop Ö, Musyoka TM, Barozi V (2022) Allostery and missense mutations as intermittently linked promising aspects of modern computational drug discovery. J Mol Biol 434:167610

WHO 2024 Malaria report. https://www.who.int/newsroom/fact-sheets/detail/malaria. Accessed online on 7 Aug 2025

Zühlke L, Karthikeyan G, Engel ME et al (2016) Clinical outcomes in 3343 children and adults with rheumatic heart disease from 14 low and middle income countries: 2-year follow-up of the global rheumatic heart disease registry (the REMEDY study). Circulation. https://doi.org/10.1161/CIRCULATIONAHA.116.024769

Zühlke L, Lawrenson J, Comitis G et al (2019) Congenital heart disease in low- and lower-middle-income countries: current status and new opportunities. Curr Cardiol Rep 21:163. https://doi.org/10.1007/s11886-019-1248-z

Zühlke L, Mayosi BM (2013) Echocardiographic screening for subclinical rheumatic heart disease remains a research tool pending studies of impact on prognosis. Curr Cardiol Rep 15:343

Evaluation of Ultrasonic-Assisted Aqueous Two-Phase Extraction of Rutin from the Leaves of *Moringa oleifera* Lam. Through Response Surface Methodology

Dakalo Lorraine Ndou, Ashwell Rungano Ndhlala, Nikita Tawanda Tavengwa, and Ntakadzeni Edwin Madala

Abstract

Rutin, a natural flavonol glycoside prevalent in various fruits and vegetables, is known to possess diverse pharmacological properties. Its extraction from plant materials has been explored through various methods, yielding different outcomes. In the present investigation, UA-ATPE was employed to extract rutin from the leaves of *Moringa oleifera*, utilizing a 25% saturated salt/ethanol solution to create an aqueous two-phase system. Different salts, such as $MgSO_4$, $(NH_4)_2SO_4$, and $NaCl$, were used, and the optimization focused on ultrasonic time and ultrasonic temperature. A CCD approach was implemented to determine the optimal experimental conditions. The various ATPE systems resulted in distinct RSMs, displaying a linear fit for $MgSO_4$/ethanol and $(NH_4)_2SO_4$/ethanol systems, and a quadratic fit for the $NaCl$/ethanol system. The optimal rutin extraction conditions were identified as 25 °C and 22.5 min for the $(NH_4)_2SO_4$/ethanol and $NaCl$/ethanol systems. Based on MRM using UHPLC-qTOF-MS, rutin concentrations at these conditions were determined to be 170.30 µg/L and 240 µg/L for the $(NH_4)_2SO_4$/ethanol and $NaCl$/ethanol systems, respectively. The $NaCl$/ethanol system thus exhibited the best performance in rutin extraction. The study also highlighted the significant impact of temperature on rutin extraction and concluded that low temperatures favour the extraction of rutin from the leaves of *M. oleifera*.

D. L. Ndou · N. T. Tavengwa
Department of Chemistry, Faculty of Science, Engineering and Agriculture, University of Venda, Thohoyandou, South Africa

A. R. Ndhlala
Green Biotechnologies Research Centre, Department of Plant Production, Soil Science and Agricultural Engineering, University of Limpopo, Sovenga, South Africa

N. E. Madala (✉)
Department of Biochemistry and Microbiology, Faculty of Science, Engineering and Agriculture, University of Venda, Thohoyandou, South Africa
e-mail: ntaka.madala@univen.ac.za

Keywords

Aqueous two-phase extraction · Rutin · Ultrasonic · Central composite design · Response surface model

© The Author(s) 2026
A. Shonhai et al. (eds.), *Advances in Biochemistry and Molecular Biology to meet Africa's Needs*, Advances in Experimental Medicine and Biology 1507, https://doi.org/10.1007/978-3-032-24254-9_2

2.1 Introduction

Moringa oleifera Lam is known for its nutritious pods and edible leaves and is used as food and medicine. It is also known to contain a great number of bioactive compounds. The leaves are the most used parts of the plant as they are rich in vitamins, carotenoids, polyphenols, phenolic acids, flavonoids, alkaloids, glucosinolates, isothiocyanates, tannins, and saponins (Paikra et al. 2017; Vergara-Jimenez et al. 2017). These bioactive compounds are responsible for the pharmaceutical properties of *M. oleifera* (Falowo et al. 2018; Lin et al. 2018). Rutinoside-bearing flavonoids such as kaempferol rutinoside, quercetin rutinoside (rutin), and isorhamnetin rutinoside are an important class of bioactive flavonoids. The disaccharide, i.e., rutinoside, enables these flavonoids to be highly bioavailable. Rutin is an attractive phytochemical because of its pharmacological properties and is thus an important flavonoid in the pharmaceutical industry (Chua 2013; Peng et al. 2018; Tursynbolat et al. 2019).

Rutin is a flavonol glycoside that has been reported to present clinically relevant functions, beneficial in preventing diseases and protecting genome stability (Gullon et al. 2017). This flavonoid exhibits other favourable properties such as antioxidant (Enogieru et al. 2018), anti-inflammatory (Peng et al. 2018), antiallergic (Rahman et al. 2021), cardioprotective (Fei et al. 2019), neuroprotective (Yang et al. 2019), antidiabetic (Tursynbolat et al. 2019), and anticancer activities (Negahdari et al. 2021; Molnar et al. 2018). More than 130 medicinal products in the market contain rutin, either alone or in combination with other active ingredients (Chua 2013; Gullon et al. 2017; Molnar et al. 2018).

Rutin is generally extracted using various organic solvents such as methanol, ethanol, acetone, isopropanol, ethyl acetate, diethyl ether, and water-solvent mixtures (Gullon et al. 2017; Huang et al. 2016; Vetrova et al. 2017; Kim and Lim 2019). Conventional extraction methods are simple to use, but they present drawbacks such as poor efficiency, high solvent consumption, and long extraction times (Izza et al. 2018; Chou et al. 2020). Non-conventional extraction methods, on the other hand, use dedicated processing aids/energy inputs to improve the extraction efficiency and/or selectivity. Some of these methods employed in the extraction of rutin include deep eutectic solvents-based ultrasound-assisted extraction (DES-UAE) (Ali et al. 2019), ultrasound-assisted extraction (UAE) (Banozic et al. 2019), microwave-assisted extraction (MAE) (Dobrincic et al. 2020), and pipette-tip micro-solid phase extraction (Ndou et al. 2024). These methods offer superior extraction efficiency in terms of cost, yield, extraction time, and/or selectivity. The drawback with these methods, however, is that scaling up can be difficult because of the costs of the equipment and installations (Kim and Lim 2019; Huang et al. 2013; Tatke and Rajan 2014; Barba et al. 2016). One method of extraction to consider is aqueous two-phase extraction (ATPE).

ATPE is a non-conventional extraction method that is formed from two or more phase-forming substances in water (e.g., two polymers, a polymer and a salt/sugar) (Zhou et al. 2018). It is a liquid-liquid fractionation technique that has been used in the separation, extraction, purification, and enrichment of some biomolecules (Jiang et al. 2019). ATPE allows the use of environmentally friendly phase-forming components such as alcohol, salts, and sugars, and can achieve rapid separation at room temperature which contradicts conventional extraction methods that often use toxic solvents, long extraction times, and high temperatures (Chong and Brooks 2021). The alcohol/salt ATPE system is advantageous because of the easy recovery, low environmental impact, high selectivity, low viscosity, fast separation effect, and low cost (Toledo et al. 2019). The most common alcohols used are short-chain alcohols such as methanol, ethanol, and 2-propanol, and the most common salts are those with cations such as Na^+, K^+, Ca^{2+}, and NH_4^+ and with anions such as Cl^-, SO_4^{2-}, CO_3^{2-}, and NO_3^{2-} (Gomis et al. 2021). Ethanol has been reported to extract weakly polar compounds such as flavonoids, saponins, and alkaloids, while 2-propanol has been reported to extract lipid-soluble compounds (Xi et al. 2023). The combination of ATPE with another method, such as the ultrasonic method, enhances the extraction efficiency (Xi et al. 2023). ATPE-based ultrasonic-assisted extraction reduces the consumption of solvents, labour, and energy, and destructs the structure of plant cell walls by acoustic cavitation that passes through the solvent, thereby enhancing the recovery of bioactive compounds (Toledo et al. 2019; Gomis et al. 2021; Xi et al. 2023).

The primary objective of this study was to enhance the efficiency of rutin extraction from the leaves of *Moringa oleifera* by employing ultrasonic-assisted aqueous two-phase extraction (UA-ATPE). Notably, this is the first study to attempt to concentrate rutin extraction from the leaves of *M. oleifera* using a salt/ethanol UA-ATPE configuration. The salts examined herein for ATPE were NaCl, $MgSO_4$, and $(NH_4)_2SO_4$. Two key extraction variables, namely ultrasonic time and ultrasonic temperature, were systematically investigated. To achieve optimization, a central composite design (CCD) approach was implemented. The outcomes were analysed using response surface methodology (RSM) using LC-MS data. RSM was chosen for its advantages of shorter processing time, predictive response capabilities, and cost-effectiveness. This study thus contributes novel insights into the targeted rutin extraction process from the leaves of *M. oleifera* using the salt/ethanol UA-ATPE method.

2.2 Materials and Methods

2.2.1 Chemicals and Reagents

Ammonium sulphate $((NH_4)_2SO_4)$, magnesium sulphate $(MgSO_4)$, sodium chloride (NaCl), and rutin hydrate were purchased from Sigma--Aldrich (Johannesburg, South Africa). Methanol (99% CP) and ethanol (99% CP) were purchased from Associated Chemical Enterprises (Johannesburg, South Africa). Ultra-pure water (0.005 μS, 18 mΩ) using a Direct-Q 5UV distiller (Massachusetts, United States of America) was used for the preparation of the salt solutions. The extractions were performed on a Scientech

Ultrasonic Cleaner and a DIAB MX-RL-Pro dragon shaker. Chromatographic separation of the metabolites in the extracts was done using a reverse-phase Shim-pack Velox C18 column (2.1 × 100 mm, 2.7 μm) with a serial number 227-32009-03 (Columbia, USA). The UPLC was connected to a Shimadzu 9030 LC-qTOF-MS detector (Shimadzu, Kyoto). The solvents used for the chromatographic runs were methanol and formic acid, which were purchased from Romil Pure Chemistry (Cambridge, UK).

2.2.2 Chromatographic and Mass Spectrometry Conditions

Rutin was separated using a Shimpack C18, 2.1 × 100 mm, 2.7 μm column from Shimadzu (Honeydew, South Africa). The column was maintained at 40 °C at a flow rate of 0.4 mL/min· and the injection volume of 3 μL was used. Mobile phase A was 0.1% formic acid in ultra-high purity water (v/v), and mobile phase B was 0.1% (v/v) formic acid in methanol.

A UPLC-QTOF-MS 9030 mass spectrometer (Shimadzu, Japan) was used for all mass spectral measurements. The mass spectrometer was equipped with an electrospray interface (ESI) operating in positive mode. ESI parameters were optimized for rutin by direct infusion of standard solutions into the mass spectrometer. The mass spectrometer was operated in the multi-reaction monitoring (MRM) mode to confirm the identity of rutin. This was achieved by selecting specific precursor to product ion transitions for each rutin based on MRM transitions. High-purity nitrogen (N_2) was used as the nebulizing and drying gas. The optimum parameters were as follows: drying gas temperature, 250 °C; drying gas flow, 10 L/min; and collision energy, 30–60 eV. For the chromatographic separation, a Shimadzu 9030 LC instrument (Shimadzu, Japan) was used. The instrument consisted of an autosampler, a thermostated column compartment, and a binary pump. LabSolutions software was used to control the LC-MS/MS instrument and for data acquisition, and the mass range was *m/z* 100–1000.

2.2.3 Preparation of ATPE

A salt (ammonium sulphate, magnesium sulphate, and sodium chloride) (8 g) was dissolved in water (32 mL) to form a 25% saturated solution. The saturated salt solution (20 mL) was mixed with *M. oleifera* powder (1 g). The resulting mixture was placed in an ultrasonic bath, and the reaction was performed at various temperatures and times. After the elapsed time, ethanol (20 mL) was added to the mixture, resulting in an aqueous two-phase system. The ethanol phase was filtered and analysed using UHPLC-qTOF-MS. The extractions were performed in triplicate.

2.2.4 Preparation of Rutin Standards

The stock standard solution of rutin was prepared in methanol at a concentration of 500 mg/L using HPLC-grade methanol. Seven standard working solutions were prepared from the stock solution through serial dilutions. The rutin standards were quantified based on scheduled multiple reaction monitoring (MRM).

2.2.5 Statistical Analysis

The central composite design response surface model (CCD RSM) was fitted to experimental data to obtain the relationship between factors and optimize the response of Z (rutin concentration) in relation to A (time) and B (temperature),

using Design Expert 13. A Two-level full factorial CCD was designed, and a total of 9 experimental runs (including 3 repetitions) were designed for all the salts. This included numerical factors such as time (10, 20, 22.5, 25, and 35 min) and temperature (25, 39, 42.5, and 46 °C). Model parameters and model significance were determined at $p < 0.05$. The fitness of the models was determined by evaluating the coefficient of regression (R^2) obtained from the analysis of variance (ANOVA). The model fit generated the response surface that defined the behavior of the response variable (i.e., concentration of rutin).

2.3 Results and Discussion

CCD was performed to determine the optimum ATPE conditions using an ultrasonic bath for the extraction of rutin from the leaves of *M. oleifera*. The results for the $MgSO_4$/ethanol, $(NH_4)_2SO_4$/ethanol, and NaCl/ethanol ATPE systems are shown in Table 2.1. The highest concentration of rutin extracted was 240 µg/L with conditions of 25 °C for 22.5 min using the NaCl/ethanol ATPE system. Table 2.1 shows that with the $MgSO_4$/ethanol system, the amount of rutin extracted increased with an increase in temperature and decreased with an increase in time. However, increasing the temperature to 46 °C resulted in a decrease in the amount of rutin extracted. In the $(NH_4)_2SO_4$/ethanol system, it was observed that an increase in temperature and time resulted in a decrease in the amount of rutin extracted. Therefore, the highest concentration of rutin was

Table 2.1 CCD experiments for ATPE optimization and concentration of rutin extracted using different salts

	Extraction factors		Rutin concentration extracted (µg/L)			
Run order	Time (min)	Temperature (°C)	$MgSO_4$	$(NH_4)_2SO_4$	NaCl	Standard order
1	25.0	46.0	169.09	157.25	204.41	3
2	25.0	39.0	165.17	154.43	240.99	2
3	22.5	42.5	168.68	154.94	233.77	9
4	10.0	42.5	188.74	162.79	230.86	4
5	22.5	25.0	145.63	170.30	240.00	6
6	22.5	42.5	173.37	152.28	234.26	8
7	35.0	42.5	150.16	152.48	216.60	5
8	22.5	42.5	141.45	159.36	220.32	7
9	20.0	39.0	173.09	157.37	239.95	1

extracted at 25 °C with a 22.5-min extraction time. Table 2.1 shows that the NaCl/ethanol system was the most effective in the extraction of rutin from the leaves of *M. oleifera*. It was noted that an increase in temperature resulted in a decrease in the concentration of rutin extracted, whereas time did not have a significant effect on the concentration of rutin extracted.

The use of the salts in the UA-ATPE is to allow partitioning of the ethanol from water, wherein the ethanol phase is enriched with the metabolite of interest; in this case, rutin. The addition of salts in this extraction method minimizes the water solubility of ethanol, in a process referred to as the salting-out effect. Upon introduction of ethanol, the precipitated metabolites move from the aqueous phase to the ethanol phase (Xi et al. 2023; Mokgehle et al. 2021). The observation that the NaCl/ethanol UA-ATPE system was the best-performing system could be attributed to the salting-out effect. It can thus be concluded that NaCl formed stronger bonds with the water molecules than $MgSO_4$ and $(NH_4)_2SO_4$, which resulted in a higher recovery of rutin from the aqueous phase. NaCl has also been reported to be the most common salt added to an ATPE system (Ng et al. 2021).

Furthermore, in a study by Shiran et al. (2020), the effect of cations on the salting-out effect was studied. In this study, a polymer-salt ATPE system was formed with polyethylene glycol (PEG) as the polymer and Na_2SO_4 and $MgSO_4$ as the salts. It was observed that Na_2SO_4 had a more salting-out effect than $MgSO_4$, and this observation was attributed to the cation effect. It was observed that the salt with multiple valence cations (Mg^{2+}) interacted with the ether oxygen of PEG, which resulted in salting-in. This led to the conclusion that the more valence of the cations, the greater the salting-in ability. Therefore, the Na^+ cation led to a more salting-out effect than the Mg^{2+} cation.

However, according to the Hofmeister series (Xie and Gao 2013), it would be expected that the $(NH_4)_2SO_4$/ethanol ATPE system would be the best performing of the three. It should be noted that $(NH_4)_2SO_4$ is more acidic than $MgSO_4$ and NaCl. Rutin, on the other hand, is more stable in an acidic medium (Ran et al. 2017). Therefore, during the salting-out process, less rutin is partitioned to the ethanol phase because of its high affinity for the acidic $(NH_4)_2SO_4$ aqueous phase. This then results in a low concentration of rutin in the ethanol phase. Another possible explanation is the hydration energy of the cations. Hydration energy increases with an increase in the cation's charge (Andersson and Stipp 2014), meaning Mg^{2+} would have the highest hydration energy among the three cations. Based on this study's observations, it can be concluded that higher hydration energy corresponds to lower extraction efficiency. However, NH_4^+ and Na^+ have the same charge, so the ion size must be considered. NH_4^+ is larger than Na^+, and hydration energy decreases with increasing ionic size. According to a study by Parsons et al. (2011), the Hofmeister series can be reversed when ionic size is considered. Therefore, it is possible that the Hofmeister series was reversed in this study, resulting in the order $Na^+ > NH_4^+$ and consequently the high extraction efficiency of the NaCl/ethanol system.

2.3.1 Statistical Data

The regression models were obtained by fitting the model obtained from the different salts to the experimental data. The adequacy of the models was assessed by model statistics, and the results are shown in Tables 2.2, 2.3, and 2.4 for the $MgSO_4$/ethanol, $(NH_4)_2SO_4$/ethanol, and NaCl/ethanol systems, respectively.

The model significance was evaluated by probability values (p-values), while the evidence of goodness of fit was assessed by R^2 values for the experimental values. From the results, a higher Fisher's F-test value and lower p-values indicate the relative significance of each term. The suitability of the models was assessed through the lack of fit. If the lack of fit values are non-significant ($p > 0.05$), then it implies that the model is valid, reliable, and precise. The lack of fit values for the $MgSO_4$/ethanol, $(NH_4)_2SO_4$/ethanol, and NaCl/ethanol systems were determined as 0.9392, 0.5596, and 0.7091, respectively. This indicates the validity, reliability, and

Table 2.2 Fit statistics of a linear model based on ANOVA for the $MgSO_4$/ethanol system

Source	Sum of squares	df	Mean square	F-value	p-value	Significance
Model	1080.36	2	540.18	4.11	0.0751	Not significant
A-Time	770.37	1	770.37	5.86	0.0518	–
B-Temperature	356.67	1	356.67	2.71	0.1505	–
Residual	788.24	6	131.37	–	–	–
Lack of fit	194.40	4	48.60	0.1637	0.9392	Not significant
Pure error	593.84	2	296.92	–	–	–
Cor Total	1868.60	8	–	–	–	–

df Degrees of freedom, – Not applicable

Table 2.3 Fit statistics of a linear model based on ANOVA for $(NH_4)_2SO_4$/ethanol system

Source	Sum of squares	df	Mean square	F-value	p-value	Significance
Model	192.5	2	96.23	8.28	0.0188	Significant
A-Time	49.46	1	49.46	4.25	0.0848	–
B-temperature	134.9	1	134.9	11.6	0.0144	–
Residual	69.77	6	11.63	–	–	–
Lack of fit	44.15	4	11.04	0.86	0.5996	Not significant
Pure error	25.62	2	12.81	–	–	–
Cor Total	262.2	8	–	–	–	–

df Degrees of freedom, – Not applicable

Table 2.4 Fit statistics of a reduced quadratic model based on ANOVA for NaCl/ethanol system

Source	Sum of squares	df	Mean square	F-value	p-value	Significance
Model	1000.8	2	500.4	11.05	0.0097	Significant
B-temperature	863.10	1	863.1	19.07	0.0047	–
B^2	563.58	1	563.6	12.45	0.0124	–
Residual	271.62	6	45.27	–	–	–
Lack of fit	146.48	4	36.62	0.585	0.7091	Not significant
Pure error	125.13	2	62.57	–	–	–
Cor Total	1272.4	8	–	–	–	–

df Degrees of freedom, – Not applicable

precision of the models. The fitting of the models revealed that the independent factors, namely ultrasonic temperature and ultrasonic time, exerted an influence on the rutin content in the response.

A model is considered insignificant if it has a p-value greater than 0.500. In this study, the model fitted to the data according to ANOVA for the $MgSO_4$/ethanol ATPE system was observed to be a linear fit with a p-value of 0.0751, which thus implied that the fit is not significant. According to Table 2.2, both temperature and time were regarded as insignificant terms with p-values of 0.1505 and 0.0518, respectively. These values ($p > 0.05$) imply that the model is not significant and is thus deemed not valid

(Shokrollahi et al. 2020). The lack of fit F-value was 4.11, which is not significant relative to the pure error (Uzoejinwa et al. 2019). The non-significant lack of fit is desirable because the intention is for the model to fit. The goodness of fit, R^2, was found to be 0.5782.

For the $(NH_4)_2SO_4$/ethanol ATPE system, the model fitted to the data according to ANOVA was observed to be a linear fit (Table 2.3). The p-value was observed to be 0.0188, proving that the model fit is significant. Time was observed to be an insignificant term with a p-value of 0.0848, while temperature had a p-value of 0.0144. This indicated that temperature was a significant term in this model. This model showed a goodness of fit of $R^2 = 0.7339$.

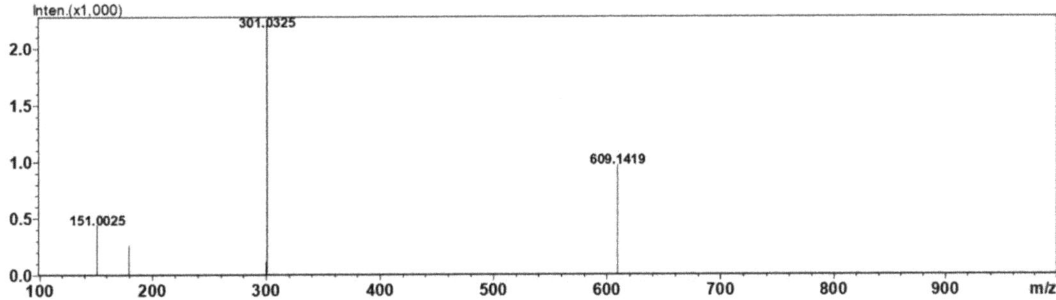

Fig. 2.1 A representation of (**a**) the fragmentation pathway of rutin and (**b**) tandem mass spectrum generated from a single ion monitoring of rutin

For the NaCl/ethanol ATPE system, the model fitted to the data according to ANOVA was observed to be a quadratic fit (Table 2.4). The p-value was found to be 0.0097, which shows that the model fit is significant. Temperature appeared to be the significant fit in this model, with a p-value of 0.0047, and the B^2 term had a p-value of 0.0124. The goodness of fit for this model was observed to be 0.7865.

2.3.2 Identification and Quantification of Rutin Using UHPLC-qTOF-MS

The presence of rutin was confirmed using the UHPLC-qTOF-MS. Figure 2.1a represents the rutin fragmentation pathway wherein it loses the sugar, resulting in the quercetin aglycone. The corresponding fragmentation pattern of rutin (measured = m/z 609.1419; calculated = m/z 609.1461) in tandem MS experiments (MS/MS) is represented in Fig. 2.1b. In this process, rutin underwent fragmentation, resulting in its flavonol aglycone, quercetin, at m/z 301.0325, attributed to the loss of the rutinoside sugar. This confirmed the presence of rutin in the extracts of the leaves of *M. oleifera* and confirmed the extraction of rutin using UA-ATPE. These results are consistent with those reported by Wojdylo and Nowicka (2019) and Fu et al. (2020) on the fragmentation of rutin. These studies reported on the loss of the rutinoside sugar, which resulted in quercetin as the major fragment ion.

2.3.3 Parameter Effects and Response Surface Models

Figure 2.2 is the Pareto chart showing the influence of ultrasonic temperature and ultrasonic time on the extraction of rutin from the leaves of *M. oleifera* using three salt/ethanol systems. It was noted that ultrasonic temperature had a

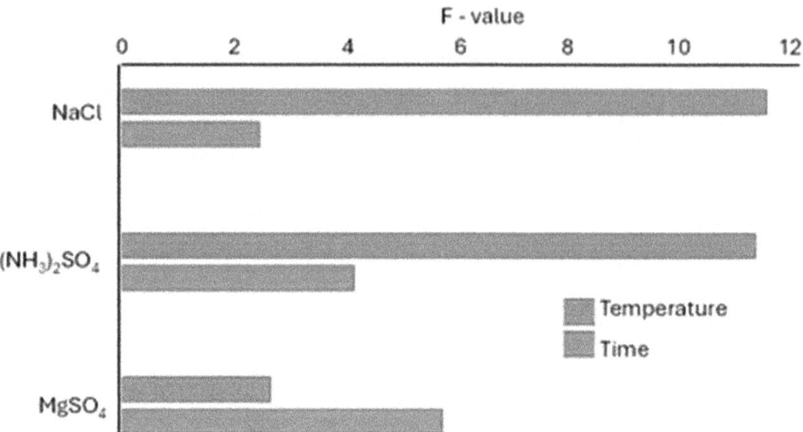

Fig. 2.2 Pareto chart showing the effects of ultrasonic time and ultrasonic temperature on the extraction of rutin from the leaves of *M. oleifera* in an ATPE using different salts

significant effect on the amount of rutin extracted, whereas ultrasonic time did not have much effect on the amount of rutin extracted. The decrease in the amount of rutin extracted from the leaves of *M. oleifera* with an increase in temperature is rather in contrast with several previous results on the extraction of rutin from different plant materials such as *Satureja montana L.* (Jakovlijevic et al. 2020), *Ruta graveolens L.* (Molnar et al. 2018), buckwheat (Kim and Lim 2019), and lemon by-products (Papoutsis et al. 2018) using DES, subcritical water, and aqueous ultrasound-assisted extraction methods This could be because each plant requires specific extraction conditions for the optimum recovery of its compounds (Cardona et al. 2017; Shafi et al. 2019). It is therefore necessary to validate the extraction method for phenolic compounds to avoid enzymatic oxidation, which results in loss of phenol function and antioxidant potential (Shafi et al. 2019). On the other hand, a decrease in the amount of rutin extracted with increasing temperature has been reported in *Lycium barbarum* L. fruits, where a temperature of 25 °C was used as the optimum temperature for the extraction of rutin (Ali et al. 2019). The differences in optimal temperatures could be because different polyphenols show different sensitivity to heat treatment depending on their structures. Flavonoids, for example, have been reported to be more sensitive to thermal degradation than phenolic acids. Elevated temperatures usually improve the extraction yield and shorten the extraction time.

However, intense temperatures or prolonged exposure to elevated temperatures can cause degradation of the thermally sensitive compounds and thus result in poor extraction yields (Dobrincic et al. 2020). The extraction yield of rutin from different plant materials is also dependent on the method of extraction. In this study, the ultrasonic power combined with the increase in temperature is most likely the reason for the observed results. In a review by Dzah et al. (2020), it was stated that the effects of ultrasound-assisted extractions and temperature on the extraction yield of polyphenolic compounds were also dependent on the plant material.

It was observed that ultrasonic temperature had a significant effect on the extraction of rutin when using the NaCl/ethanol and $(NH_4)_2SO_4$/ethanol systems. However, when the $MgSO_4$/ethanol system was used, time was observed to be a more significant term than temperature. These observations agree with the fit statistics detailed above in Tables 2.2, 2.3 and 2.4.

The response surface plots in Fig. 2.3 show the effect of ultrasonic time and ultrasonic temperature in the extraction of rutin from the leaves *M. oleifera* in the salt/ethanol ATPE-systems. Figure 2.3a represents the response surface plot for the $MgSO_4$/ethanol UA-ATPE system. This plot shows that the concentration of rutin extracted increases with an increase in temperature and decreases with a decrease in time. The $(NH_4)_2SO_4$/ethanol UA-ATPE system is represented in Fig. 2.3b, wherein it was observed that

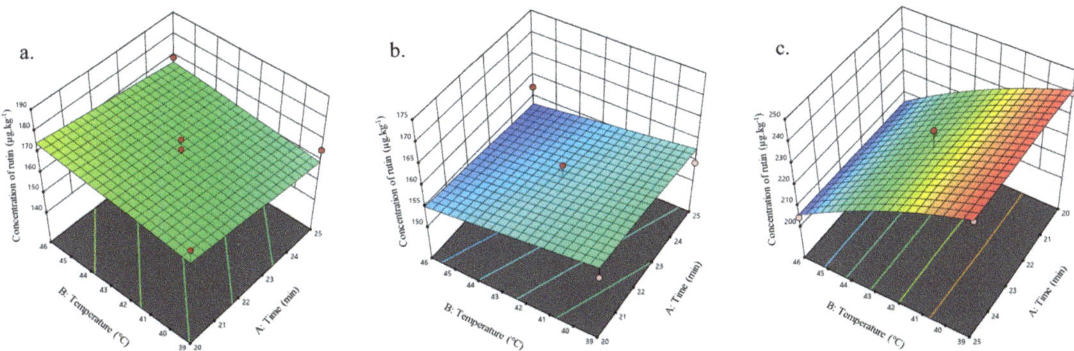

Fig. 2.3 Response surface plots on the effect of ultrasonic time and ultrasonic temperature on the extraction of rutin from the leaves of *M. oleifera* for the (**a**) MgSO$_4$/ethano (**b**) (NH$_4$)$_2$SO$_4$/ethanol and (**c**) NaCl/ethanol ATPE systems

the concentration of rutin extracted decreased with an increase in time and temperature. Figure 2.3c represents the NaCl/ethanol UA-ATPE system, wherein it was observed that the concentration of rutin decreased with an increase in temperature. It was also observed that time did not have a significant effect on the concentration of rutin extracted.

2.4   Conclusion

The extraction of rutin from the leaves of *Moringa oleifera* through three salt/ethanol UA-ATPE systems was performed. The salts investigated in the salt/ethanol ATPE systems were NaCl, MgSO$_4$, and (NH$_4$)$_2$SO$_4$. Ultrasonic temperature and ultrasonic time were optimized, and the maximum concentration of rutin was extracted through the NaCl/ethanol UA-ATPE system at 25 °C. It was observed in this study that an increase in ultrasonic temperature and ultrasonic time resulted in a decrease in the extraction of rutin from the leaves of *M. oleifera* extracts. However, based on the ANOVA statistics, it was noted that the ultrasonic temperature was a significant factor ($p < 0.05$) and the ultrasonic time was an insignificant factor ($p > 0.05$) for the NaCl/ethanol and (NH$_4$)$_2$SO$_4$/ethanol UA-ATPE systems. However, for the ethanol/MgSO$_4$ UA-ATPE system, time had a more significant effect than temperature on the extraction of rutin. The response surface models confirmed the effect of tempera-

ture and time on the extraction of rutin using the three salt/ethanol UA-ATPE systems. From this study, it was observed that the extraction of rutin from the leaves of *M. oleifera* is favourable at low temperature. This could be because rutin is sensitive to high temperatures and therefore degrades at these elevated temperatures, resulting in low extraction yield. This observation coincides with what was reported in a study by Vo et al. (2024). It was reported that an increase in temperature resulted in the degradation of flavonoids, causing the total flavonoid content (TFC) to decrease. In the same study, ultrasonic time was observed to not have any significant effect on the TFC. The UA-ATPE method propelled by the salting-out technique has proven to be time and energy-efficient for the extraction of rutin from the leaves of *M. oleifera* Lam. In a study by Gao et al. (2021), where UA-ATPE was used for the extraction of anthocyanins from blueberry residue, the authors reported this method as efficient, energy-saving, and green and could be a good replacement for the traditional organic solvent extraction method (Gao et al. 2021). Studies by Dong et al. (2015) and Tan et al. (2022) reported UA-ATPE to be a promising technique for the extraction, enrichment, and purification of bioactive compounds from plants (Dong et al. 2015; Tan et al. 2022). According to another study by Mokgehle et al. (2021), the salting-out method was reported to be an efficient approach for the extraction of metabolites from plants, thus making the extraction process less tedious.

Owing to its multiple bioactivities, the extraction of rutin is therefore imperative, and because of its presence in many herbal medicines, the findings of the current study can form a basis on which traditional healers can use to optimize its extraction during their preparation, instead of using just pure water, as is the case currently.

Acknowledgements The authors would like to acknowledge the LC-MS facility at the University of Venda. The Executive Dean, Professor Natasha Potgieter, is acknowledged for providing funding for the LC-MS instrument maintenance.

Data Availability Statement The data that support the findings of this study are available from the corresponding author upon reasonable request.

Compliance with Ethical Standards

Funding This study was funded by the National Research Foundation (NRF) (grant number 118447).

Conflicts of Interest Ndou DL declares that she has no conflicts of interest. Ndhlala AR declares that he has no conflicts of interest. Tavengwa NT declares that he has no conflicts of interest. Madala NE declares that he has no conflicts of interest.

Ethical Approval This chapter does not contain any studies with human participants or animals performed by any of the authors.

References

Ali MC, Chen J, Zhang H et al (2019) Effective extraction of flavonoids from *Lycium barbarum* L. fruits by deep eutectic solvents-based ultrasound-assisted extraction. Talanta 203:16–22. https://doi.org/10.1016/j.talanta.2019.05.012

Andersson MP, Stipp SLS (2014) Predicting hydration energies for multivalent ions. J Comput Chem 35:2070–2075. https://doi.org/10.1002/jcc.23733

Banozic M, Banjari I, Jakovljevic M et al (2019) Optimization of ultrasound-assisted extraction of some bioactive compounds from tobacco waste. Molecules 24(8):1611. https://doi.org/10.3390/molecules24081611

Barba FJ, Zhenzhou Z, Koubaa M et al (2016) Green alternative methods for the extraction of antioxidant bioactive compounds from winery wastes and by-products: a review. Trends Food Sci Technol 49:96–109. https://doi.org/10.1016/j.tifs.2016.01.006

Cardona MI, Toro RM, Costa GM et al (2017) Influence of extraction process on antioxidant activity and rutin content in *Physalis peruviana* calyses extract. J Appl Pharm Sci 7(6):164–168. https://doi.org/10.7324/JAPS.2017.70623

Chong KY, Brooks MSL (2021) Effects of recycling on the aqueous two-phase extraction of bioactives from haskap leaves. Sep Purif Technol 255:117755. https://doi.org/10.1016/j.seppur.2020.117755

Chou SC, Nasir HM, Mohd-Setapar SH et al (2020) A glimpse into the extraction methods of active compounds from plants. Crit Rev Anal Chem 52(4):667–696. https://doi.org/10.1080/10408347.2020.1820851

Chua LS (2013) A review on plant-based rutin extraction methods and its pharmacological activities. J Ethnopharmacol 150(3):805–817. https://doi.org/10.1016/j.jep.2013.10.036

Dobrincic A, Repajic M, Garofulic IE et al (2020) Comparison of different extraction methods for the recovery of olive leaves polyphenols. Processes 8(9):1008. https://doi.org/10.3390/pr8091008

Dong B, Yuan X, Zhao Q et al (2015) Ultrasound-assisted aqueous two-phase extraction of phenylethanoid glycosides from *Cistanche deserticola* Y. C. Ma stems. J Sep Sci 38:1194–1203. https://doi.org/10.1002/jssc.201401410

Dzah CS, Duan Y, Zhang H et al (2020) The effects of ultrasound assisted extraction on yield, antioxidant, anticancer and antimicrobial activity on polyphenol extracts: a review. Food Biosci 35:1100547. https://doi.org/10.1016/j.fbio.2020.100547

Enogieru AB, Haylett W, Hiss DC et al (2018) Rutin as a potent antioxidant: implications for neurodegenerative disorders. Oxidative Med Cell Longev 2018:6241017. https://doi.org/10.1155/2018/6241017

Falowo AB, Mukumbo FE, Idamokoro EM et al (2018) Multi-functional application of *M. oleifera* Lam. in nutrition and animal food products: a review. Food Res Int 106:317–334. https://doi.org/10.1016/j.foodres.2017.12.079

Fei J, Sun Y, Duan Y et al (2019) Low concentration of rutin treatment might alleviate the cardiotoxicity effect of pirarubicin on cardiomyocytes via activation of PI3K/AKTmTOR signalling pathway. Biosci Rep 39:BSR20190546. https://doi.org/10.1042/BSR20190546

Fu Y, Sun R, Yang J et al (2020) Characterization and quantification of phenolic constituents in peach blossom by UPLC-LTQ-Orbitrap-MS and UPLC-DAD. Nat Prod Commun 15(1):1–9. https://doi.org/10.1177/1934578X19884437

Gao Y, Ji Y, Wang F et al (2021) Optimization the extraction of anthocyanins from blueberry residue by dual-aqueous phase method and cell damage protection study. Food Sci Biotechnol 30(13):1709–1719. https://doi.org/10.1007/s10068-021-00994-w

Gomis A, Garcia-Cano J, Font A, Gomis V (2021) Operational limits in processes with water, salt, and short-chain alcohol mixtures as aqueous two-phase systems and problems in its simulation. Ind Eng Chem Res 60(6):2578–2587. https://doi.org/10.1021/acs.iecr.0c05891

Gullon B, Lu-Chau TA, Moreira MT et al (2017) Rutin: a review on extraction, identification and purification methods, biological activities and approaches to enhance its bioavailability. Trends Food Sci Technol 67:220–235. https://doi.org/10.1016/j.tifs.2017.07.008

Huang HW, Hsu CP, Yang BB, Wang CY (2013) Advances in the extraction of natural ingredients by high pressure extraction technology. Trends Food Sci Technol 33(1):54–62. https://doi.org/10.1016/j.tifs.2013.07.001

Huang Y, Feng F, Jiang J et al (2016) Green and efficient extraction of rutin from tartary buckwheat hull by using natural deep eutectic solvents. Food Chem 221:1400–1405. https://doi.org/10.1016/j.foodchem.2016.11.013

Izza N, Dewi SR, Setyanda A et al (2018) Microwave-assisted extraction of phenolic compounds from *M. oleifera* seed as anti-biofouling agents in membrane processes. MATEC Web Conf 204:03003. https://doi.org/10.1051/matecconf/201820403003

Jakovlijevic M, Vladic J, Vidovic S et al (2020) Application of deep eutectic solvents for the extraction of Rutin and Rosmarinic Acid from *Satureja montana* L. and evaluation of the extracts antiradical activity. Plants 9:153. https://doi.org/10.3390/plants9020153

Jiang B, Na J, Wang L et al (2019) Separation and enrichment of antioxidant peptides from whey protein isolate hydrolysate by aqueous two-phase extraction and aqueous two-phase flotation. Foods 8(1):34. https://doi.org/10.3390/foods8010034

Kim DS, Lim SB (2019) Subcritical water extraction of rutin from the aerial parts of common buckwheat. J Supercrit Fluids 152:1–7. https://doi.org/10.1016/j.supflu.2019.104561

Lin M, Zhang J, Chen X (2018) Bioactive flavonoids in *M. oleifera* and their health-promoting properties. J Funct Foods 47:469–479. https://doi.org/10.1016/j.jff.2018.06.011

Mokgehle T, Madala N, Gitari W, Tavengwa N (2021) Deciphering the effects of kosmotrope and chaotrope salts during aqueous two-phase extraction (ATPE) of polyphenolic compounds and glycoalkaloids from the leaves of a nutraceutical plant, *Solanum retroflexum*, with the aid of UHPLC-qTOF-MS. Appl Biol Chem 64(28):1–15. https://doi.org/10.1186/s13765-021-00603-8

Molnar M, Jakovljevic M, Jokic S (2018) Optimization of the process conditions for the extraction of Rutin from *Ruta graveolens* L. by choline chloride based deep eutectic solvents. Solvent Extr Res Dev Jpn 25(2):109–116. https://doi.org/10.15261/serdj.25.109

Ndou DL, Mtolo BP, Khwathisi A et al (2024) Development of the pipette-tip micro-solid-phase extraction for extraction of Rutin from Moringa oleifera lam. Using activated hollow carbon Nanospheres as sorbents. Int J Anal Chem 2024:2681595. https://doi.org/10.1155/2024/2681595

Negahdari R, Bohlouli S, Sharifi S et al (2021) Therapeutic benefits of rutin and its nanoformulations. Phytother Res 35:1719–1738. https://doi.org/10.1002/ptr.6904

Ng HS, Kee PE, Yim HS et al (2021) Characterization of alcohol/salt aqueous two-phase system for optimal separation of gallic acids. J Biosci Bioeng 131(5):537–542. https://doi.org/10.1016/j.jbiosc.2021.01.004

Paikra BK, Dhongade HKJ, Gidwani B (2017) Phytochemistry and pharmacology of *M. oleifera* Lam. J Pharmacopunct 20(3):194–200. https://doi.org/10.3831/KPI.2017.20.022

Papoutsis K, Pristijono P, Golding JB et al (2018) Optimizing a sustainable ultrasound-assisted extraction method for the recovery of polyphenols from lemon by-products: comparison with hot water and organic solvent extractions. Eur Food Res Technol 244:1353–1365. https://doi.org/10.1007/s00217-018-3049-9

Parsons DF, Bostrom M, Nostro PL, Ninham BW (2011) Hofmeister effects: interplay of hydration, nonelectrostatic potentials, and ion size. Phys Chem Chem Phys 13:12352–12367. https://doi.org/10.1039/C1CP20538B

Peng F, Xu P, Zhao B-Y et al (2018) The application of deep eutectic solvent on the extraction and in vitro antioxidant activity of rutin from *Sophora japonica* bud. J Food Sci Technol 55(6):2326–2333. https://doi.org/10.1007/s13197-018-3151-9

Rahman F, Tabrez S, Ali R et al (2021) Molecular docking analysis of rutin reveals possible inhibition of SARS-CoV-2 vital proteins. J Tradit Complement Med 11(2):173–179. https://doi.org/10.1016/j.jtcme.2021.01.006

Ran F, Liu H, Wang X, Guo Y (2017) A novel molybdenum disulfide nanosheet self-assembled flower-like monolithic sorbent for solid-phase extraction with high efficiency and long service life. J Chromatogr A 1507:18–24. https://doi.org/10.1016/j.chroma.2017.05.018

Shafi W, Mansoor S, Jan S et al (2019) Variability in Catechin and Rutin contents and their antioxidant potential in diverse apple genotypes. Molecules 24:943. https://doi.org/10.3390/molecules24050943

Shiran HS, Baghbanbashi M, Ahsaie FG, Pazuki G (2020) Study of curcumin partitioning in polymer-salt aqueous two phase systems. J Mol Liq 303:112629. https://doi.org/10.1016/j.molliq.2020.112629

Shokrollahi M, Rezakazemi M, Younas M (2020) Producing water from saline streams using membrane distillation: modeling and optimization using CFD and design expert. Int J Energy Res 44(11):8841–8853. https://doi.org/10.1002/er.5578

Tan J, Cui P, Ge S et al (2022) Ultrasound assisted aqueous two-phase extraction of polysaccharides from *Cornus officinalis* fruit: modeling, optimization, purification, and characterization. Ultrason Sonochem 84:105966. https://doi.org/10.1016/j.ultsonch.2022.105966

Tatke P, Rajan M (2014) Comparison of conventional and novel extraction techniques for the extraction of scopoletin from *Convolvulus pluricaulis*. IJPER 48(1):27–31. https://doi.org/10.5530/ijper.48.1.5

Toledo MO, Farias FO, Igarashi-Mafra L, Mafra MR (2019) Salt effect on ethanol-based aqueous biphasic systems applied to alkaloids partition: an experimental and theoretical approach. J Chem Eng Data 64(5):2018–2026. https://doi.org/10.1021/acs.jced.8b01024

Tursynbolat S, Bakytkarim Y, Huang J, Wang L (2019) Highly sensitive simultaneous electrochemical determination of myricetin and rutin via solid phase extraction on a ternary Pt@r-GO@MWCNTs nanocomposite. J Pharm Anal 9(5):358–366. https://doi.org/10.1016/j.jpha.2019.03.009

Uzoejinwa BB, He X, Wang S et al (2019) Co-pyrolysis of macroalgae and lignocellulosic biomass. J Therm Anal Calorim 136:2001–2016. https://doi.org/10.1007/s10973-018-7834-2

Vergara-Jimenez M, Almatrafi MM, Fernandez ML (2017) Bioactive components in *M. oleifera* leaves protect against chronic disease. Antioxidants 6(4):91. https://doi.org/10.3390/antiox6040091

Vetrova EV, Maksimenko EV, Borisenko SN (2017) Extraction of Rutin and quercetin antioxidants from the buds of Sophora Japonica (*Sophora japonica* L.) by subcritical water. Russ J Phys Chem B 11(7):1202–1206. https://doi.org/10.1134/S1990793117070193

Vo TP, Ho TAT, Ha NMH et al (2024) Recovering bioactive compounds from yacon (*Smallanthus son-chifolius*) using the ultrasonic-microwave-assisted extraction technique. Appl Food Res 4:100451. https://doi.org/10.1016/j.afres.2024.100451

Wojdylo A, Nowicka P (2019) Anticholinergic effects of *Actinidia arguta* fruits and their polyphenol content determined by liquid chromatography-photodiode array detector-quadrupole time of flight-mass spectrometry (LC-MS-PDA-Q/TOF). Food Chem 271:216–223. https://doi.org/10.1016/j.foodchem.2018.07.084

Xi J, Zhou X, Wang Y, Wei S (2023) Short-chain alcohol/salt-based aqueous two-phase system as a novel solvent for extraction of plant active ingredients: a review. Trends Food Sci Technol 138:78–84. https://doi.org/10.1016/j.tifs.2023.06.001

Xie WJ, Gao YQ (2013) A simple theory for the Hofmeister series. J Phys Chem Lett 4:4247–4252. https://doi.org/10.1021/jz402072g

Yang H, Wang C, Zhang L et al (2019) Rutin alleviates hypoxia/reoxygenation-induced injury in myocardial cells by up-regulating SIRT1 expression. Chem Biol Interact 297:44–49. https://doi.org/10.1016/j.cbi.2018.10.016

Zhou S, Wu X, Huang Y et al (2018) Microwave-assisted aqueous two-phase extraction of alkaloids from *Radix Sophorae Tonkinensis* with an ethanol/Na_2HPO_4 system: process optimization, composition identification and quantification analysis. Ind Crop Prod 122:316–328. https://doi.org/10.1016/j.indcrop.2018.06.004

Exploratory Correlation of De Novo Transcriptome Data and GC-MS Volatile Essential Oils Profile of *W. salutaris* Leaves

Vuyiseka Nkqenkqa, Vanessa A. Jackson, and Richard Mundembe

Abstract

Warburgia salutaris, a medicinal plant native to southern Africa, is renowned for its essential oil, which is rich in monoterpenes and sesquiterpenes with therapeutic properties. This study explores the correlation between de novo transcriptome data and the volatile essential oil profile of *W. salutaris* leaves to elucidate the gene-metabolite relationships involved in secondary metabolite biosynthesis. Essential oil was extracted from fresh leaves using hydrodistillation and analysed via gas chromatography-mass spectrometry (GC-MS), identifying 34 volatile compounds, including monoterpenes and sesquiterpenes. To explore the genetic pathways of terpene biosynthesis, a transcriptome was generated using RNA sequencing, yielding over 79 million raw reads, with 60% usable data. The assembly produced 55,687 transcripts, of which 68,004 were coding sequences. Functional annotation through the Kyoto Encyclopedia of Genes and Genomes (KEGG) database identified genes encoding mono- and sesquiterpene synthases. However, not all metabolites in the essential oil were represented by corresponding genes, as some terpenes act as precursors. Nine of the 34 identified compounds were directly or indirectly linked to transcriptome data. This first comparative study of *W. salutaris* essential oil and de novo transcriptome data provides valuable insights into the interplay between gene transcription and metabolite production, highlighting the complexity of plant metabolism.

V. Nkqenkqa (✉)
Department of Environmental Health and Occupational Studies, Faculty of Applied Sciences, Cape Peninsula University of Technology, Cape Town, South Africa

Department of Biotechnology and Consumer Science, Faculty of Applied Sciences, Cape Peninsula University of Technology, Cape Town, South Africa
e-mail: nkqenkqav@cput.ac.za

V. A. Jackson · R. Mundembe
Department of Biotechnology and Consumer Science, Faculty of Applied Sciences, Cape Peninsula University of Technology, Cape Town, South Africa
e-mail: jacksonva@cput.ac.za; mundember@cput.ac.za

Keywords

Warburgia salutaris · Chemical composition · Terpenes · Transcriptome

3.1 Introduction

Warburgia salutaris (Pepper bark) is a tree species that belongs to the Canellaceae family. The tree is found in eastern and southern African countries. In South Africa, it is distributed in

© The Author(s) 2026
A. Shonhai et al. (eds.), *Advances in Biochemistry and Molecular Biology to meet Africa´s Needs*, Advances in Experimental Medicine and Biology 1507, https://doi.org/10.1007/978-3-032-24254-9_3

KwaZulu-Natal, Limpopo and Mpumalanga provinces (Harvey-Brown et al. 2022). Among other uses, the plant is known for its medicinal value and is used to treat various human and animal ailments as detailed in comprehensive reviews by Maroyi (2014), Leonard and Viljoen (2015) and Meddows-Taylor and Ramadwa (2025). *Warburgia salutaris* exhibits secondary metabolites from a class of terpenoids, which contribute to the plant's therapeutic value (Abuto et al. 2016).

Terpenoids comprise one of the most diverse groups of plant secondary metabolites, including monoterpenes (C_{10}), sesquiterpenes (C_{15}), diterpenes (C_{20}), triterpenes (C_{30}), tetraterpenes (C_{40}) and polyterpenes as well as steroids (C_{27}), which are derived from triterpenes (Wink 2015). More than 40,000 different structures of terpenoids have been identified (Yu and Utsumi 2009), and over 80,000 terpene compounds have been produced from different living organisms (Christianson 2017). Terpenoids are highly diverse in chemical structure, exhibiting hundreds of different carbon skeletons. However, they share a common feature of biosynthesis, and their classification is based on the number of common five-carbon isoprene units in their skeletal structure (Yu and Utsumi 2009).

The classes of terpenes of interest in this study are monoterpenes and sesquiterpenes. These classes include numerous volatile essential oil compounds (Feng-qi et al. 2017) that plants emit in response to herbivores, serving as direct defences by deterring or harming herbivores upon ingestion, or as indirect defences by attracting their natural enemies through induced volatiles (Kollner et al. 2008). The volatile compounds also play a crucial role in influencing pollinators and seed dispersers, either attracting or repelling them (Kollner et al. 2008). Furthermore, the volatile compounds from plants are known to exhibit antimicrobial properties. Khumalo et al. (2019) tested the antimicrobial activity of volatile essential oil compounds from *W. salutaris* and found them to be effective against skin pathogens, *Pseudomonas aeruginosa* ATCC 743971 and *Staphylococcus aureus* ATCC 25923, and respiratory tract pathogens *Klebsiella pneumoniae*

ATCC 13883 and *Moraxella catarrhalis* ATCC 23246. This is probably why the plant is traditionally used to treat various ailments and illnesses caused by these disease agents. To understand the biosynthesis of the volatile compounds found in *W. salutaris*, it is important to examine the role of terpene synthases (TPSs), the enzymes that drive their formation.

Many genes are involved in the biosynthesis of terpenoids. The unique backbone of the terpenoids is produced under the catalysis of different kinds of terpenoid synthases (TPSs), including monoterpene and sesquiterpene synthases. Terpene synthases are the primary enzymes responsible for catalysing the formation of monoterpenes (C_{10}), sesquiterpenes (C_{15}) or diterpenes (C_{20}) from the substrates geranyl diphosphate (GPP), farnesyl diphosphate (FPP) or geranylgeranyl diphosphate (GGPP), respectively. It is generally recognised that the mevalonate (MVA) pathway in the cytosol is responsible for the synthesis of sesquiterpenes, phytosterols and ubiquinone. In contrast, monoterpenes, gibberellins, abscisic acid, carotenoids and the prenyl moiety of chlorophylls, plastoquinone and tocopherol are produced in plastid-localised 2-C-methyl-d-erythritol 4-phosphate (MEP) pathway (Lange and Ahkami 2013).

In both MVA and MEP pathways, the upstream enzymes catalyse the synthesis of IPP and DMAPP, which are the precursors of all terpenoids (Bergman et al. 2024). Downstream, terpene synthases (TPSs) play a critical function in the synthesis of diverse terpenoids. Other important enzymes are involved in the biosynthesis of terpenes, including the prenyltransferases that condense IPP and DMAPP to produce larger prenyl diphosphates as well as terpene-modifying enzymes such as cytochrome P450 monooxygenases (CYPs) (Wang et al. 2025). The regulation of terpenoid biosynthesis in plants is intricate, likely resulting from the diverse roles of terpenoids during plant development and in response to biotic and abiotic factors (Hu et al. 2024).

Elucidation of genes coding for these synthases will lead to a better understanding of the pathways and open possibilities for biotechnological manipulation. Transcriptomic approaches are more comprehensive in analysing gene expression. A transcriptome is defined as a complete set of RNA transcripts in a cell produced under the specified conditions (McGettigan 2013). This high-throughput approach allows the detection of lowly expressed transcripts and identifies new genes, exons and transcript isoforms (McGettigan 2013).

Little genomic work has been done on the genus *Warburgia*. Wang et al. (2015) utilised the Illumina sequencing instrument to generate tran-scriptomic and genomic data from *W. ugandensis* and genomic data from *W. salutaris*. The de novo transcriptome data derived from the bark of *W. ugandensis* included genes encoding enzymes in terpenoid and polyunsaturated fatty acid biosyn-thetic pathways. Except for the transcriptome derived from *W. ugandensis* [sequence read archive (SRA) SRP056452 on NCBI with acces-sion SRX970743] by Wang et al. (2015), there is no other transcriptome data available on the pub-licly available data from the genus *Warburgia* from NCBI (as of 10 January 2025).

The objectives of the study were to identify the volatile compounds present in *W. salutaris* using GC-MS and to discover, identify and char-acterise genes that encode enzymes involved in the terpenoid biosynthesis pathways, specifically monoterpene and sesquiterpene.

3.2 Materials and Methods

3.2.1 Sample Collection, Hydrodistillation and GC-MS Analysis

Fresh mature leaves of *W. salutaris* [accession 176/04 (P3)] were collected from Kirstenbosch Botanical Garden—South African National Botanical Institute (SANBI) in Cape Town. Upon collection, for transcriptomic analyses, the leaves were immediately placed in liquid nitrogen, while those for GC-MS analyses were kept at ambient temperature and analysed on the same day.

The Clevenger-type apparatus was used for hydrodistillation of volatile essential oils. The oils were analysed using GC-MS at Central Analytical Facilities (CAF) at Stellenbosch University. Briefly, the sample was diluted with hexane (1:10), and 1 µl was injected into the GC operated at a 50:1 split ratio. The separation of essential oils was performed on a gas chromato-graph (6890N, Agilent Technologies Network) coupled to an Agilent Technologies inert XL EI/CI Mass Selective Detector (MSD) (5975B, Agilent Technologies Inc., Palo Alto, CA). The GC-MS system was coupled to a CTC Analytics PAL autosampler. Separation of the essential oils was performed in a non-polar ZB-5Ms (30 m, 0.25 mm ID, 0.25 µm film thickness) capillary column. The mass spectrometer was operated under electron impact (EI) mode at an ionisation energy of 70 eV, scanning from 35 to 500 m/z. Helium was used as a carrier gas at a flow rate of 1 mL/min.

3.3 Whole Transcriptome Sequencing

3.3.1 Total RNA Extraction and cDNA Library Preparation

Total RNA was extracted from a lysate with CTAB buffer using the NucleoSpin® RNA Plant kit (Macherey-Nagel, Düren, Germany). The RNA integrity (RINe) was assessed on the TapeStation 4150, using the RNA ScreenTape assay (Agilent Technologies, Waldbronn, Germany) according to the protocol, G2991-90021 Rev. B. The concentration of the RNA was determined by fluorometry on the Qubit 4 using the Qubit RNA high sensitivity assay kit (ThermoFisher Scientific). To assess the level of DNA contamination, the Qubit 1 x dsDNA high-sensitivity assay was used (ThermoFisher Scientific). The quality and quan-tity of RNA yielded were measured using spec-trophotometry. Upon enrichment for mRNA, the cDNA library was constructed on the Ion Torrent™ S5™ system according to the protocol, MAN0010654 REV D.0.

3.3.2 Sequence Analysis, Assembly and Bioinformatic Analysis

The sequencing was performed on the Ion Torrent™ GeneStudio™ S5 Prime System using sequencing solutions and reagents according to the protocol, MAN0010851 REVF.0. Clean reads were assembled into transcripts using the SPAdes assembler. BUSCO analysis on Galaxy was conducted to obtain an indication of the extent to which the data represent the plant's transcriptome. Augustus for gene prediction was also run on the transcripts. The Kyoto Encyclopedia of Genes and Genomes (KEGG) database was used for functional annotation in the form of KEGG Orthology (KO) assignment using the BlastKOALA and GhostKOALA annotation by BLASTP and GHOSTX searches, respectively.

3.4 Results

The chemical profile of the essential oils from mature leaves of *W. salutaris* was analysed using the gas chromatography-mass spectrometry (GC-MS) technique. A total of 34 compounds were detected, of which only one (peak 28) is unknown, and the respective retention times and relative abundances are shown in Table 3.1. The essential oils extracted consist of terpenes, specifically monoterpenes and sesquiterpenes. The compounds detected, including the unknown, accounted for 94.79% of the total oil content.

Monoterpenes were the first to be eluted on the column, followed by sesquiterpenes based on retention times in minutes, as shown in Fig. 3.1. Eight monoterpene compounds were found in the essential oils of *W. salutaris,* as depicted from peaks 1–8 out of the 34 peaks (Fig. 3.1). Nonetheless, the monoterpene β-pinene (15,96%) was the major compound found in the essential oil extract, followed by (*E*)-β-ocimene (13,26%) and (*Z*)-ocimene (9,86%) (Fig. 3.1). This explains why the relative abundance of monoterpenes is almost half of the volume of the essential oils (42.95%). Sesquiterpenes, on the other hand, were more diverse, making up 26 of the 34 compounds. This, however, accounted for 51.84% of

relative abundance. The major compounds for sesquiterpenes were (*E*)-β-farnesene (8,59%), followed by β-caryophyllene (7,78%), β-cubebene (7,60%) and nerolidol (5,90%).

Previous studies have demonstrated that the hydrodistillation of plant essential oils and their analysis using GC-MS reveal that monoterpenes and sesquiterpenes are the major constituents of the oil content. Lawal et al. (2014) obtained a significantly lower number of compounds (23) in the essential oil of *W. salutaris* compared to the current study. The difference could be attributed to the use of air-dried leaves instead of fresh leaves and the release of certain volatile compounds before the hydrodistillation process. The major oil constituents obtained by Lawal et al. (2014) using GC-MS were myrcene (27.5%), limonene (16.9%), and (*E*)-β-ocimene (11.1%) and (*Z*)-β-ocimene (9.5%) amongst monoterpenes, while α-humulene (5.6%) and β-caryophyllene (4.7%) were the only prominent sesquiterpenes identified in the oil. These findings by Lawal et al. (2014) are consistent with those of Leonard et al. (2020), even though fresh material was used for hydrodistillation, where myrcene (0.6–65.3%), (*E*)-β-ocimene (not detected (nd)–56.9%) and (*Z*)-β-ocimene (nd–19.1%) were the dominant monoterpenes. On the other hand, sesquiterpenes were present at lower levels than monoterpenes, with the most abundant sesquiterpenes identified to be β-caryophyllene (nd–9.0%), α-humulene (nd–8.3%), germacrene D (0.1–6.9%) and (*E*)-α-farnesene (0.3–6.3%) (Leonard et al. 2020). The profile of oil composition in the current study is similar to the previous studies, where monoterpenes were present in greater abundance than sesquiterpenes.

3.4.1 Total RNA Quality, Transcriptome Sequencing and De Novo Assembly

The RNA quality control (QC) for RNA integrity (RINe) cut-off for RNASeq from plant leaves (or green material) is >5.0. The *W. salutaris* sample had good-quality RNA with an RINe of 5.2. The

Table 3.1 GC-MS analysis of essential oils composition from leaves of *W. salutaris*

Terpene class	No. of peaks	Retention time (min)	Library identification	Relative abundance of compounds (area%)
Monoterpenes	1	10,6147	β-Pinene	15,96
	2	12,2216	(Z)-Ocimene	9,86
	3	12,6021	(E)-β-Ocimene	13,26
	4	14,3159	Linalool L	2,65
	5	14,363	3-methylene-6-hepten-2-one	0,10
	6	15,1371	Allo-ocimene	0,26
	7	17,452	4,5-epoxy-1-isopropyl-4-methyl-1-cyclohexene	0,24
	8	17,656	7,7-dimethyl-2-methoxy norborn-2-ene	0,62
Sesquiterpenes	9	22,1527	α-Copaene[a]	0,67
	10	22,3438	β-Bourbonene	0,18
	11	22,5043	Germacrene-D[a]	0,33
	12	22,5364	β-Elemene	0,19
	13	23,3101	β-Caryophyllene[a]	7,78
	14	24,2641	(E)-β-Farnesene	8,59
	15	24,8737	β-Cubebene	7,60
	16	25,1674	α-Bergamotene	2,05
	17	25,2895	α-Muurolene[a]	0,23
	18	25,4941	(E,E)-α-Farnesene	1,42
	19	25,6254	α-Amorphene[a]	0,23
	20	25,767	δ-Cadinene*	1,44
	21	26,8916	Nerolidol	5,90
	22	27,1716	(+) Spathulenol	0,74
	23	27,2537	Caryophyllene oxide	1,12
	24	27,9112	Liguhodgsonal	0,38
	25	28,2293	Longifolenaldehyde	0,29
	26	28,6649	δ-Cadinene[a]	0,42
	27	28,9585	T-Cadinol	0,65
	28	29,377	Unknown	3,80
	29	29,8288	Farnesol 2[a]	0,41
	30	30,3359	Farnesol Isomer A[a]	0,37
	31	30,8345	(3S,4R,5R,6R)-4,5-Bis(hydroxymethyl)-3,6-dimethylcyclohexene	0,67
	32	31,0115	(+/−)-(1RS,4aRS,8aRS)-decahydro-5,5,8a-trimethyl-2-methylene-1-naphthylmethanol	1,30
	33	31,4231	Drimenol	4,75
	34	31,9704	7,11-Epoxyisogermacrone	0,33

[a]Compounds that are directly or indirectly linked to the transcriptome data

concentrations of RNA and DNA, measured using the Qubit fluorometric quantification, were 994 ng/μL and 60.2 ng/μL, respectively. The RNA concentration was high (1410 ng/μL) and free of impurities as indicated by a 260/280 ratio of 2.18 and a 260/230 ratio of 2.26. Less than 10% of the extracted material was DNA. The test passed the QC threshold, and the total amount of RNA isolated was more than 5000 ng, which was sufficient for the ribosomal-subunit depletion step during library construction. The quality and quantity of RNA were sufficient for library construction and sequencing.

The cDNA library from the leaf of *W. salutaris* was obtained at a concentration of 114 nM, which is considered sufficient for template preparation and enrichment. The Ion Torrent™ Prime produced more than 79 million raw reads, of which 60% were usable (Fig. 3.2). The total size of the clean bases generated was 8.45 GB, with a

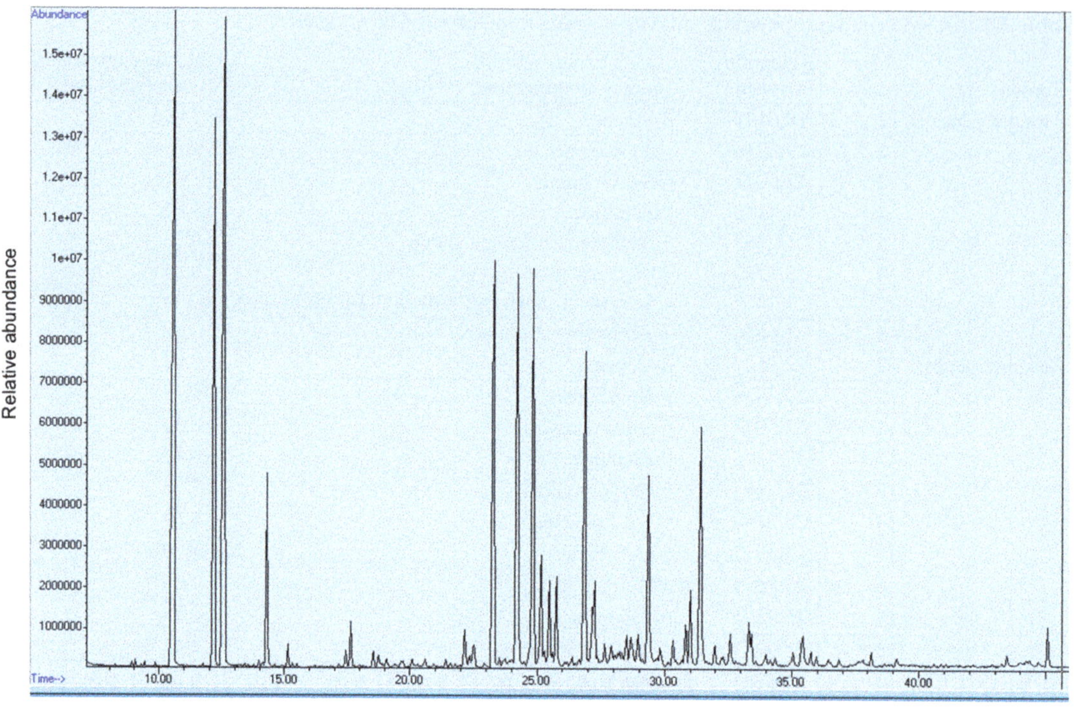

Fig. 3.1 GC-MS chromatogram of *W. salutaris* leaf essential oils. The numbers correspond to the peaks for the compounds detected, as shown in Table 3.1

mean read length of 106 bp and a median of 100 bp (Fig. 3.2). Table 3.2 shows a summary of the assembly statistics of the transcriptome, where usable reads were assembled into a total of 55,687 transcripts, of which 20,545 were more than or equal to 1 Kbp with a total size of at least 30 Mbp. Furthermore, 10 transcripts were more than or equal to 5 Kbp with a total size of more than 59 Kbp. The largest contig was 8138 bp and the GC content was 42.86%, with a N50 of 1085 bp. These results showed that the throughput and sequencing quality were sufficiently high for further bioinformatics analyses.

3.4.2 Functional Annotation and Classification of Transcripts

A web-based analysis platform known as Galaxy. eu was used to analyse the *W. salutaris* transcriptomic data. The transcripts were run using the Augustus command to predict gene sequences.

The 55,687 transcripts corresponded to 68,004 coding sequences, each of which had a corresponding protein sequence. The sequences were searched against the KEGG pathway database to better understand the molecular interaction, reaction, and relation network for pathways. The KEGG database was also used for functional annotation in the form of KEGG Orthology (KO) assignment using the BlastKOALA and GhostKOALA annotation by BLASTP and GHOSTX searches, respectively. The automatic server by BLASTP searches for sequence similarity against the reduced database, while GHOSTX searches against a nonredundant set of KEGG genes (Kanehisa et al. 2016).

All the pathways that are found on the KEGG database were represented by some sequence/s from the transcriptome. However, not all the sequences from the transcriptome were annotated for identification. Our interest was focused on volatile essential oil compounds, specifically monoterpenes and sesquiterpenes. The elucidation of biosynthetic pathways is essential for

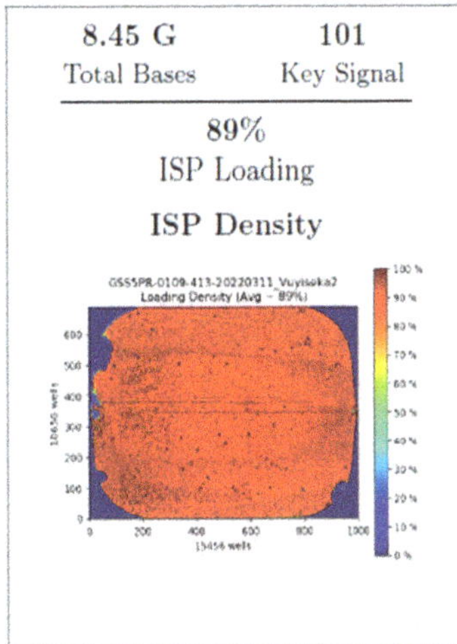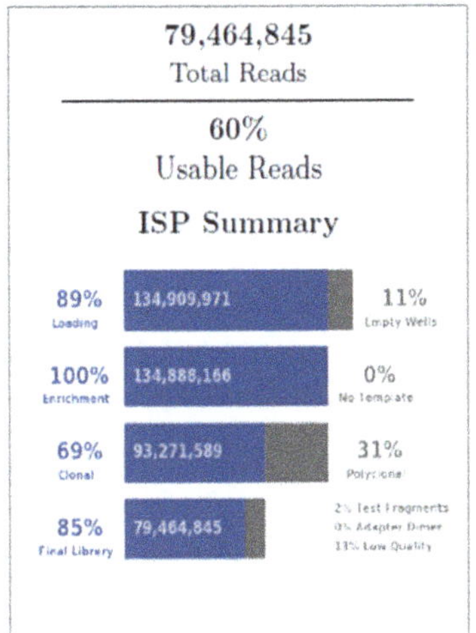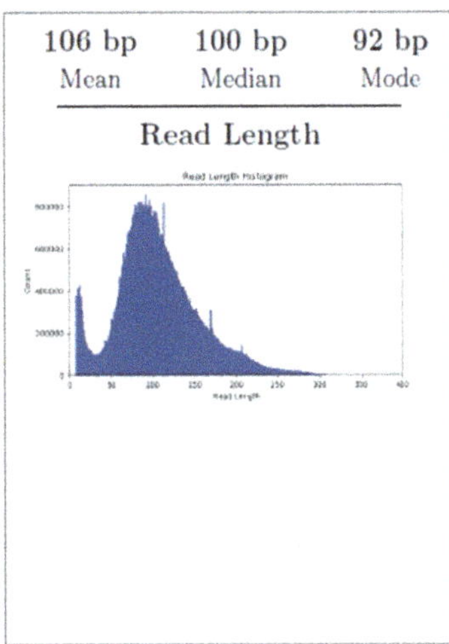

Fig. 3.2 Run summary and statistics of the sequencing and assembly of the *W. salutaris* transcriptome

Table 3.2 Assembly statistics of the *W. salutaris* transcriptome

Contigs	Transcripts
# contigs (> = 0 bp)	55687
# contigs (> = 1000 bp)	20545
# contigs (> = 5000 bp)	10
# contigs (> = 10,000 bp)	0
# contigs (> = 25,000 bp)	0
# contigs (> = 50,000 bp)	0
Total length (> = 0 bp)	55600484
Total length (> = 1000 bp)	30880305
Total length (> = 5000 bp)	59184
Total length (> = 10,000 bp)	0
Total length (> = 25,000 bp)	0
Total length (> = 50,000 bp)	0
# contigs	55356
Largest contig	8138
Total length	55435655
GC (%)	42.86
N50	1085
N75	757
L50	17506
L75	32868

All statistics are based on contigs of size > = 500 bp, unless otherwise noted (e.g., "# contigs (>= 0 bp)" and "Total length (>= 0 bp)" include all contigs)

obtaining expressed genes, which are the candidates for key pathway enzymes in plant products.

Of the 68,004 coding and protein sequences produced, only 32.2% (21920) were annotated using the KEGG database. As shown in Fig. 3.3, the functional categories of the annotated sequences include Protein families: genetic information processing (4672), Genetic information processing (4273), Carbohydrate metabolism (1774), Protein families: metabolism (1531), Environmental information processing (1464), Protein families: signalling and cellular processes (1366), Cellular processes (1312), Organismal systems (793), Lipid metabolism (643), Amino acid metabolism (631), Energy metabolism (518), Metabolism of cofactors and vitamins (481), Human diseases (436), Glycan biosynthesis and metabolism (365), Metabolism of terpenoids and polyketides (263), Nucleotide metabolism (262), Biosynthesis of other secondary metabolites (157), Unclassified: signalling and cellular processes (140), Metabolism of other amino acids (123), Unclassified: genetic information processing (44), 617 sequences were under unclassified metabolism and 52 sequences were unclassified.

3.4.3 Genes Related to Terpenoid Biosynthesis in *W. salutaris*

Using the de novo assembled transcriptome, KEGG functional categories revealed that the terpene backbone biosynthesis was represented by

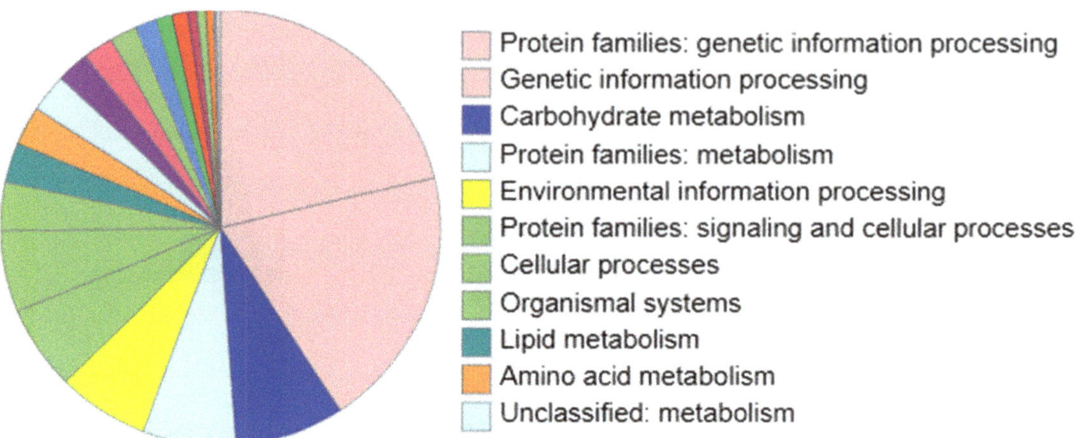

Fig. 3.3 Functional categories of the annotated sequences

31 genes from both the MVA (cytosol) and MEP (plastids) pathways (Fig. 3.4). These genes encode enzymes involved in the biosynthesis of key intermediates of the terpene pathway. The expressed key intermediate (upstream) genes align with the production of specific metabolites that produce specific terpene compounds. These 31 transcripts that map onto the terpenoid backbone de novo biosynthetic pathway are important and effectively validate the transcriptome. Our specific interest in this investigation was in specific metabolites that accumulate towards the end of the pathways, and to see if mRNA transcripts for their synthases are detectable as part of the transcriptome.

The number of genes annotated in terpenoid backbone biosynthesis in *W. ugandensis* by Wang et al. (2015) was higher (53) than those found in the current study (31). As noted by Wang et al. (2015), the low number of genes annotated seems reasonable for a medicinal plant due to its complex secondary metabolites. Also, the relatively low number of annotated genes in this plant species is probably due to insufficient publicly available plant transcriptome and genome data, especially on the Canellaceae family.

For the downstream terpenoid pathways, the monoterpenoid biosynthesis pathway was represented by two genes coding for 8-hydroxygeraniol dehydrogenase (K23232, EC:1.1.1.324) and linalool 8-monooxygenase (K05525, EC:1.14.14.84). 8-hydroxygeraniol dehydroge-

nase uses 8-hydroxygeraniol as its substrate and converts it to 8-oxogeranial, a precursor for other monoterpene products, which is ultimately used as a substrate for iridoid synthase, forming the iridoid ring of iridoids (Kries et al. 2016). This substrate is an important natural compound of pharmacological importance (Kamatou and Viljoen 2008). Linalool 8-monooxygenase catalyses the conversion of linalool into various compounds such as 8-hydroxylinalool and 8-oxolinalool, which have important fragrances and pharmacological applications (Kamatou and Viljoen 2008).

In addition, four other genes were annotated to the sesquiterpene biosynthesis pathway enzymes, NAD+-dependent farnesol dehydrogenase (K15891, EC:1.1.1.354), (-)-germacrene D synthase (K15803, EC:4.2.3.75), premnaspirodiene oxygenase (K15472, EC:1.14.14.151) and α-humulene synthase/beta-caryophyllene synthase (K14184, EC:4.2.3.104, 4.2.3.57). The enzyme NAD+-dependent farnesol dehydrogenase catalyses the conversion of farnesol into farnesal, a metabolite that is used in pest control and insecticide development (Mayoral et al. 2009). The enantiomer (-)-germacrene D synthase catalyses the general precursor of sesquiterpenes, farnesyl diphosphate, to produce (-)-germacrene D, a flavour compound that can act as a plant defence chemical, insect deterrent, pheromone or as an antibiotic (Schmidt et al. 1999). Premnaspirodiene oxygenase catalyses

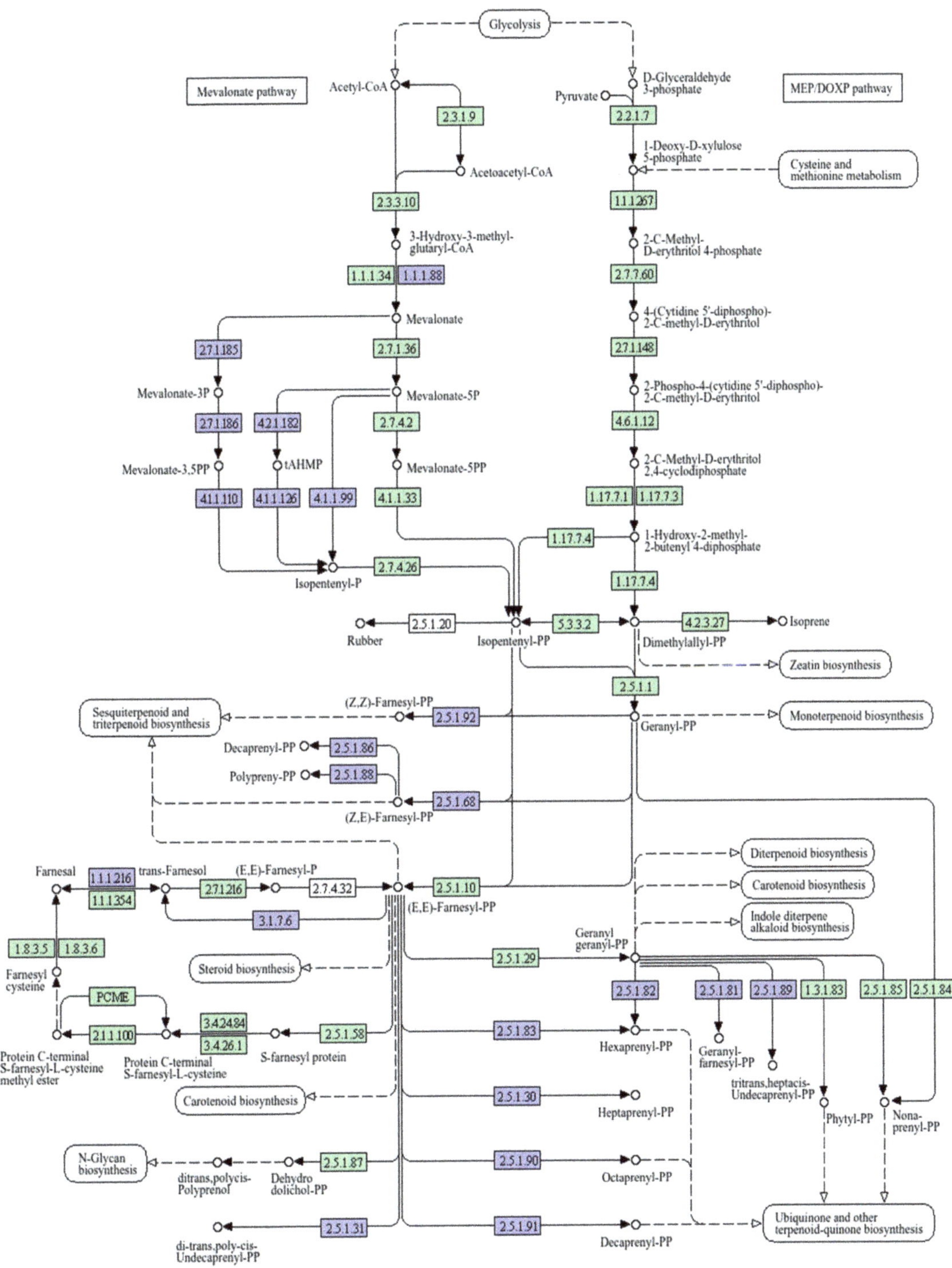

Fig. 3.4 Terpenoid backbone biosynthesis from the KEGG pathway. Organism-specific pathway: green boxes (34) are hyperlinked to GENES entries by converting K numbers (KO identifiers) to gene identifiers in the reference pathway, indicating the presence of genes in the genome and the completeness of the pathway. Boxes and arrows show standard KEGG annotations. EC 1.17.7.4 is repeated, 2.5.1.84 and 2.5.1.85 both are K05356 (SPS), and 1.8.3.5 and 1.8.3.6 both are K05906 (PCYOX1). This leaves a total of 31 genes in the terpenoid backbone biosynthesis

the conversion of premnaspirodiene to solavetivone. Solavetivone is a potent antifungal alexin with potential applications in agriculture for plant defence, pest repellent and pesticide formulations (Takahashi et al. 2007). It also has potential value in pharmacology due to its anti-inflammatory and antimicrobial action and has been investigated for anti-cancer activity (Takahashi et al. 2007). Solavetivone may also potentially be used as a natural preservative in foods and cosmetics. Alpha humulene synthase catalyses the conversion of farnesyl diphosphate to α-humulene, which has an earthy, woody and slightly spicy aroma (de Lacerda Leite et al. 2021). It exhibits anti-inflammatory and antimicrobial properties and has been investigated for its potential anti-cancer effects (de Lacerda Leite et al. 2021). Other classes of terpenoids, such as diterpenoid and triterpenoid biosynthesis, were also represented. In many instances, sequences that appeared to be full-length genes had isoforms or truncated gene segments that lacked nucleotide sequences corresponding to the start or stop codons.

As noted in the current study, only six genes coding for enzymes in the downstream terpenoid pathway were annotated, and these explain nine metabolites. When correlating the transcriptome data with GC-MS data, the number of annotated transcripts is significantly lower than the number of compounds detected by GC-MS. This could be attributed to the fact that some monoterpene and sesquiterpene compounds are known to be precursors of other terpene products into which they spontaneously rearrange. For example, Schmidt et al. (1998) noted that a sesquiterpene compound like germacrene D is a precursor for some other sesquiterpenes. In such cases, precursor germacrene D may not accumulate to significant levels and, as such, may fail to be detected by GC-MS analysis. A study conducted by Bülow and König (2000) on the rearrangement of germacrene D through acid catalysis, light, and heat produced other sesquiterpene rearranged products such as cadinene, muurolene, amorphene and selinene. In other species, however, such as *Solidago*, germacrene D is not a metabolic pathway intermediate and will accumulate as a final product (Schmidt et al. 1998).

In this paper, we report nine sesquiterpene compounds linked directly or indirectly to the transcriptome data. Where GC-MS fails to correlate with transcriptome data, the reason might be due to other phenomena, such as spontaneous chemical rearrangement of molecules, or molecular genetic phenomena such as messenger RNA instability.

3.5 Conclusion

This work demonstrates a correlation between the GC-MS data and transcriptome data. Nine out of 34 compounds were directly or indirectly linked to the de novo transcriptome data. Although the number of compounds with transcripts was low, it is understood that not all corresponding transcripts for some metabolites may be detected because of various chemical and genetic phenomena. The transcriptome was of good quality with more than 55 thousand transcripts, which produced more than 68 thousand coding and protein sequences, 31 of which were terpene backbone biosynthesis genes. Downstream from these, we had 6 genes: 8-hydroxygeraniol dehydrogenase (EC:1.1.1.324) and linalool 8-monooxygenase (EC:1.14.14.84], as well as NAD+-dependent farnesol dehydrogenase (EC:1.1.1.354), (-)-germacrene D synthase (EC:4.2.3.75), premnaspirodiene oxygenase (EC:1.14.14.151) and α-humulene synthase/beta-caryophyllene synthase (EC:4.2.3.104, 4.2.3.57) and these catalyse reactions that produce bioactive monoterpenes and sequiterpenes compounds. The platform demonstrated here will contribute towards easier cloning and overexpression of these compounds in heterologous systems.

Conflict of Interest Author VN declares that she has no conflict of interest. Author VAJ declares that she has no conflict of interest. Author RM declares that he has no conflict of interest.

Ethical Approval This chapter does not contain any studies with human participants or animals performed by any of the authors.

Funding This study was funded by the Council for Scientific and Industrial Research (CSIR) of South Africa, the Kirstenbosch National Botanical Garden—SANBI provided the plant material, and Stellenbosch University—CAF facilitated sequencing, transcriptome assembly, and GC-MS analysis.

References

Abuto JO, Muchugi A, Mburu D et al (2016) Variation in antimicrobial activity of *Warburgia ugandensis* extracts from different populations across the Kenyan Rift Valley. J Microbiol Res 6(3):55–64

Bergman ME, Kortbeek RW, Gutensohn et al (2024) Plant terpenoid biosynthetic network and its multiple layers of regulation. Prog Lipid Res 95:101287

Bülow N, König WA (2000) The role of germacrene D as a precursor in sesquiterpene biosynthesis: investigations of acid catalyzed, photochemically and thermally induced rearrangements. Phytochemistry 55(2):141–168

Christianson DW (2017) Structural and chemical biology of terpenoid cyclases. Chem Rev 117(17):11570–11648

de Lacerda Leite GM, de Oliveira Barbosa M, Lopes MJP et al (2021) Pharmacological and toxicological activities of α-humulene and its isomers: a systematic review. TIFS 115:255–274

Feng-qi L, Ning-ning F, Jing-jiang Z et al (2017) Functional characterization of (*E*)-β-caryophyllene synthase from lima bean and its up-regulation by spider mites and alamethicin. J Integr Agric 16(10):2231–2238

Harvey-Brown Y, Mhlongo NN, Raimondo D et al (2022) Warburgia salutaris (G. Bertol.) Chiov. National Assessment: red list of South African Plants version. Accessed on 2025/07/21Supplementary materials

Hu Y, Zheng T, Dong J et al (2024) Regulation of the main terpenoids biosynthesis and accumulation in fruit trees. Hortic Plant J. https://doi.org/10.1016/j.hpj.2024.08.002

Kamatou GP, Viljoen AM (2008) Linalool–a review of a biologically active compound of commercial importance. Nat Prod Commun 3(7):1934578X0800300727

Kanehisa M, Sato Y, Kawashima M et al (2016) KEGG as a reference resource for gene and protein annotation. Nucleic Acids Res 44(D1):D457–D462

Khumalo GP, Sadgrove NJ, Van Vuuren S et al (2019) Antimicrobial activity of volatile and non-volatile isolated compounds and extracts from the bark and leaves of *Warburgia salutaris* (Canellaceae) against skin and respiratory pathogens. S Afr J Bot 122:547–550

Kollner TG, Held M, Lenk C et al (2008) A maize (E)-β-caryophyllene synthase implicated in indirect defense responses against herbivores is not expressed in most American maize varieties. Plant Cell 20(2):482–494

Kries H, Caputi L, Stevenson CE et al (2016) Structural determinants of reductive terpene cyclization in iridoid biosynthesis. Nat Chem Biol 12(1):6–8

Lange BM, Ahkami A (2013) Metabolic engineering of plant monoterpenes, sesquiterpenes and diterpenes—current status and future opportunities. Plant Biotechnol J 11:169–196

Lawal OA, Ogunwande IA, Opoku AR et al (2014) Chemical composition and antibacterial activities of essential oil of *Warburgia salutaris* (BertolF.) Chiov. from South Africa. J Biol Act Prod Nat 4(4):272–277

Leonard C, Kamatou G, van Vuuren S et al (2020) Essential oil variation within *Warburgia salutaris*– a coveted Ethnomedicinal aromatic tree. Chem Biodivers 17(11):e2000542

Leonard CM, Viljoen AM (2015) Warburgia: a comprehensive review of the botany, traditional uses and phytochemistry. J Ethnopharmacol 165:260–285

Maroyi A (2014) The genus *Warburgia*: a review of its traditional uses and pharmacology. Pharm Biol 52(3):378–391

Mayoral JG, Nouzova M, Navare A et al (2009) NADP+-dependent farnesol dehydrogenase, a corpora allata enzyme involved in juvenile hormone synthesis. PNAS 106(50):21091–21096

McGettigan PA (2013) Transcriptomics in the RNA-seq era. Curr Opin Chem Biol 17(1):4–11

Meddows-Taylor S, Ramadwa TE (2025) A comprehensive review of the traditional uses, pharmacological activity and phytochemistry of *Warburgia salutaris* in southern Africa. S Afr J Bot 179:134–146

Schmidt CO, Bouwmeester HJ, de Kraker JW (1998) Biosynthesis of (+)-and (−)-germacrene D in Solidago canadensis: isolation and characterization of two enantioselective germacrene D synthases. Angew Chem Int Ed 37(10):1400–1402

Schmidt CO, Bouwmeester HJ, Franke S (1999) Mechanism of the biosynthesis of sesquiterpene enantiomers (+)- and (−)-germacrene D in *Solidago canadensis*. Chirality 11:353–362

Takahashi S, Yeo YS, Zhao Y (2007) Functional characterization of premnaspirodiene oxygenase, a cytochrome P450 catalyzing regio-and stereo-specific hydroxylations of diverse sesquiterpene substrates. J Biol Chem 282(43):31744–31754

Wang X, Zhou C, Yang X (2015) *De novo* transcriptome analysis of *Warburgia ugandensis* to identify genes involved in terpenoids and unsaturated fatty acids biosynthesis. PLoS One 10(8):e0135724

Wang Q, Jiang J, Liang Y et al (2025) Expansion and functional divergence of terpene synthase genes in angiosperms: a driving force of terpene diversity. Hortic Res 12(1):uhae272

Wink M (2015) Modes of action of herbal medicines and plant secondary metabolites. Medicine 2:251–286

Yu F, Utsumi R (2009) Diversity, regulation, and genetic manipulation of plant mono- and sesquiterpenoid biosynthesis. Cell Mol Life Sci 66:3043–3052

Malaria Public Health Status: Global Context and Update on Gulf Cooperation Council Countries

4

Tanveer Ahmad (iD), Bushra A. Alhammadi,
Shaikha Y. Almaazmi (iD), Sahar Arafa (iD),
Gregory L. Blatch (iD), Tanima Dutta (iD),
Jason E. Gestwicki (iD), Robert A. Keyzers (iD),
Aisha Meskiri, Ahmed Sharafeldin,
Addmore Shonhai (iD), and Harpreet Singh (iD)

Abstract

From 2000 to 2019, the global malaria death toll fell from 864,000 to 576,000 deaths; however, progress on reducing this toll has since slowed, with the COVID-19 pandemic contributing to an increase in the mortality rate since 2020. The emergence of widespread resistance in the major malaria parasite (*Plasmodium falciparum*) to first-line treatment drugs has highlighted the need for new antimalarials and smarter drug delivery strategies. Nevertheless, recent successes of the first malaria vaccines (RTS,S/AS01 and R21/MM) have renewed the promise of widely applicable vaccines in the future. Since 2015, the Eastern Mediterranean Region has experi-

Supplementary Information The online version contains supplementary material available at https://doi.org/10.1007/978-3-032-24254-9_4.

T. Ahmad · B. A. Alhammadi · S. Y. Almaazmi ·
S. Arafa · A. Sharafeldin
Faculty of Health Sciences, Higher Colleges of Technology, Sharjah, United Arab Emirates

G. L. Blatch (✉)
Biomedical Biotechnology Research Unit, Department of Biochemistry, Microbiology and Bioinformatics, Rhodes University, Makhanda, South Africa

Faculty of Health Sciences, Higher Colleges of Technology, Sharjah, United Arab Emirates

The Vice Chancellery, The University of Notre Dame Australia, Fremantle, Australia
e-mail: g.blatch@ru.ac.za

T. Dutta
Department of Diagnostic Genomics, Pathwest, QEII Medical Centre, Nedlands, WA, Australia

J. E. Gestwicki
Department of Pharmaceutical Chemistry and the Institute for Neurodegenerative Diseases, University of California San Francisco, San Francisco, CA, USA

R. A. Keyzers
Centre for Biodiscovery & School of Chemical and Physical Sciences, Victoria University of Wellington, Wellington, New Zealand

A. Meskiri
Faculty of Health Sciences, Higher Colleges of Technology, Abu Dhabi, United Arab Emirates

A. Shonhai
Department of Biochemistry and Microbiology, Faculty of Science, Engineering and Agriculture, University of Venda, Thohoyandou, South Africa

H. Singh
Department of Bioinformatics, Hans Raj Mahila Maha Vidyalaya, Jalandhar, Punjab, India

© The Author(s) 2026
A. Shonhai et al. (eds.), *Advances in Biochemistry and Molecular Biology to meet Africa´s Needs*, Advances in Experimental Medicine and Biology 1507,
https://doi.org/10.1007/978-3-032-24254-9_4

enced an overall increase in malaria cases and deaths. In the Arabian Peninsula, while most of the Gulf Cooperation Council (GCC) countries have been declared free of indigenous malaria (Bahrain, Kuwait, Qatar, and the United Arab Emirates), malaria is still endemic in two GCC countries (Saudi Arabia and Oman) and their neighbors (Yemen). Furthermore, a number of factors threaten malaria control in this region, especially anti-malarial drug resistance, the emergence of highly invasive mosquito species, increasing average temperatures, and imported malaria. In this review, we assess the current worldwide status of malaria before providing a critical evaluation of the malaria situation in the GCC countries.

Keywords

Plasmodium falciparum · malaria · antimalarial drugs · RTS,S/A01 and R21/MM vaccines · Gulf Cooperation Council Countries

4.1   Global Status of Malaria Pre- and Post-COVID-19

Malaria in humans is caused by species of the *Plasmodium* genus, with the most devastating and virulent species being *P. falciparum,* a unicellular protozoan parasite capable of intracellular existence. Over the 2000–2019 period, the global malaria death toll declined from 864,000 to 576,000 deaths (World Malaria Report 2023a, World Health Organization, WHO). However, this trend was reversed in 2020, with 631,000 deaths recorded worldwide (a 10% increase compared to the number of deaths recorded in 2019), before levelling at around 610,000 deaths over the 2021–2022 period (World Malaria Report 2023a, WHO). There were 63,000 deaths reported due to disruptions to essential malaria services during the COVID-19 pandemic (2019–2021) (World Malaria Report 2022c, WHO). While there have been reviews on the status of malaria in the Eastern Mediterranean Region (EMR; Fig. 4.1a, b) based primarily on data prior to the

pandemic (Al-Awadhi et al. 2021; Iqbal et al. 2021), this review represents a significant post-COVID-19 update on the status of malaria in this region, with a particular focus on the Gulf Cooperation Council (GCC) countries (Bahrain, Kuwait, Qatar, Oman, Saudi Arabia, and the United Arab Emirates, UAE). Clinical data has been retrieved from publicly available databases for the GCC countries, and systematically collated (Table 4.1) and critically analyzed, thereby providing the most up-to-date snapshot of malaria in these countries pre- and post-COVID-19. Hence, in this review we assess the current worldwide status of malaria in terms of public health, prevention and treatment, before providing an Arabian regional perspective, with a critical evaluation of the malaria situation in the GCC countries.

In the EMR as defined by WHO (Fig. 4.1a, b), annual cases of malaria decreased during 2000–2015, from 6.9 million to 4.3 million, before steadily increasing, reaching 8.3 million in 2022, a 92% increase (World Malaria Report 2023a, WHO). Mortality due to malaria in the EMR also decreased from 13,600 to 7500 deaths during 2000–2014, but then increased during 2014–2022 to reach 15,900 deaths, a more than 100% increase. Interestingly, publicly available data retrieved from Our World in Data ("Malaria" by Max Roser and Hannah Ritchie, for the period 2000–2021;   ourworldindata.org/malaria; Fig. 4.1a, b), representing estimates of malaria deaths by the Institute of Health Metrics and Evaluation (IHME), were found to be slightly different from the WHO data. While the IHME data show the same overall trends in mortality as the WHO data, the number of deaths was somewhat higher (Fig. 4.1a, b). While a number of countries in the EMR appear to be free of indigenous malaria, there is still the threat of imported malaria (Al-Awadhi et al. 2021). In the Arabian Peninsula, most of the GCC countries have been declared free of indigenous malaria by the WHO (Bahrain, Kuwait, Qatar, and the UAE; World Malaria Report 2022c, WHO); however, imported malaria accounts for a significant number of deaths in this region, and malaria is still endemic in two GCC countries (Saudi Arabia and Oman)

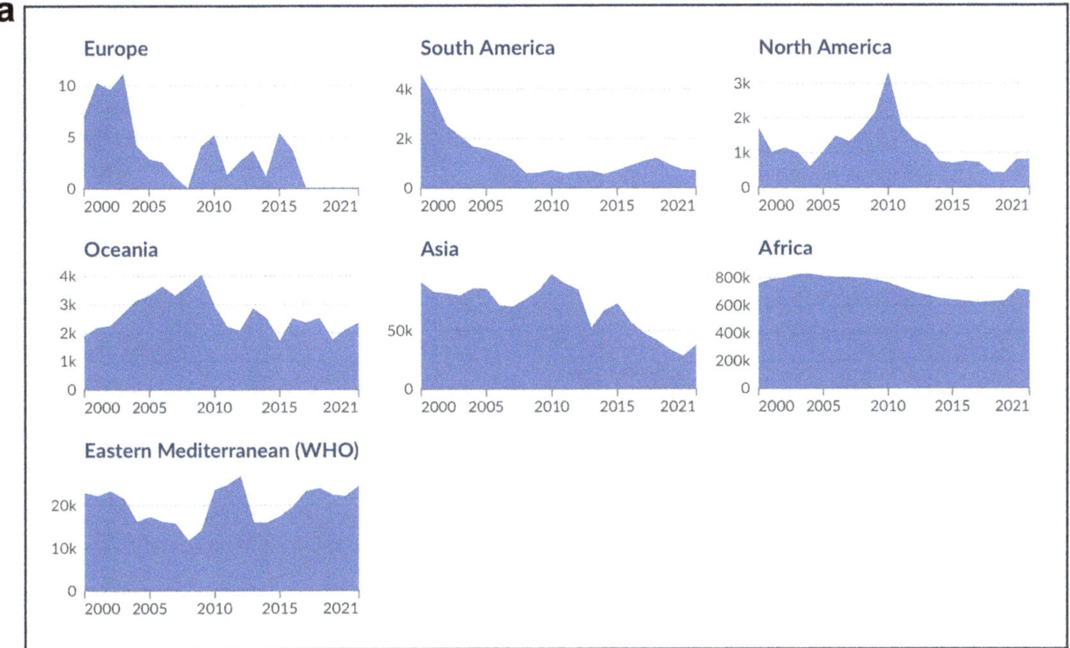

Fig. 4.1 Malaria Status globally, in the Eastern Mediterranean Region (EMR; as defined by the World Health Organization, WHO) and the Gulf Cooperation Council Countries (GCC). The annual number of deaths (2000–2021) from malaria across all ages and both sexes for the major regions of the world (**a**), and the EMR countries (with GCC countries highlighted in yellow) (**b**). The data were retrieved online from Our World in Data ("Malaria" by Max Roser and Hannah Ritchie, ourworldindata.org/malaria; accessed 17 June 2024), and represent estimates of malaria deaths by the Institute of Health Metrics and Evaluation (IHME), which are notably higher than those of the WHO; however, the trends are the same. The difference between the IHME and WHO data are due to differences in the methodologies used for data sourcing and data processing (Yoon et al. 2018). (**c**) The number of newly diagnosed cases of malaria (2020) in the GCC Countries (Bahrain, Kuwait, Qatar, Oman, Saudi Arabia, and the UAE) within the context of neighboring countries of the EMR. (The data were obtained online from the Malaria Atlas Project (data.malariaatlas.org, and rendered using the Mapbox platform, www.mapbox.com/about/maps; accessed 17 July 2023))

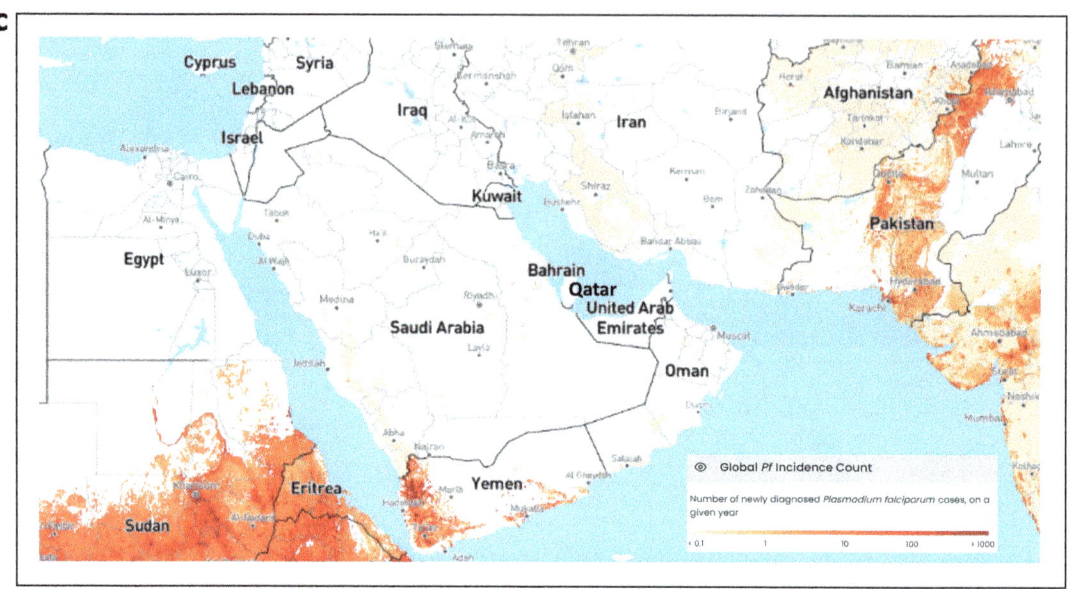

Fig. 4.1 (continued)

and its neighbors (Yemen) (Al-Awadhi et al. 2021) (Fig. 4.1c).

There are a number of factors that are threatening malaria control measures, including but not limited to: antimalarial drug resistance (Balikagala et al. 2021); false negative diagnostic test results due to *pfhrp2/3* gene deletions (Feleke et al. 2021); insecticide resistance in mosquitoes to widely used insecticides (pyrethroids, organochlorines, organophosphates and carbamates) (Riveron et al. 2018; Hancock et al. 2020; Venkatesan 2024); and the emergence of various highly invasive mosquito species, in particular *Anopheles stephensi* (Ahmed et al. 2021). The potential synergistic negative impact of the emergence of insecticide resistance in invasive species of mosquitoes is of grave concern. Insecticide resistance is acquired mainly through target resistance (e.g. mutant voltage-gated sodium channels, VGSCs, with reduced binding to pyrethroids), metabolic resistance (e.g. increase in detoxification enzymes such as cytochrome P450 monooxygenases, esterases, carboxylesterases and glutathione *S*-transferases), and penetration resistance (e.g. increase in structural protein involved in cuticle thickness, such as CPR63, AsCPF1 and AaCPR100A) (Barnes et al. 2017;

Riveron et al. 2018; Kouamo et al. 2021). Also of major concern are the emerging signs of antimalarial drug resistance in *P. falciparum* in Africa and Southeast Asia, particularly to first-line drug treatment (artemisinin resistance, which manifests as delayed parasite clearance) and other drugs (antifolates, naphthoquinones, antibiotics like clindamycin and doxycycline and 4-aminoquinolines) (Balikagala et al. 2021). Furthermore, there are only a few novel targets for the development of new antimalarial drugs (Shibeshi et al. 2020), highlighting the need for the identification of further drug targets for antimalarial drug discovery. Elucidation of the structural properties of the protein-protein interactions required for the survival and pathology of the parasite in humans is critical to the continuous development of novel antimalarial drugs.

Similar to antimalarial drug development, advancement in effective malaria vaccine development has also been challenged by the pleiotropic nature of the parasite. Despite such hurdles, in 2021, the first malaria vaccine to be approved and licensed for widespread use was released (RTS,S/AS01; Mosquirix™; Laurens 2020; Beeson et al. 2022). Furthermore, very recently, a second malaria vaccine, R21/Matrix-M™ (R21/

Table 4.1 Status of malaria in the GCC countries

Country	Status[a]	Clinical cases[b]															*Plasmodium* spp.[c]	Origin[d]
		Imported					Indigenous					Total						
		'18	'19	'20	'21	'22	'18	'19	'20	'21	'22	'18	'19	'20	'21	'22		
Bahrain[e]	Malaria Free (2012)	53	60	15	NR	NR	0	0	0	NR	NR	53	60	15	NR	NR	Pv, Pf, Pm, Po, Mx	Africa, India, Nepal, Pakistan, Philippines
Kuwait[f]	Malaria Free (1979)	293	175	NR	NR	NR	0	0	NR	NR	NR	293	175	NR	NR	NR	Pv, Pf, Mx	Afghanistan, Africa, India, Pakistan
Oman[g]	Endemic (low)	891	1326	276	172	257	30	15	0	0	1	921	1338	276	172	259	Pv, Pf	Africa, India
Qatar[h]	Malaria Free (2012)	NR	NR	NR	NR	NR	NR	NR	NR	NR	NR	NR	NR	NR	NR	NR	Pv, Pf, Mx	Africa, India
Saudi Arabia[i]	Endemic (low)	2517	2029	3453	2470	NR	61	38	83	0	0	2711	2152	3658	2616	NR	Pv, Pf, Mx	Africa, India, Pakistan
UAE[j]	Malaria Free (2007)	3238	2881	915	1014	NR	0	0	0	0	0	3238	2881	915	1014	NR	Pv, Pf, Mx	Africa, India, Pakistan

[a]Countries and Territories Certified Malaria-Free by WHO, 2023 (www.who.int/teams/global-malaria-programme/elimination/countries-and-territories-certified-malaria-free-by-who)

[b]NR-No record in the public domain; note, due to unassigned cases, the total number of cases sometimes exceeds the sum of imported and indigenous cases

[c]The major *Plasmodium* species causing malaria (see text for details); the abbreviations are as follows: *P. vivax* (Pv); *P. falciparum* (Pf); *P. malariae* (Pm); *P. ovalae* (Po); and Mixed, Pv and Pf (Mx)

[d]The major countries or regions for imported malaria cases (see text for the details)

[e]Bahrain: Bahrain Ministry of Health Statistics (www.moh.gov.bh/Ministry/Statistics); in 2018 six out of 53 malaria cases were Bahrain nationals

[f]Kuwait: Iqbal et al. (2020) and Sher and Latif (2023); in 2019 one mortality was reported ("WHO Mortality Database portal"; platform.who.int/mortality/countries/country-details/MDB/kuwait)

[g]Oman: Ministry of Health, Oman. Annual Health Report 2022 (www.moh.gov.om/en/web/statistics/annual-reports)

[h]Qatar: There appear to be no recent records of malaria cases in the public domain (see text for details of historical information up to 2016)

[i]Saudi Arabia: WHO Saudi Arabia-Health Indicators (data.humdata.org/dataset/who-data-for-saudi-arabia); World Malaria Report 2023, WHO (www.who.int/teams/global-malaria-programme/reports)

[j]UAE: Ministry of Health and Prevention, Infectious Disease Data in the UAE 2021–2022 (mohap.gov.ae/en/open-data/mohap-open-data); The zero indigenous cases for 2022 were captured from the World Malaria Report 2023, WHO (www.who.int/teams/global-malaria-programme/reports)

MM), was recommended for release (Siddiqui et al. 2023). While these vaccines hold much promise, they are designed for children, and hence further research and development is needed to produce malaria vaccines with broader reach at the population level. These malaria vaccines are critically evaluated later in the section on "Antimalarial drugs and vaccines."

Given the recent increase in case and mortality rates, and the emerging threats to the control, elimination and ultimately eradication of malaria in regions of the globe with high burden, there is a dire need for a globally integrated approach to this infectious disease (Venkatesan 2024). Despite the negative impact of COVID-19 on the public health status of malaria, the pandemic has shown what can be done when a disease is viewed as a global threat. Indeed, there is an urgent need for a universal effort in applying more strategic approaches towards the eradication of malaria. In the following sections we evaluate the global status of malaria in terms of prevention and treatment, before focusing on the GCC countries.

4.2 Malaria Prevention

Since mosquitoes act as vectors, one of the key preventative methods to control the spread of malaria is to reduce the vector population. Several approaches are used to control the malaria vector population, including drainage of swamps, application of pesticides and use of surface coverings (such as oil) to kill mosquito larvae present in standing water bodies. Indoor residual sprays (IRS) and long-lasting insecticidal nets (LLINs) are still the primary approaches to malaria prevention (Corrêa et al. 2019). Despite its effectiveness as a malaria control tool, the use of dichlorodiphenyltrichloroethane (DDT) is controversial because of the persistence of the chemical in the environment (Bouwman et al. 2011). The main insecticides used to control malaria vectors include pyrethroids, organochlorines, carbamates, and organophosphates (Killeen et al. 2017). Of these, pyrethroids pose the least public health concerns as they are bio-degradable (Casida et al. 1975; Cycoń and Piotrowska-Seget

2016). However, as discussed in the previous section, the consistent use of insecticides leads to resistance to the chemicals by the mosquito vector. The recommended approach to mitigate against mounting insecticide resistance is to make use of IRS or combine IRS with LLINs (West et al. 2014). An alternative approach proposed to prevent insecticide resistance is to mix the active ingredients, and this has been reported to be effective when applied to mosquito bed nets (Pennetier et al. 2008). An additional advantage of this approach is that the active ingredients interact synergistically when applied in combination, such that low doses are highly effective (Pennetier et al. 2008).

Preventative approaches to malaria also involve a combination of vaccines and chemopreventive approaches, none of which are effective in isolation. Several chemoprevention approaches for malaria have been stipulated in global guidelines, amongst them: intermittent preventive treatments during pregnancy (IPTp), and for infants (IPTi); seasonal malaria chemoprevention (SMC); and for emergency situations, mass drug administration (MDA) (Plowe 2022). Effective chemopreventive approaches are associated not only with reduced disease burden but also with a reduced incidence of drug resistance.

4.3 Antimalarial Drugs and Vaccines

4.3.1 Antimalarial Drug Discovery and Combination Therapies

Resistance has emerged in *P. falciparum* to artemisinin (and its derivatives), the globally recommended first-line therapy for malaria. Artemisinin and its derivatives (e.g., artesunate, artemether and arteether), synthesized by the plant *Artemisia annua*, are able to rapidly clear the parasite by inhibiting protein synthesis. Studies have shown that mutations in the *P. falciparum* Kelch13-encoding gene result in reduced levels of the essential protein PfKelch13, leading to artemisinin resistance (Siddiqui et al. 2021; Paloque et al. 2022). A mutation in PfKelch13 enables the early ring stage of the parasite to avoid going through cell cycle arrest and continue its intra-erythrocytic development when artemisinin is no longer available at effective concentrations (Mok et al. 2015, 2021). Resistance has been shown to develop in six Southeast Asian countries, including Cambodia, Myanmar, Thailand, Vietnam, Laos, and southern China (Yunnan and Guangxi) (Balikagala et al. 2021), and it poses a serious threat to other global regions. This emerging drug resistance has put a spotlight on the fact that there are only a few validated antimalarial drug targets (Shibeshi et al. 2020). One approach to overcoming resistance to artemisinin and other antimalarials has been combination drug therapy using complementary or synergistic drugs that act on distinct targets. Indeed, global guidelines have been released and implemented, involving the use of combinations of antimalarial drugs with distinct targets and mechanisms of action (WHO Guidelines for Malaria 2022a). However, this approach has had mixed success, highlighting that for significant improvement towards treatments where drug resistance is minimized, new antimalarials for combination drug therapies should ideally be derived from novel chemical scaffolds that have novel modes of action (Luo et al. 2023). Furthermore, while there are a number of promising candidate compounds feeding into next-generation antimalarial treatments, many target only the asexual, erythrocytic stage of the parasite's life cycle. Thus, one logical approach is to identify novel chemotypes targeting the other parasitic stages or multiple stages, a topic that has been extensively reviewed (Luo et al. 2023). Broadly, contemporary approaches involve the identification of a protein that is required for the growth or survival of the parasite, followed by the identification of a chemical series that inhibits the target. A key to antimalarial drug discovery is the subsequent medicinal chemistry efforts, which are designed to enhance binding to the parasite protein while sparing the closest human orthologs. To illustrate this concept, the malarial heat shock proteins (HSPs) are a case in point. The HSPs are a family of molecular chaperones whose expression is upregulated under stress (e.g. heat, low oxygen levels, pH varia-

tions, and nutrient deprivation), and which function to prevent the accumulation of toxic levels of misfolded and aggregated proteins under such conditions (Jee 2016; Edkins and Boshoff 2021). The major families of HSPs, including HSP10, HSP40, HSP60, HSP70, HSP90, and HSP100, are highly conserved from prokaryotes to eukaryotes, forming an intricate network critical for ensuring protein homeostasis (proteostasis). Consistent with this role, most HSPs have been found to be essential for the growth, survival, viability, and differentiation at all stages of the life cycle of kinetoplastid and apicomplexan parasites, including the malaria parasite (Shonhai et al. 2011; Zhang et al. 2018). Many HSPs are also essential for the pathogenesis and virulence of the malaria parasite (Blatch 2022). These observations have led to the suggestion that the *P. falciparum* HSPs (PfHSPs), especially PfHSP40 (Daniyan et al. 2016; Daniyan and Blatch 2017; Dutta et al. 2021a; Almaazmi et al. 2022), PfHSP70 (Przyborski et al. 2015; Batinovic et al. 2017; Zininga et al. 2017a, b; Chen et al. 2018; Day et al. 2019; Cortés et al. 2020; Barth et al. 2022; Mrozek et al. 2022), PfHSP90 (Banumathy et al. 2003; Bayih et al. 2016; Wang et al. 2016; Posfai et al. 2018; Shahinas et al. 2013; Seraphim et al. 2019; Stofberg et al. 2021) and their complexes (Botha et al. 2011; Cockburn et al. 2011; Cockburn et al. 2014; Dutta et al. 2021b; Almaazmi et al. 2023; Singh et al. 2023), could be prospective antimalarial drug targets. Unlike many other targets that have been explored, inhibition of these proteins could have broad effects on parasite survival at multiple stages, forming a potentially important addition to the antimalarial armamentarium.

In the preceding paragraphs, we outlined progress towards developing HSPs and their complexes as new targets for antimalarial drug discovery. Ideally, optimized versions of these single-agent compounds might be combined with both emerging and existing drugs. Combination therapies have become a promising tool for reducing burdens on the path to elimination (Luo et al. 2023). Mass drug administration (MDA) is the contemporaneous delivery of a full course of antimalarial medicine at a population level, regardless of whether individuals are infected with malaria or not, before or during high malaria transmission seasons (Eisele 2019; WHO Guidelines for Malaria 2022a). According to the WHO, MDA use is recommended in settings approaching elimination, where vector control and monitoring are effective, treatment is accessible, and re-infection is minimal (WHO Guidelines for Malaria 2022a). For example, programmatic MDA (pMDA) campaigns with dihydroartemisinin-piperaquine (DHAP) were conducted in the Southern Province of Zambia as a malaria elimination tool (Fraser et al. 2020). Interrupted time-series results with comparison groups indicated that two rounds of pMDA with DHAP over one year with moderate coverage resulted in a significant short-term reduction of malaria transmission in low transmission areas, but not eradication (Fraser et al. 2020). In recent years, MDA intervention using mosquitocidal drugs has been used in Gambia, combined with antimalarial treatments with ivermectin and DHAP, in an area of moderate malaria, where there is high coverage of insecticide-treated nets and indoor residual spraying. The implementation of MDA was found to be both safe and well-tolerated with a high coverage of standard control interventions, resulting in significant reductions in malaria prevalence, and to possibly promote malaria elimination where there was high coverage of vector control interventions (Dabira et al. 2022).

4.3.2 Antimalarial Vaccines and Their Role in Mass Drug Administration

What is the status of antimalarial vaccines, and their role in antimalarial MDA? The first malaria vaccine to be released, RTS,S/AS01, is recommended for young children (5–6 months old, with a booster at 2 years old) living in moderate-to-high malaria transmission locations (RTS,S Clinical Trials Partnership 2015; Beeson et al. 2022). RTS,S/AS01 is a genetically engineered fusion protein-based vaccine (also referred to as a subunit vaccine) that targets the pre-erythrocytic

stage of the *P. falciparum* life cycle. The fusion protein consists of the central tandem repeat region (R) and C-terminal T-cell epitopes (T) of the *P. falciparum* circumsporozoite protein, PfCSP, fused to the S antigen of the hepatitis B virus (S). This fusion protein (RTS) and the hepatitis B surface antigen, HBsAg (S), are co-produced in a yeast (*Saccharomyces cerevisiae*) expression system, and formulated with the adjuvant AS01 (Zavala 2022). RTS,S/AS01 clinical trials in malaria-endemic African countries reported only a modest decrease in malaria cases for test versus control groups (clinical efficacy), with ~55% efficacy in young children (5–17 months), and ~31–26% efficacy in infants (6–12 weeks) (Beeson et al. 2022; Zavala 2022). Nevertheless, this vaccine appears to be safe, readily delivered, and already making a notable positive impact on protection from malaria among young children at the community level (Beeson et al. 2022). Indeed, the vaccine is reported to have already reached approximately 2 million children in three countries in Africa (Ghana, Kenya and Malawi). However, despite the promising potential of RTS,S/AS01, it does not meet the WHO threshold clinical efficacy (75%), and it offers only modest, short-lived and strain-specific protection, requiring four doses plus a booster shot within 18 months of first administration (RTS,S Clinical Trials Partnership 2015; Neafsey et al. 2015; Laurens 2020). The recently approved second malaria vaccine, R21/Matrix-M™ (R21/MM), offers greater clinical efficacy, having exceeded the WHO efficacy threshold. R21/MM is also a subunit-based vaccine targeting the pre-erythrocytic stage of the parasite lifecycle, and designed for the prevention of malaria in young children (Datoo et al., 2021). This vaccine comprises R21, a PfCSP central tandem repeat region fused to the N-terminus of the HBsAg, co-produced with HBsAg in a yeast (*Hansenula polymorpha*) expression system, and formulated in the adjuvant Matrix-M (MM). R21/MM clinical trials in malaria-endemic African countries have demonstrated 74% (low dose) and 77% (high dose) efficacy in young children (5 and 17 months) (Datoo et al. 2022; Siddiqui et al. 2023). While there are now two malaria vaccines for young children, one released and licenced (RTS,S/AS01), and one approved for release (R21/MM), there is still currently no malaria vaccine available for older children, adolescents or adults at risk of malaria. Mass drug administration is proposed to interrupt transmission permanently if combined with mass vaccination (Tun et al. 2017). For example, the RTS,S/AS01 vaccine was shown to be safe and immunogenic for healthy adult Thai volunteers without adverse effects when administered in combination with DHAP and primaquine (von Seidlein et al. 2020). The results of these interventions support the combined use of multiple antimalarial drugs in concert with malaria vaccines to accelerate the elimination of malaria. However, further large-scale intervention-research studies are necessary to improve the application of vaccine-drug strategies, as well as innovative approaches to next-generation vaccines to produce malaria vaccines with greater reach across the life span.

4.4 Malaria in the GCC Countries

4.4.1 The History of Malaria in the Region

The GCC countries are all located in the Arabian Peninsula (Fig. 4.1c), and for more than 30 years they have shared several common characteristics such as a dry, hot climate, high income generated from an oil-based economy leading to rapid development, and modern healthcare and infrastructure (Al-Awadhi et al. 2021). In addition, these countries have large and dynamic expatriate populations feeding into their workforce. In most of the GCC countries, the expatriate population outnumbers its citizens (Iqbal et al. 2021). The majority of the expatriates are from malaria-endemic developing countries within Africa (e.g. Ethiopia, Nigeria and Sudan; Table 4.1) and south/southeast Asia (e.g. Afghanistan, Bangladesh, India, Nepal, Pakistan, Philippines and Sri Lanka; Table 4.1) (Al-Awadhi et al. 2021; Iqbal et al. 2021). On an annual basis, the GCC countries welcome millions of new expatriates and expatriates returning from visits to their home countries, thereby increasing the risk of imported malaria entering the region (Iqbal et al. 2021). Since the 1950s, the GCC countries have increased their investments in malaria control, leading to the interruption of local malaria transmission. This successfully eliminated malaria (malaria-free status) in all the GCC countries, apart from Oman and Saudi Arabia (Al-Rumhi et al. 2020; Table 4.1). Oman still suffers from periodic outbreaks of malaria, while in Saudi Arabia, isolated clusters of indigenous malaria prevail. Hence, the GCC countries have shifted policy toward a malaria-free Arabian Peninsula, with an emphasis on approaches that prevent reintroduction of malaria, as detailed in the country-by-country section below (Regional Malaria Action Plan 2016–2020: Towards a Malaria-Free Region, WHO). None of the GCC countries is among the 25 high malaria burden countries, and since 2010 there have been zero, or close to zero, annually reported malarial deaths in this region (Our World in Data, IHME 2022; World Malaria Report 2022c, WHO; Fig. 4.1b).

4.4.2 Malaria Status by Country

Bahrain According to the WHO, Bahrain was declared malaria-free in 2012 (Table 4.1). However, the Bahrain Ministry of Health reported the national eradication of malaria in 1979 (Al-Awadhi et al. 2021). There have been limited research reports dedicated to the status of malaria in Bahrain (Ismaeel et al. 2004); however, a number of articles reviewing malaria epidemiology in the Eastern Mediterranean Region have included Bahrain (Al-Awadhi et al. 2021; Iqbal et al. 2021). Furthermore, since the year 2000, the Bahrain Ministry of Health has published health statistics, which include epidemiological data on malaria cases (Table 4.1). The previous reports indicated malaria cases of 1572 for the 1992–2001 period (Ismaeel et al. 2004) and 133 in 2017 (Iqbal et al. 2021). Over the 2018–2020 period, the trend of declining malaria cases has continued, with only 15 cases reported in 2020

(Table 4.1). As reported previously, all malaria cases appear to be imported among expatriates originating mostly from Pakistan and India, with the predominant causative agent being *P. vivax*. The last time Bahrain nationals presented with malaria was 2018 (6 out of a total of 53 malaria cases; Table 4.1); however, these are most likely citizens returning from travel in malaria-endemic countries.

Kuwait The WHO declared Kuwait malaria-free in 1979 (Table 4.1). Ever since malaria was declared to be eradicated from Kuwait in 1963, no indigenous cases have been reported. However, several studies reported that despite precautions, imported malaria cases have been identified among expatriate families travelling from regions on the Indian subcontinent, Southeast Asia, and Africa where malaria is endemic (Iqbal et al. 2000, 2003, 2020; Sher and Latif 2023). Nevertheless, the number of malaria cases has steadily decreased from 1400 (in 1993) to less than 200 (in 2019) due to a highly effective malaria control program managed by the Ministry of Health, Kuwait (Sher and Latif 2023; Table 4.1). While the majority of early reports (1985 to 2000) of malaria cases were single infections with *P. falciparum* and *P. vivax*, recent studies have revealed that the majority of current malaria cases in Kuwait were caused by mixed infections of both *P. falciparum* and *P. vivax* parasites, and that nearly all these mixed-infection cases were among travelers from India, the largest expatriate group in Kuwait (Iqbal et al. 2020; Sher and Latif 2023).

Oman Oman is still officially considered a malaria-endemic country, and while malaria has been a major public health issue since the 1960s, national eradication programs have considerably reduced the number of indigenous cases, which are now approaching zero on an ongoing basis in recent years (Table 4.1). In 1975, a malaria training program was initiated (Snow et al. 2013). This program had multiple key objectives and measures to reduce malaria cases in Oman, such as the establishment of malaria training centers, identification of vector sources and vector control

using larvivorous fish, *Aphanius dispar*, and continuous tracking of larval control and the widespread use of larvicides. Furthermore, during the 1973–1979 period, in specific high-endemic areas, weekly chloroquine prophylaxis treatment was offered to school children. Despite these measures, Oman remained endemic with the number of malaria cases escalating to almost 33,000 in 1990 (Hassan 2017). In 1991, the Omani Ministry of Health launched the National Malaria Eradication Programme (NMEP), with the aim of reducing parasite incidence to 1/10,000 population by the year 2000, by robust vector control, and early identification, management and curing of active cases (Simon et al. 2017). Over the 1994–2004 period, indigenous cases decreased from 4415 to zero, with no local transmission reported between 2004 and 2006 (Simon et al. 2017; Al Mukhaini et al. 2023). However, since 2007, due to the continuous existence of the vector, local transmission has been reported in small outbreaks (Simon et al. 2017; Al Mukhaini et al. 2023). In 2016, malaria control efforts were further increased in Oman by enhancing malaria surveillance as a fundamental approach for better monitoring and assessment of the effectiveness of interventions (Global Technical Strategy for Malaria 2016–2030, 2021 update, WHO). Analysis of publicly available malaria infection data for the 2018–2022 period (Table 4.1), revealed that the annual number of imported malaria cases was significant but steadily dropping, while the number of indigenous malaria cases was virtually zero for the last 3 years (Table 4.1). Interestingly, the predominant malaria species has significantly changed over the years, somewhat mirroring the changes in the origin of imported cases. In the 1988–1999 period, the majority of malaria cases were due to infection by *P. falciparum* and were attributed to imported cases from East Africa. Over the 2000–2017 period, there was a shift to predominantly *P. vivax* cases due to an increasing imported workforce from India (Al Mukhaini et al. 2023). However, since 2018 *P. falciparum* cases have again predominated, most likely due to an influx of immigrants from East Africa (Al Mukhaini et al. 2023). In 2022, a total of 259 cases were recorded (with only one indigenous case), mostly resulting from infection by *P. vivax* (52.5%), followed by *P. falciparum* (42.5%) (Table 4.1).

Qatar In 1970, Qatar eliminated malaria transmission, with the WHO declaring it malaria-free in 2012 (Table 4.1). A number of studies (Al-Kuwari 2009; Farag et al. 2018; Al-Rumhi et al. 2020) and recent reviews of the malaria situation in Qatar (Al-Awadhi et al. 2021; Iqbal et al. 2021) have reported that all malaria cases have been travel-related, largely due to expatriates who have visited their home countries, which are malaria-endemic, but also Qatari nationals who have visited malaria-endemic countries. The two predominant malaria vectors in Qatar were *An. stephensi* and *An. Multicolor*, which were eliminated by malaria elimination programs and intense vector control (Al-Kuwari 2009). A study employing molecular diagnostic techniques reported that between 2013 and 2016, most cases of imported malaria were due to *P. vivax*, while *P. falciparum* and mixed *P. falciparum/P. vivax* infections occurred far less frequently (Al-Rumhi et al. 2020). *P. vivax* infections originated in the Indian subcontinent, whereas *P. falciparum* infections primarily originated from Africa. The study also demonstrated a high genetic variability of *P. falciparum* strains and the incidence of mutations in the PfKelch13-encoding gene, and other genes linked to drug resistance in malaria parasites. Scientists fear that there is a risk of the re-establishment of drug-resistant malaria in Qatar and other GCC nations due to the increasing introduction of multidrug-resistant *P. falciparum* (Al-Rumhi et al. 2020).

Saudi Arabia Saudi Arabia is still officially classified as endemic for malaria. In 1998, Saudi Arabia faced a major malaria outbreak, but due to intensive federal malaria control programs, the number of reported indigenous malaria cases sharply declined (Shibl et al. 2012; Coleman et al. 2014; El Hassan et al. 2015). Over the past two decades, extensive measures have been introduced to reduce indigenous malaria. Analysis of publicly available data for the 2018–2022 period

(Table 4.1) revealed that while the number of imported malaria cases per annum was significant, the number of indigenous malaria cases has been relatively low (Table 4.1). Indeed, by 2021, Saudi Arabia achieved its inaugural year of reporting zero indigenous malaria cases, which was maintained through 2022 (Table 4.1). The majority of reported indigenous malaria cases appear to originate from endemic countries bordering Saudi Arabia, which were not part of the GCC (El Hassan et al. 2015; Alshahrani et al. 2016, 2019; Al Zahrani et al. 2018). A study conducted in southwestern Saudi Arabia reported that the vast majority of cases were due to *P. falciparum* (98%), with very few cases caused by mixed *P. falciparum/P. vivax* infection (1.85%), and no detection of *P. vivax* cases (Bin Dajem 2015). Another study investigated the type of malaria parasite in western Saudi Arabia (the Makkah region), where millions of immigrants annually visit to perform the Islamic pilgrimage, Hajj, and reported that 95% of malaria cases were among non-Saudi nationals. Most of the malaria cases were caused by *P. falciparum* (67%) and *P. vivax* (32%), with the majority (62%) of cases originating from immigrants traveling from India, Pakistan and Nigeria (Memish et al. 2014). Antimalarial drug resistance alleles have been reported in Saudi Arabia, especially mutations associated with chloroquine resistance and antifolate drug resistance, suggesting that improvements were needed in future antimalarial drug regimens (Soliman et al. 2018; Dafalla et al. 2020; Madkhali et al. 2020).

United Arab Emirates (UAE) Since 2007, the UAE has been declared by the WHO as malaria-free (Table 4.1). However, imported malaria cases have been reported annually among working expatriates or returning travelers from endemic countries (Al-Awadhi et al. 2021; Table 4.1). A study performed in a Dubai hospital between 2008 and 2010 reported 629 malaria cases, where most of the patients (90.1%) were from Pakistan or India, 7% were from sub-Saharan Africa and none of the reported cases were Emirati citizens. The study showed that most of the cases (78%) were caused by *P. vivax* followed by *P. falciparum* and only 2% of infections were caused by mixed *P. vivax/P. falciparum* (Nilles et al. 2014; Al-Awadhi et al. 2021). The UAE Ministry of Health and Prevention has implemented national strategies to maintain a malaria-free status, including early malaria detection and treatment of imported cases, workforce training in malaria prevention and control, and enhanced vector control. These measures appear to be effective, as publicly available data for a five-year period (2018–2022; Table 4.1), indicated that the number of imported malaria cases has been steadily decreasing year-on-year, while the number of indigenous malaria cases has been zero for the entire period (Table 4.1). Interestingly, the majority of imported malaria cases are reported from the Emirates of Abu Dhabi and Dubai (Table S1), which is perhaps not surprising given that they are major cities, channeling millions of travelers into the UAE every year through their international airports. Strategic preventative measures could be implemented, such as compulsory pre-screening of travelers on route to the UAE prior to departure from malaria-endemic countries. National strategies such as these, informed by tried and tested COVID-19 infection control procedures, could potentially result in a substantial reduction in future imported malaria cases.

4.4.3 Impact of Mosquito Vector Insecticide Resistance in the Region

Over the last two decades, insecticide resistance in the highly invasive malaria vector, *Anopheles stephensi*, has spread widely across most endemic countries, posing a substantial threat to the successful control of malaria (Riveron et al. 2018; World Malaria Report 2023a, WHO). This is of major concern to the GCC countries, given they are under constant risk of the reintroduction of *Anopheles* species (including species previously undetected in the region), particularly from neighboring countries with compromised malaria vector control measures (Al-Eryani et al. 2023).

The WHO has released two "Vector Alerts" (WHO Vector Alerts 2019, 2022b), highlighting the threat of the spread of *An. stephensi* into regions where it has not previously been detected, notably countries in the Horn of Africa and Yemen (Malaria Threats Map, Global Malaria Programme, World Health Organization, WHO 2024; apps.who.int/malaria/maps/threats/; accessed 20 March 2024; Fig. 4.2). The presence of *An. stephensi* was not reported in Yemen until 2021, although it may have been resident but not detected (Al-Eryani et al. 2023). Furthermore, the vector was found to be resistant to multiple insecticides (pyrethroids, organochlorines, organophosphates and carbamates) (Fig. 4.2). Yemen has recently become the center of increasingly acute conflict and upheaval, resulting in the displacement of segments of the population into make-shift urban dwellings, which has inadvertently created conditions favorable to the spread of the malaria vector (Allan et al. 2023). This has seriously compromised the capacity of Yemen to control the spread of malaria, and hence its proximity to the GCC countries poses a major threat to the spread and re-establishment of malaria in this largely malaria-free region. In particular, the threat of the spread of insecticide-resistant and invasive mosquito species across the Horn of Africa and through Yemen to the GCC countries requires that stricter cross-border vector surveillance and control measures be established in the Arabian Peninsula.

4.4.4 Impact of Climate Change on Malaria Cases in the Region

The Arabian Peninsula is highly vulnerable to the negative impacts of climate change, particularly the adverse effects on the control of infectious diseases (Waha et al. 2017). The spread of malaria relies on a strong ecosystem between humans, the environment and infected mosquitoes, and the degree of malaria distribution and seasonal activity are highly sensitive to climatic factors (Costello et al. 2009). Given that mosquitoes are poikilothermic, variation in temperature, humidity and rainfall can significantly affect mosquito

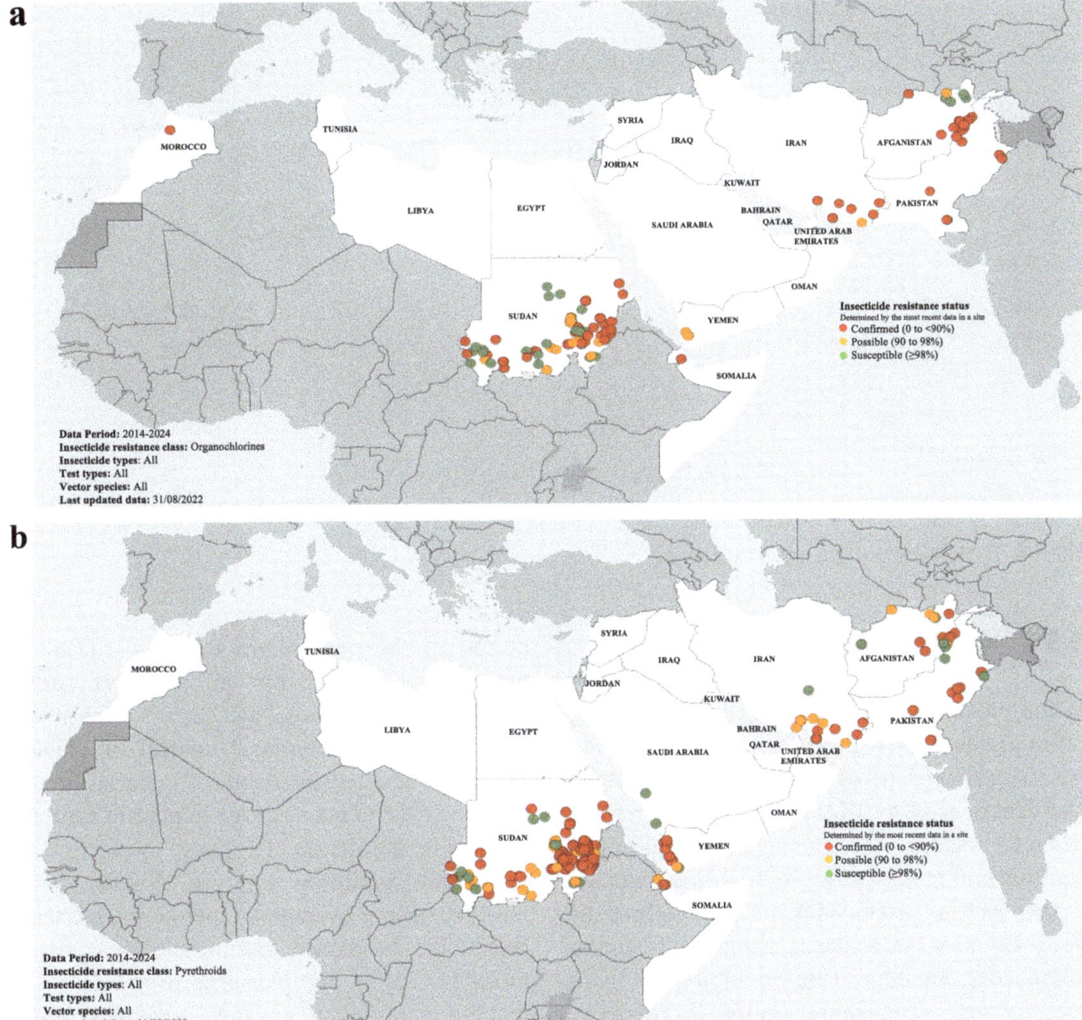

Fig. 4.2 Maps of the status of mosquito vector insecticide resistance in the Eastern Mediterranean Region (EMR). The maps present the status of resistance to organochlorines (**a**), pyrethroids (**b**) and carbamates (**c**). All three maps depict the most recent insecticide resistance data for the period 2014–2024. (The data was sourced from Malaria Threats Map, Global Malaria Programme, World Health Organization, WHO 2024 (apps.who.int/malaria/maps/threats/; accessed 20 March 2024). © WHO 2024)

prevalence, successful development of the malaria parasite within the vector, and malaria transmission (Carrington et al. 2013; Shapiro et al. 2017; Mordecai et al. 2019). From a risk framework perspective, there is a high likelihood that climate change will cause an increase in malaria infections not only in endemic areas, but also in current malaria-free regions. Modelling of potential future worldwide climate change scenarios indicates conditions in tropical highlands could favor the spread of malaria, socioeconomic factors notwithstanding (Caminade et al. 2014). In the next 10 years, GCC countries are predicted to suffer from extreme heat waves (Salimi and Al-Ghamdi 2020) and severe water constraints (Bayram and Öztürk 2021). There are data gaps regarding climate dynamics and disease populations across the GCC, reflecting an underrepresentation in global health forecast research for this region (Rawson et al. 2023), and highlighting the urgent need for such public health research in the near future.

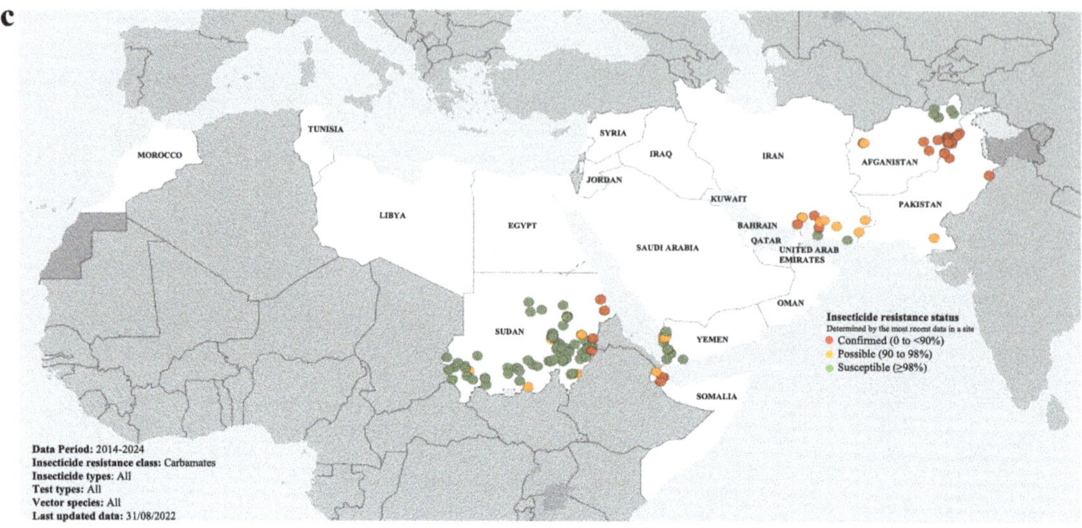

Fig. 4.2 (continued)

4.5 Conclusion

While global malaria cases and deaths fell significantly during the 2000–2015 period, progress has since slowed, particularly in the EMR, which has experienced an overall increase in malaria cases and deaths. The emergence of widespread resistance in *P. falciparum* to first-line treatment drugs, such as artemisinin, has highlighted the need for new antimalarial drugs and smarter approaches, including better combination therapies and mass drug administration. Furthermore, the threat of the spread of insecticide-resistant and invasive mosquitoes from endemic to non-endemic regions requires that stricter cross-border vector surveillance and control measures be established. The recent successes of the first malaria vaccine (RTS, S/AS01), and the approval of a second (R21/MM) have renewed the promise of widely applicable vaccines, which provide long-term protection and are affordable and accessible in malaria-endemic areas. In the Arabian Peninsula, while most of the GCC countries have been declared free of indigenous malaria (Bahrain, Kuwait, Qatar, and the United Arab Emirates), malaria is still endemic in two GCC countries (Saudi Arabia and Oman) and their neighbors (Yemen). The GCC region is poised in the near future to move towards a malaria-free status for all member states. Preventative measures have proven highly successful in population-level control of malaria, while the targeted treatment of malaria-infected patients will continue to be an important approach for preventing deaths. However, evolving environmental conditions such as climate change with increasing average temperatures, increased insecticide resistance and invasiveness of the mosquito vector, and increased resistance of the malaria parasite to first-line treatment drugs, mean that malaria should still be considered a major threat to public health in the GCC region. Therefore, further strengthening of current malaria-control frameworks, with well-developed prevention, surveillance, monitoring, diagnosis and treatment regimes, will be required before the entire GCC region can achieve malaria-free status. A region is only as safe as its "weakest link," and hence, as the world learnt from the COVID-19 pandemic, borderless, collective intent can also lead to robust solutions to public health threats such as malaria.

Author Contributions All authors listed have made a direct and intellectual contribution to the work and approved the manuscript for publication.

Funding This study was funded by a Higher Colleges of Technology (UAE) Interdisciplinary Research Grant (grant number 213471).

Conflict of Interest Author TA declares that he has no conflict of interest. Author BAA declares that she has no conflict of interest. Author SYA declares that she has no conflict of interest. Author SA declares that she has no conflict of interest. Author GLB declares that he has no conflict of interest. Author TD declares that she has no conflict of interest. Author JEG declares that he has no conflict of interest. Author RAK declares that he has no conflict of interest. Author AM declares that she has no conflict of interest. Author AS (Ahmed Sharafeldin) declares that he has no conflict of interest. Author AS (Addmore Shonhai) declares that he has no conflict of interest. Author HS declares that he has no conflict of interest.

Ethical Approval This chapter is a review of previously published accounts; as such, no animal or human studies were performed.

References

Ahmed A, Khogali R, Elnour MAB, Nakao R, Salim B, Elhaj H (2021) Emergence of the invasive malaria vector *Anopheles stephensi* in Khartoum State, Central Sudan. Parasites Vectors 14(1):511

Al Mukhaini SK, Mohammed OA, Gerbers S, Al Awaidy ST (2023) The progress towards national malaria elimination: The experience of Oman. Oman Med J 38(3):e500

Al Zahrani MH, Omar AI, Abdoon AMO, Ibrahim AA, Alhogail A, Elmubarak M, Elamin YE, Al Helal MA, Alshahrani AM, Abdelgader TM et al (2018) Cross-border movement, economic development and malaria elimination in the Kingdom of Saudi Arabia. BMC Med 16(1):98

Al-Awadhi M, Ahmad S, Iqbal J (2021) Current status and the epidemiology of malaria in the Middle East region and beyond. Microorganisms 9(2):338

Al-Eryani SM, Irish SR, Carter TE, Lenhart A, Aljasari A, Montoya LF, Hutin YJ (2023) Public health impact of the spread of *Anopheles stephensi* in the WHO Eastern Mediterranean Region countries in Horn of Africa and Yemen: need for integrated vector surveillance and control. Malar J 22(1):187

Al-Kuwari MG (2009) Epidemiology of imported malaria in Qatar. J Travel Med 16(2):119–122

Allan R, Weetman D, Sauskojus H, Budge S, Hawail TB, Baheshm Y (2023) Confirmation of the presence of *Anopheles stephensi* among internally displaced people's camps and host communities in Aden city, Yemen. Malaria J 22(1):1

Almaazmi SY, Singh H, Dutta T, Blatch GL (2022) Exported J domain proteins of the human malaria parasite. Front Mol Biosci 9:978663

Almaazmi SY, Kaur RP, Singh H, Blatch GL (2023) The *Plasmodium falciparum* exported J domain proteins fine-tune human and malarial Hsp70s: pathological exploitation of proteostasis machinery. Front Mol Biosci 10:1216192

Al-Rumhi A, Al-Hashami Z, Al-Hamidhi S, Gadalla A, Naeem R, Ranford-Cartwright L et al (2020) Influx of diverse, drug resistant and transmissible *Plasmodium falciparum* into a malaria-free setting in Qatar. BMC Infect Dis 20:1–10

Alshahrani AM, Abdelgader TM, Saeed I, Al-Akhshami A, Al-Ghamdi M, Al-Zahrani MH, El Hassan I, Kyalo D, Snow RW (2016) The changing malaria landscape in Aseer region, Kingdom of Saudi Arabia: 2000–2015. Malar J 15(1):538

Alshahrani AM, Abdelgader TM, Saeed I, Al-Akhshami A, Al-Ghamdi M, Al-Zahrani MH, Snow RW (2019) Risk associated with malaria infection in Tihama Qahtan, Aseer Region, Kingdom of Saudi Arabia: 2006–2007. Malaria Control Elimin 5(2):144

Balikagala B, Fukuda N, Ikeda M, Katuro OT, Tachibana SI, Yamauchi M, Opio W, Emoto S, Anywar DA, Kimura E, Palacpac NMQ, Odongo-Aginya EI, Ogwang M, Horii T, Mita T (2021) Evidence of Artemisinin-resistant malaria in Africa. N Engl J Med 385(13):1163–1171

Banumathy G, Singh V, Pavithra SR, Tatu U (2003) Heat shock protein 90 function is essential for Plasmodium falciparum growth in human erythrocytes. J Biol Chem 278(20):18336–18345

Barnes KG, Irving H, Chiumia M, Mzilahowa T, Coleman M, Hemingway J, Wondji CS (2017) Restriction to gene flow is associated with changes in the molecular basis of pyrethroid resistance in the malaria vector *Anopheles funestus*. Proc Natl Acad Sci 114(2):286–291

Barth J, Schach T, Przyborski JM (2022) HSP70 and their co-chaperones in the human malaria parasite *P. falciparum* and their potential as drug targets. Front Mol Biosci 9:968248

Batinovic S, McHugh E, Chisholm SA, Matthews K, Liu B, Dumont L, Charnaud SC, Schneider MP, Gilson PR, de Koning-Ward TF, Dixon MW (2017) An exported protein-interacting complex involved in the trafficking of virulence determinants in *Plasmodium*-infected erythrocytes. Nat Commun 8(1):16044

Bayih AG, Folefoc A, Mohon AN et al (2016) In vitro and in vivo anti-malarial activity of novel harmine-analog heat shock protein 90 inhibitors: a possible partner for artemisinin. Malar J 15(1):579

Bayram H, Öztürk AB (2021) Global climate change, desertification, and its consequences in Turkey and the Middle East. In: Climate Change and Global Public Health. Humana Press, New York, NY, pp 445–458

Beeson JG, Kurtovic L, Valim C, Asante KP, Boyle MJ, Mathanga D, Dobano C, Moncunill G (2022) The

RTS,S malaria vaccine: Current impact and foundation for the future. Sci Transl Med 14(671):eabo6646

Bin Dajem SM (2015) Molecular investigation of mixed malaria infections in southwest Saudi Arabia. Saudi Med J 36(2):248–251

Blatch GL (2022) *Plasmodium falciparum* Molecular Chaperones: Guardians of the Malaria Parasite Proteome and Renovators of the Host Proteome. Front Cell Dev Biol 10:921739

Botha M, Chiang AN, Needham PG, Stephens LL, Hoppe HC, Külzer S, Przyborski JM, Lingelbach K, Wipf P, Brodsky JL, Shonhai A, Blatch GL (2011) *Plasmodium falciparum* encodes a single cytosolic type I Hsp40 that functionally interacts with Hsp70 and is upregulated by heat shock. Cell Stress Chaperones 16(4):389–401

Bouwman H, van den Berg H, Kylin H (2011) DDT and malaria prevention: addressing the paradox. Environ Health Perspect 119(6):744–747

Caminade C, Kovats S, Rocklov J, Tompkins AM, Morse AP, Colón-González FJ, Stenlund H, Martens P, Lloyd SJ (2014) Impact of climate change on global malaria distribution. Proc Natl Acad Sci USA 111(9):3286–3291

Carrington LB, Armijos MV, Lambrechts L, Barker CM, Scott TW (2013) Effects of fluctuating daily temperatures at critical thermal extremes on *Aedes aegypti* life-history traits. PLoS One 8(3):e58824

Casida JE, Ueda K, Gaughan LC, Jao LT, Soderlund DM (1975) Structure-biodegradability relationships in pyrethroid insecticides. Arch Environ Contam Toxicol 3(4):491–500

Chen Y, Murillo-Solano C, Kirkpatrick MG, Antoshchenko T, Park HW, Pizarro JC (2018) Repurposing drugs to target the malaria parasite unfolding protein response. Sci Rep 8(1):1–12

Cockburn IL, Pesce ER, Pryzborski JM, Davies-Coleman MT, Clark PG, Keyzers RA, Stephens LL, Blatch GL (2011) Screening for small molecule modulators of Hsp70 chaperone activity using protein aggregation suppression assays: inhibition of the plasmodial chaperone PfHsp70-1. Biol Chem 392(5):431–438

Cockburn IL, Boshoff A, Pesce ER, Blatch GL (2014) Selective modulation of plasmodial Hsp70s by small molecules with antimalarial activity. Biol Chem 395(11):1353–1362

Coleman M, Al-Zahrani MH, Coleman M, Hemingway J, Omar A, Stanton MC, Thomsen EK, Alsheikh AA, Alhakeem RF, McCall PJ et al (2014) A country on the verge of malaria elimination–the Kingdom of Saudi Arabia. PLoS One 9(9):e105980:1–e105980:8

Corrêa APSA, Galardo AKR, Lima LA et al (2019) Efficacy of insecticides used in indoor residual spraying for malaria control: an experimental trial on various surfaces in a "test house". Malar J 18(1):345

Cortés GT, Wiser MF, Gómez-Alegría CJ (2020) Identification of *Plasmodium falciparum* HSP70-2 as a resident of the *Plasmodium* export compartment. Heliyon 6(6):e04037

Costello A, Abbas M, Allen A, Ball S, Bell S, Bellamy R, Friel S, Groce N, Johnson A, Kett M, Lee M (2009) Managing the health effects of climate change: lancet and University College London Institute for Global Health Commission. Lancet 373(9676):1693–1733

Cycoń M, Piotrowska-Seget Z (2016) Pyrethroid-degrading microorganisms and their potential for the bioremediation of contaminated soils: a Review. Front Microbiol 7:1463

Dabira ED, Soumare HM, Conteh B, Ceesay F, Ndiath MO, Bradley J, Mohammed N, Kandeh B, Smit MR, Slater H, Peeters Grietens K, Broekhuizen H, Bousema T, Drakeley C, Lindsay SW, Achan J, D'Alessandro U (2022) Mass drug administration of ivermectin and dihydroartemisinin–piperaquine against malaria in settings with high coverage of standard control interventions: a cluster-randomised controlled trial in The Gambia. Lancet Infect Dis 22(4):519–528

Dafalla OM, Alzahrani M, Sahli A, Al Helal MA, Alhazmi MM, Noureldin EM, Mohamed WS, Hamid TB, Abdelhaleem AA, Hobani YA et al (2020) Kelch 13-propeller polymorphisms in *Plasmodium falciparum* from Jazan region, southwest Saudi Arabia. Malar J 19(1):397

Daniyan MO, Blatch GL (2017) Plasmodial Hsp40s: New Avenues for Antimalarial Drug Discovery. Curr Pharm Des 23(30):4555–4570

Daniyan MO, Boshoff A, Prinsloo E, Pesce ER, Blatch GL (2016) The malarial exported PFA0660w is an Hsp40 co-chaperone of PfHsp70-x. PLoS One 11(2):e0148517

Datoo MS, Natama HM, Somé A, Bellamy D, Traoré O, Rouamba T, Tahita MC, Ido NFA, Yameogo P, Valia D, Millogo A, Ouedraogo F, Soma R, Sawadogo S, Sorgho F, Derra K, Rouamba E, Ramos-Lopez F, Cairns M, Provstgaard-Morys S et al (2022) Efficacy and immunogenicity of R21/Matrix-M vaccine against clinical malaria after 2 years' follow-up in children in Burkina Faso: a phase 1/2b randomised controlled trial. Lancet Infect Dis 22(12):1728–1736

Day J, Passecker A, Beck HP, Vakonakis I (2019) The *Plasmodium falciparum* Hsp70-x chaperone assists the heat stress response of the malaria parasite. FASEB J 33(12):14611

Dutta T, Pesce ER, Maier AG, Blatch GL (2021a) Role of the J Domain Protein Family in the Survival and Pathogenesis of *Plasmodium falciparum*. Adv Exp Med Biol 1340:97–123

Dutta T, Singh H, Gestwicki JE, Blatch GL (2021b) Exported plasmodial J domain protein, PFE0055c, and PfHsp70-x form a specific co-chaperone-chaperone partnership. Cell Stress Chaperones 26(2):355–366

Edkins AL, Boshoff A (2021) General Structural and Functional Features of Molecular Chaperones. Adv Exp Med Biol 1340:11–73

Eisele TP (2019) Mass drug administration can be a valuable addition to the malaria elimination toolbox. Malar J 18(1):281

El Hassan IM, Sahly A, Alzahrani MH, Alhakeem RF, Alhelal M, Alhogail A, Alsheikh AA, Assiri AM,

ElGamri TB, Faragalla IA et al (2015) Progress toward malaria elimination in Jazan Province, Kingdom of Saudi Arabia: 2000–2014. Malar J 14:444

Farag E, Bansal D, Chehab MAH, Al-Dahshan A, Bala M, Ganesan N, Al Abdulla YA, Al Thani M, Sultan AA, Al-Romaihi H (2018) Epidemiology of Malaria in the State of Qatar, 2008-2015. Mediterr J Hematol Infect Dis 10(1):e2018050

Feleke SM, Reichert EN, Mohammed H, Brhane BG, Mekete K, Mamo H, Petros B, Solomon H, Abate E, Hennelly C, Denton M, Keeler C, Hathaway NJ, Juliano JJ, Bailey JA, Rogier E, Cunningham J, Aydemir O, Parr JB (2021) *Plasmodium falciparum* is evolving to escape malaria rapid diagnostic tests in Ethiopia. Nat Microbiol 6(10):1289–1299

Fraser M, Miller JM, Silumbe K, Hainsworth M, Mudenda M, Hamainza B, Moonga H, Chizema Kawesha E, Mercer LD, Bennett A, Schneider K, Slater HC, Eisele TP, Guinovart C (2020) Evaluating the impact of programmatic mass drug administration for malaria in Zambia using routine incidence data. J Infect Dis 225(8):1415–1423

Hancock PA, Hendriks CJM, Tangena JA, Gibson H, Hemingway J, Coleman M, Gething PW, Cameron E, Bhatt S, Moyes CL (2020) Mapping trends in insecticide resistance phenotypes in African malaria vectors. PLoS Biol 18(6):e3000633

Hassan KS (2017) World Malaria Day: Our story with malaria in Oman. Sultan Qaboos Univ Med J 17(2):e133–e134

IHME (2022) Our World in Data. Institute of Health Metrics and Evaluation (IHME). www.ourworldindata.org/malaria. Accessed 17 July 2023

Iqbal J, Sher A, Hira PR, Al-Anezi AA (2000) Risk of introduction of drug-resistant malaria in a non-endemic country, Kuwait: A real threat? Med Princ Pract 9:125–130

Iqbal J, Hira PR, Al-Ali F, Sher A (2003) Imported malaria in Kuwait (1985-2000). J Travel Med 10(6):324–329

Iqbal J, Al-Awadhi M, Ahmad S (2020) Decreasing trend of imported malaria cases but increasing influx of mixed *P. falciparum* and *P. vivax* infections in malaria-free Kuwait. PLoS One 15(12):e0243617

Iqbal J, Ahmad S, Sher A, Al-Awadhi M (2021) Current epidemiological characteristics of imported malaria, vector control status and malaria elimination prospects in the Gulf Cooperation Council (GCC) countries. Microorganisms 9(7):1431

Ismaeel AY, Senok AC, Jassim Al-Khaja KA, Botta GA (2004) Status of malaria in the Kingdom of Bahrain: a 10-year review. J Travel Med 11(2):97–101

Jee H (2016) Size dependent classification of heat shock proteins: a mini-review. J Exerc Rehabil 12(4):255–259

Killeen GF, Masalu JP, Chinula D, Fotakis EA, Kavishe DR, Malone D, Okumu F (2017) Control of malaria vector mosquitoes by insecticide-treated combinations of window screens and eave baffles. Emerg Infect Dis 23(5):782–789

Kouamo MFM, Ibrahim SS, Hearn J, Riveron JM, Kusimo M, Tchouakui M, Ebai T, Tchapga W, Wondji MJ, Irving H, Boudjeko T, Boyom FF, Wondji CS (2021) Genome-wide transcriptional analysis and functional validation linked a cluster of epsilon glutathione S-transferases with insecticide resistance in the major malaria vector *Anopheles funestus* across Africa. Genes 12(4):561

Laurens MB (2020) RTS,S/AS01 vaccine (Mosquirix™): an overview. Hum Vaccin Immunother 16(3):480–489

Luo AP, Giannangelo C, Siddiqui G, Creek DJ (2023) Promising antimalarial hits from phenotypic screens: a review of recently-described multi-stage actives and their modes of action. Front Cell Infect Microbiol 13:1308193

Madkhali AM, Al-Mekhlafi HM, Atroosh WM, Ghzwani AH, Zain KA, Abdulhaq AA, Ghailan KY, Anwar AA, Eisa ZM (2020) Increased prevalence of *pfdhfr* and *pfdhps* mutations associated with sulfadoxine-pyrimethamine resistance in *Plasmodium falciparum* isolates from Jazan Region, Southwestern Saudi Arabia: important implications for malaria treatment policy. Malar J 19(1):446

Memish ZA, Alzahrani M, Alhakeem RF, Bamgboye EA, Smadi HN (2014) Toward malaria eradication in Saudi Arabia: evidence from 4-year surveillance in Makkah. Ann Saudi Med 34(2):153–158

Mok S, Ashley EA, Ferreira PE, Zhu L, Lin Z, Yeo T, Chotivanich K, Imwong M, Pukrittayakamee S, Dhorda M, Nguon C, Lim P, Amaratunga C, Suon S, Hien TT, Htut Y, Faiz MA, Onyamboko MA, Mayxay M, Newton PN et al (2015) Drug resistance. Population transcriptomics of human malaria parasites reveals the mechanism of artemisinin resistance. Science 347(6220):431–435

Mok S, Stokes BH, Gnädig NF, Ross LS, Yeo T, Amaratunga C, Allman E, Solyakov L, Bottrill AR, Tripathi J, Fairhurst RM, Llinás M, Bozdech Z, Tobin AB, Fidock DA (2021) Artemisinin-resistant K13 mutations rewire *Plasmodium falciparum's* intra-erythrocytic metabolic program to enhance survival. Nat Commun 12(1):530

Mordecai EA, Caldwell JM, Grossman MK, Lippi CA, Johnson LR, Neira M, Rohr JR, Ryan SJ, Savage V, Shocket MS, Sippy R (2019) Thermal biology of mosquito-borne disease. Ecol Lett 22(10):1690–1708

Mrozek A, Antoshchenko T, Chen Y, Zepeda-Velázquez C, Smil D, Kumar N, Lu H, Park HW (2022) A non-traditional crystal-based compound screening method targeting the ATP binding site of GRP78 for identification of novel nucleoside analogues. Front Mol Biosci 9:956095

Neafsey DE, Juraska M, Bedford T, Benkeser D, Valim C, Griggs A, Lievens M, Abdulla S, Adjei S, Agbenyega T, Agnandji ST, Aide P, Anderson S, Ansong D, Aponte JJ, Asante KP, Bejon P, Birkett AJ, Bruls M et al (2015) Genetic diversity and protective efficacy of the RTS,S/AS01 malaria vaccine. N Engl J Med 373(21):2025–2037

Nilles EJ, Alosert M, Mohtasham MA, Saif M, Sulaiman L, Seliem RM, Kotlyar S, Dziura JD, Al-Najjar FJ (2014) Epidemiological and clinical characteristics of imported malaria in The United Arab Emirates. J Travel Med 21(3):201–206

Paloque L, Coppée R, Stokes BH, Gnädig NF, Niaré K, Augereau JM, Fidock DA, Clain J, Benoit-Vical F (2022) Mutation in the *Plasmodium falciparum* BTB/POZ domain of K13 protein confers artemisinin resistance. Antimicrob Agents Chemother 66(1):e0132021

Pennetier C, Costantini C, Corbel V, Licciardi S, Dabiré RK, Lapied B, Chandre F, Hougard JM (2008) Mixture for controlling insecticide-resistant malaria vectors. Emerg Infect Dis 14(11):1707–1714

Plowe CV (2022) Malaria chemoprevention and drug resistance: a review of the literature and policy implications. Malar J 21(1):104

Posfai D, Eubanks AL, Keim AI, Lu KY, Wang GZ, Hughes PF, Kato N, Haystead TA, Derbyshire ER (2018) Identification of Hsp90 Inhibitors with Anti-*Plasmodium* Activity. Antimicrob Agents Chemother 62(4):e01799–e01717

Przyborski JM, Diehl M, Blatch GL (2015) Plasmodial HSP70s are functionally adapted to the malaria parasite life cycle. Front Mol Biosci 2:34

Rawson T, Doohan P, Hauck K, Murray KA, Ferguson N (2023) Climate change and communicable diseases in the Gulf Cooperation Council (GCC) countries. Epidemics 42:100667

Riveron JM, Tchouakui M, Mugenzi L, Menze BD, Chiang MC, Wondji CS (2018) Insecticide resistance in malaria vectors: an update at a global scale. In: Towards malaria elimination-a leap forward. IntechOpen. https://doi.org/10.5772/intechopen.78375

RTS,S Clinical Trials Partnership (2015) Efficacy and safety of RTS,S/AS01 malaria vaccine with or without a booster dose in infants and children in Africa: final results of a phase 3, individually randomised, controlled trial. Lancet 386(9988):31–45

Salimi M, Al-Ghamdi SG (2020) Climate change impacts on critical urban infrastructure and urban resiliency strategies for the Middle East. Sustain Cities Soc 54:101948

Seraphim TV, Chakafana G, Shonhai A, Houry WA (2019) *Plasmodium falciparum* R2TP complex: driver of parasite Hsp90 function. Biophys Rev 11(6):1007–1015

Shahinas D, Folefoc A, Taldone T, Chiosis G, Crandall I, Pillai DR (2013) A purine analog synergizes with chloroquine (CQ) by targeting *Plasmodium falciparum* Hsp90 (PfHsp90). PLoS One 8(9):e75446

Shapiro LL, Whitehead SA, Thomas MB (2017) Quantifying the effects of temperature on mosquito and parasite traits that determine the transmission potential of human malaria. PLoS Biol 15(10):e2003489

Sher A, Latif SA (2023) Increasing trend of mixed infections of *Plasmodium falciparum* and *Plasmodium vivax* among imported malaria cases in Kuwait, a non-endemic country. Open Access Library Journal 10:e9733

Shibeshi MA, Kifle ZD, Atnafie SA (2020) Antimalarial drug resistance and novel targets for antimalarial drug discovery. Infect Drug Resist 13:4047–4060

Shibl A, Senok A, Memish Z (2012) Infectious diseases in the Arabian Peninsula and Egypt. Clin Microbiol Infect 18(11):1068–1080

Shonhai A, Maier AG, Przyborski JM, Blatch GL (2011) Intracellular protozoan parasites of humans: the role of molecular chaperones in development and pathogenesis. Protein Pept Lett 18(2):143–157

Siddiqui FA, Liang X, Cui L (2021) *Plasmodium falciparum* resistance to ACTs: emergence, mechanisms, and outlook. Int J Parasitol. Drugs Drug Resist 16:102–118

Siddiqui AJ, Bhardwaj J, Saxena J, Jahan S, Snoussi M, Bardakci F, Badraoui R, Adnan M (2023) A critical review on human malaria and schistosomiasis vaccines: current state, recent advancements, and developments. Vaccine 11(4):792

Simon B, Sow F, Al Mukhaini SK, Al-Abri S, Ali OAM, Bonnot G, Bienvenu AL, Petersen E, Picot S (2017) An outbreak of locally acquired *Plasmodium vivax* malaria among migrant workers in Oman. Parasite 24:25

Singh H, Almaazmi SY, Dutta T, Keyzers RA, Blatch GL (2023) In silico identification of modulators of J domain protein-Hsp70 interactions in *Plasmodium falciparum*: a drug repurposing strategy against malaria. Front Mol Biosci 10:1158912

Snow RW, Amratia P, Zamani G, Mundia CW, Noor AM, Memish ZA, Al Zahrani MH, Al Jasari A, Fikri M, Atta H (2013) The malaria transition on the Arabian Peninsula: progress toward a malaria-free region between 1960-2010. Adv Parasitol 82:205–251

Soliman RH, Garcia-Aranda P, Elzagawy SM, Hussein BE, Mayah WW, Martin Ramirez A, Ta-Tang TH, Rubio JM (2018) Imported and autochthonous malaria in West Saudi Arabia: results from a reference hospital. Malar J 17(1):286

Stofberg ML, Caillet C, de Villiers M, Zininga T (2021) Inhibitors of the *Plasmodium falciparum* Hsp90 towards selective antimalarial drug design: the past, present and future. Cells 10(11):2849

Tun STT, von Seidlein L, Pongvongsa T, Mayxay M, Saralamba S, Kyaw SS, Chanthavilay P, Celhay O, Nguyen TD, Tran TN, Parker DM, Boni MF, Dondorp AM, White LJ (2017) Towards malaria elimination in Savannakhet, Lao PDR: mathematical modelling driven strategy design. Malar J 16(1):483

Venkatesan P (2024) The 2023 WHO World Malaria Report. Lancet Microbe 5(3):e214

von Seidlein L, Hanboonkunupakarn B, Jittamala P, Pongsuwan P, Chotivanich K, Tarning J, Hoglund RM, Winterberg M, Mukaka M, Peerawaranun P, Sirithiranont P, Doran Z, Ockenhouse CF, Ivinson K, Lee C, Birkett AJ, Kaslow DC, Singhasivanon P, Day NPJ, Dondorp AM, White NJ, Pukrittayakamee S (2020) Combining antimalarial drugs and vaccine for malaria elimination campaigns: a randomized safety and immunogenicity trial of RTS,S/AS01 administered with dihydroartemisinin, piperaquine, and primaquine in healthy Thai adult volunteers. Hum Vaccin Immunother 16(1):33–41

Waha K, Krummenauer L, Adams S, Aich V, Baarsch F, Coumou D, Fader M, Hoff H, Jobbins G, Marcus R, Mengel M (2017) Climate change impacts in the Middle East and Northern Africa (MENA) region and their implications for vulnerable population groups. Reg Environ Chang 17:1623–1638

Wang T, Mäser P, Picard D (2016) Inhibition of *Plasmodium falciparum* Hsp90 contributes to the antimalarial activities of aminoalcohol-carbazoles. J Med Chem 59(13):6344–6352

West PA, Protopopoff N, Wright A, Kivaju Z, Tigererwa R, Mosha FW, Kisinza W, Rowland M, Kleinschmidt I (2014) Indoor residual spraying in combination with insecticide-treated nets compared to insecticide-treated nets alone for protection against malaria: a cluster randomised trial in Tanzania. PLoS Med 11(4):e1001630

WHO (2017) Regional Malaria Action Plan 2016–2020: towards a Malaria-Free Region. World Health Organization, Regional Office for the Eastern Mediterranean. applications.emro.who.int/docs/EMROPUB_2017_EN_19546.pdf. Accessed 17 July 2023

WHO (2019) Vector alert: *Anopheles stephensi* invasion and spread: Horn of Africa, the Republic of the Sudan and surrounding geographical areas, and Sri Lanka: information note. World Health Organization, Geneva. iris.who.int/handle/10665/326595

WHO (2021) Global Technical Strategy for Malaria 2016–2030, 2021 Update. World Health Organization, Geneva. www.who.int/publications/i/item/9789240031357. Accessed 17 July 2023

WHO (2022a) Guidelines for Malaria. World Health Organization, Geneva. reliefweb.int/report/world/who-guidelines-malaria-3-june-2022. Accessed 17 July 2023

WHO (2022b) Vector alert: *Anopheles stephensi* invasion and spread in Africa and Sri Lanka: information note. World Health Organization, Geneva. www.who.int/publications/i/item/9789240067714

WHO (2022c) World Malaria Report. World Health Organization, Geneva. www.who.int/teams/global-malaria-programme/reports. Accessed 17 July 2023

WHO (2023a) World Malaria Report. World Health Organization, Geneva. www.who.int/teams/global-malaria-programme/reports. Accessed 20 Mar 2024

WHO (2023b) Countries and Territories Certified Malaria-Free by WHO. World Health Organization, Geneva. www.who.int/teams/global-malaria-programme/elimination/countries-and-territories-certified-malaria-free-by-who. Accessed 17 July 2023

Yoon SJ, Kim YE, Kim EJ (2018) Why They Are Different: Based on the Burden of Disease Research of WHO and Institute for Health Metrics and Evaluation. Biomed Res Int 2018:7236194

Zavala F (2022) RTS,S: The first malaria vaccine. J Clin Invest 132:e156588

Zhang M, Wang C, Otto TD, Oberstaller J, Liao X, Adapa SR, Udenze K, Bronner IF, Casandra D, Mayho M, Brown J (2018) Uncovering the essential genes of the human malaria parasite *Plasmodium falciparum* by saturation mutagenesis. Science 360(6388):eaap7847

Zininga T, Pooe OJ, Makhado PB, Ramatsui L, Prinsloo E, Achilonu I, Dirr H, Shonhai A (2017a) Polymyxin B inhibits the chaperone activity of *Plasmodium falciparum* Hsp70. Cell Stress Chaperones 22(5):707–715

Zininga T, Ramatsui L, Makhado P, Makumire S, Achilinou I, Hoppe H, Dirr H, Shonhai A (2017b) (−)-Epigallocatechin-3-Gallate inhibits the chaperone activity of *Plasmodium falciparum* Hsp70 chaperones and abrogates their association with functional partners. Molecules 22(12):2139

5

Tendamudzimu Harmfree Dongola
and Addmore Shonhai

Abstract

About 15 to 20% of the total Hsp70 is released to the extracellular space of some tumour cells, where it associates with cholesterol-rich microdomains located on the cell surface. Extracellular Hsp70 is known to recruit natural killer cells, leading to the concomitant release of granzyme B (GrB). This leads to the selective destruction of the tumour cells. A 14-mer TKD motif (KDNNLLGRFELSG) of human Hsp70 (hHsp70) is implicated in activating the NK cells. One of the defining features of *Plasmodium falciparum*, the most prevalent and lethal agent of malaria, is that it exports one of its Hsp70s, named PfHsp70-x, to the infected human red blood cell (RBC). In addition, hHsp70 is also present in the RBC. It has been reported that the infection of the RBC by the malaria parasite is associated with the migration of Hsp70 to the membrane of the host cell. Interestingly, GrB is elevated in severe malaria (SMA). PfHsp70-x, present on the outer leaflet of the infected RBC membrane, may stimulate NK cells to release GrB. We discuss the possible roles of Hsp70 and GrB in driving the development of SMA, and the prospects of Hsp70-GrB-mediated antimalarial therapy.

Keywords

Plasmodium falciparum · PfHsp70-x · Human Hsp70 · TKD motif · Granzyme B · Red blood cell

The main victims of malaria are children under the age of 5 and pregnant women, with over 93% of cases reported in sub-Saharan Africa (WHO 2024). Although there has been a general decline in malaria-related deaths since 2000, malaria control efforts have not yielded consistent gains over the years (Fig. 5.1). For example, between 2019 and 2021, there was a significant increase in malaria cases and deaths, and this may have been caused by the allocation of resources to the management of the COVID-19 pandemic (Fig. 5.1). Consequently, in 2021, there were 247 million cases and 619,000 deaths (WHO 2022). This represents an increase in malaria cases in Africa from 222 to 232 per 1000 individuals in 2019 and 2020, respectively (WHO 2021). It is, therefore, necessary to develop effective malaria control measures towards the eradication of the disease. The increase in malaria cases continued in 2023, reaching 263 million, an increase of 11 million

T. H. Dongola
Department of Biochemistry and Microbiology, University of Venda, Thohoyandou, South Africa

A. Shonhai (✉)
Department of Biochemistry and Microbiology, Faculty of Science, Engineering and Agriculture, University of Venda, Thohoyandou, South Africa
e-mail: addmore.shonhai@univen.ac.za

© The Author(s) 2026
A. Shonhai et al. (eds.), *Advances in Biochemistry and Molecular Biology to meet Africa´s Needs*, Advances in Experimental Medicine and Biology 1507,
https://doi.org/10.1007/978-3-032-24254-9_5

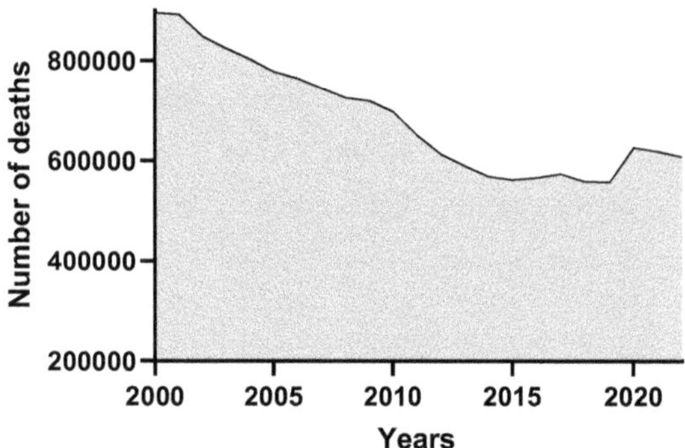

Fig. 5.1 Graph representing malaria deaths. The graph was plotted based on the World Health Organization malaria reports for the period 2000–2024

cases compared to 2022 (WHO 2024). The main factor in the growing number of malaria cases and related deaths is parasite drug resistance (Menard and Dondorp 2017), necessitating the need for new antimalarial therapies.

Malaria is caused by protozoa of the genus *Plasmodium*, of which five species (*P. vivax*, *P. falciparum*, *P. knowlesi*, *P. malariae*, and *P. ovale*) infect humans. While occasional cases of severe malaria (SMA) caused by *P. vivax* (Rahimi et al. 2014) and *P. knowlesi* (William et al. 2011) have been reported, it is *P. falciparum* that accounts for the most severe cases of the disease (Trampuz et al. 2003). The malaria parasite has two distinct life stages: a stint in the cold-blooded *Anopheles* mosquito vector followed by development within the warm-blooded human host. Sporozoites are injected into the human host during a mosquito bite. The sporozoites travel via the bloodstream, subsequently invading the hepatocytes. The sporozoites multiply asexually in the hepatocytes over the next 7–10 days, causing no symptoms. The sporozoites eventually develop into merozoites (Arama and Troye-Blomberg 2014). The merozoites are released into the blood when the schizonts rupture, infecting the red blood cells (RBCs). It is at the RBC stage that clinical malaria manifests. The breakdown of the infected RBC leads to the release of the inorganic crystal, hemozoin (Jani et al. 2008), and glucose phosphate isomerase (Srivastava et al. 1992) (GPI), both of which enter the bloodstream. These products modulate the immune system to

produce cytokines, leading to the manifestation of malaria symptoms such as periodic chills and fevers. The infection of the RBC by the parasite results in the modification of the cell surface of the host cell, leading to cyto-adherence, giving rise to complications such as cerebral malaria.

5.1 The Host Immune Response to Malaria

Upon invading the human host, *P. falciparum* undergoes two stages: an asymptomatic pre-erythrocytic (liver) stage, and a symptomatic erythrocytic stage (blood stage). The migration of sporozoites to the liver is asymptomatic; however, it is thought that some immune response is initiated at this stage (Zheng et al. 2014). Some of the key immunological responses associated with this stage include: the apoptosis of infected cells (Meslin et al. 2007), isolation and targeting of parasites in specific compartments for elimination (Yano and Kurata 2011), production of type I IFN through the induction of parasite RNA (Liehl et al. 2015), and targeting of sporozoites through light chain 3 (LC3)-mediated autophagy (Risco-Castillo et al. 2015). Parasites that evade this immune response replicate rapidly within the liver and are eventually released as several thousand merozoites that enter the circulatory system (Prado et al. 2015).

Antibodies exhibit anti-parasitic activity by binding to the iRBCs to initiate their phagocytosis (opsonization) by circulating macrophages (Dups et al. 2014). Furthermore, antibodies can prevent the invasion of RBCs by attaching to extracellular merozoites and earmarking them for clearance or lysis by the host complement system. T-cells play an essential role by both direct and indirect mechanisms. One way in which T-cells directly respond is by producing pro-inflammatory cytokines such as IFN-γ and TNF-α, thus alerting macrophages and other components of the anti-parasite immune response (Nasr et al. 2014). An indirect mechanism of T-cells involves activation of specific B-cell clones to produce anti-parasite antibodies (Kafuye-Mlwilo et al. 2012). Apart from T-cells, the host also produces natural killer (NK) cells upon encountering foreign antigens (Nagata et al. 2002). The NK cells, in turn, release granules that bind to ligands expressed on the cell surface of the infected cells (Nagata et al. 2002). NK cells target malaria-parasite iRBCs through antibody-dependent cellular cytotoxicity (ADCC), involving the release of lytic granules that contain perforin, granzyme B, and granulysin, as well as the release of IFNγ (Damelang et al. 2021; Dick and Hart 2022).

While ageing RBCs are destroyed in the spleen, the parasite iRBCs evade splenic clearance through sequestration, cytoadherence, and rosetting (Chotivanich et al. 2002; Lee et al. 2019). In the same way, the malaria parasites evade the host's complement system, constituted by a membrane attack complex (MAC) that lyses pathogens, with the evading parasites playing an important role in the development of SMA (Schmidt et al. 2015).

5.2 Hsp70 and its Role in the Development of Malaria

Hsp70 functions as a molecular chaperone, protecting cells during cellular stress (Bukau and Walker 1989; Flaherty et al. 1990; Shonhai et al. 2005). Hsp70s are typically induced in response to stress. However, some Hsp70s are constitutively expressed and are referred to as heat cognate proteins 70 (Hsc70) (Fakhari et al. 2013). Hsp70 is made up of two main domains: the N-terminal ATPase domain (NBD), whose size is 45 kDa, and the C-terminal substrate-binding domain (SBD), with a size of approximately 25 kDa (Fig. 5.2; Flaherty et al. 1990; Wang et al. 1993). The SBD is further subdivided into two components: the β-domain (SBDβ), which harbors the substrate-binding pocket, and the α-helical lid (SBDα) (Fig. 5.2; Wang et al. 1993). The NBD and SBD are joined by a flexible linker region that is important for allosteric communication (Chakafana et al. 2019a). The structure-function features of Hsp70s of *P. falciparum* and their human counterparts have been previously described (Chakafana et al. 2019b; Shonhai 2021).

Fig. 5.2 Hsp70 structural organization. (Hsp70 possesses two domains: nucleotide-binding domain (NBD) and substrate-binding domain (SBD). The latter is characterised by two subdomains: SBDβ (binds the peptide substrate) and SBDα (lid segment). The NBD and the SBD are joined by a flexible linker.

Intracellular Hsp70 is a molecular chaperone that acts as a housekeeping and stress response molecule. It is responsible for several critical functions such as protein folding, transportation, and dissociation of protein complexes (Kim et al. 2013; Saibil 2013; Shonhai 2021). To maintain homeostasis, Hsp70 collaborates with several chaperones such as Hsp40, Hsp60, Hsp90, and Hsp100 (Buchberger et al. 1996; Diamant et al. 2000; Bukau et al. 2006; Shonhai 2021). The cellular clients of Hsp70 number in several hundreds, highlighting its importance in cellular development (Nollen et al. 2001; Mayer and Bukau 2005; Ryu et al. 2020; Dongola et al. 2023; Chakafana et al. 2024). Hsp70 has been found to protect cells under heat stress conditions, acting as a molecular chaperone to shield the cell from death (Shonhai et al. 2007; Makumire et al. 2021; Dongola et al. 2024). Numerous studies have implicated Hsp70 in the suppression of apoptosis (Jäättelä et al. 1998; Zorzi and Bonvini 2011; Vostakolaei et al. 2021). Consequently, Hsp70 upregulation is associated with the progression of cancer (Albakova et al. 2020; Elmallah et al. 2020).

Conditions that mimic the fever episodes associated with malaria induce the expression of several Hsps, including Hsp70, indicating their significance in the cytoprotection of the parasite during the development of malaria (Pavithra et al. 2004; Botha et al. 2011; Gitau et al. 2012; Zininga et al. 2016). At least six Hsp70s members are represented in the genome of *P. falciparum* (Shonhai et al. 2007; Shonhai 2021). The subcellular localization and roles of these proteins have been described (Table 5.1). The roles of these proteins in the development and pathogenicity of *P. falciparum* have been previously reviewed (Shonhai

Table 5.1 Cellular distribution and functions of *P. falciparum* Hsp70s

Protein name	Localization	Functions
PfHsp70–1 (PF3D7_0818900)	Cytosol and nucleus[a,b]	Essential cytosol/nuclear localised chaperone[c–f]
PfHsp70–2 (PF3D7_0917900)	ER[d]	Import and fold proteins in the ER[d,g]
PfHsp70–3 (PF3D7–1,134,000)	Mitochondrion[h]	Import of proteins into the mitochondrion and facilitating their folding[i]
PfHsp70-x (PF3D7_0831700)	PV, host cell cytosol[j,k]	Implicated in the trafficking and folding of parasite proteins in the host iRBC; also thought to facilitate GrB uptake by parasite iRBC[g,l]
PfHsp70-y (PF3D7_1344200)	ER[m]	Essential chaperone and a possible NEF for PfHsp70-2[m,n]
PfHsp70-z (PF3D7_0708800)	Cytosol[g]	Essential for the survival of the parasite and is thought to serve as PfHsp70–1 NEF[o,p]

[a]Kumar et al. 1991, [b]Njunge et al. 2015, [c]Przyborski et al. 2015, [d]Makumire et al. 2021, Cortés et al. 2020, [e]Shonhai et al. 2005, [f]Chiang et al. 2009, [g]Shonhai et al. 2007, [h]Kumar et al. 1991, [i]Njunge et al. 2013, [j]Charnaud et al. 2017, [k]Külzer et al. 2010, [l]Cobb et al. 2017, [m]Chakafana et al. 2019a, b, [n]Kudyba et al. 2019, [o]Muralidharan et al. 2012, [p]Zininga et al. 2016

ER endoplasmic reticulum, *PV* parasitophorous vacuole, *NEF* nucleotide exchange factor

et al. 2007; Shonhai 2021). Here, our interest is largely focused on PfHsp70-x, which is exported to the iRBC (Külzer et al. 2012).

One of the defining features of *P. falciparum* is that it exports PfHsp70-x, an Hsp70 that is encoded on the parasite genome (Külzer et al. 2012). PfHsp70-x has been found to occur in complex with other malarial proteins exported to the RBC (Külzer et al. 2012; Grover et al. 2013; Zhang et al. 2017; Behl et al. 2019). Notably, *P. falciparum* is the only agent of malaria that exports Hsp70 to the iRBC. Although PfHsp70-x is not essential, knockout of its gene reduced parasite virulence (Charnaud et al. 2017). In addition, the knockdown of its gene compromised the growth of the parasite under conditions that mimic fever episodes of clinical malaria (Day et al. 2019). Altogether, these reports suggest a possible role of PfHsp70-x in augmenting the virulence of *P. falciparum*. Indeed, a previous study suggested that host chaperones, including hHsp70 and hHsp90, occur largely in the cytosol of uninfected RBC, and that these proteins migrate to the membrane of the iRBC (Banumathy et al. 2002). Studies based on cancer cells demonstrated the occurrence of Hsp70 on the outer leaflet of the membrane coupled to its externalisation (Gehrmann et al. 2008). We infer from these findings that both PfHsp70-x and its hHsp70 could be located on the outer leaflet of the iRBC membrane. In addition, the membrane-bound Hsp70 and Hsp90 were previously reported to occur in complex with knob-enriched proteins such as *P. falciparum* histidine-rich protein 1 (PfHRP1) (Banumathy et al. 2002). This further suggests that chaperones of both parasite and human origin may facilitate the presentation of knob-like structures located on the surface of the iRBC. It is most probable that PfHsp70-x complements the role of hHsp70 in this regard. That of all human malaria-causing pathogens, only *P. falciparum* expresses the Hsp70-x isoform could explain why this species is responsible for the most severe form of the disease.

5.3 The Role of PfHsp70-x and Some of Its Interactors at the Host-Parasite Interface

The malaria parasite *P. falciparum* exports approximately 10% of its proteome into the cytosol of the iRBC, where these proteins extensively remodel the host cell to support parasite survival and promote pathogenesis (Maier et al. 2008; Jonsdottir et al. 2021). This export process is mediated by the *Plasmodium* translocon of exported proteins (PTEX), a multiprotein complex embedded in the parasitophorous vacuole membrane (PVM) that facilitates the translocation of parasite proteins across the PVM and into the host cytoplasm (Hiller et al. 2004; Marti et al. 2005; de Koning-Ward et al. 2009; Bullen et al. 2012). Most exported proteins contain the PEXEL that guides them through the PTEX pathway (Hiller et al. 2004; Boddey et al. 2009). Once in the host cytosol, these proteins contribute to virulence by enabling cytoadherence, immune evasion, and nutrient uptake (Hiller et al. 2004; Marti et al. 2005; de Koning-Ward et al. 2009; Bullen et al. 2012).

A key virulence protein, *P. falciparum* erythrocyte membrane protein 1 (PfEMP1), is trafficked to the iRBC surface where it mediates adhesion to host endothelial receptors, a process essential for avoiding splenic clearance and contributing to SMA syndromes (Batinovic et al. 2017). Both PfHsp70-x and select co-chaperones of the Hsp40 family that are exported to the iRBC are proposed to facilitate the presentation of PfEMP1 on the iRBC surface (Maier et al. 2008; Külzer et al. 2012). In addition to its chaperone functions, PfHsp70-x is proposed to play a role in modulating host immune responses. A previous study suggests that members of the Hsp70 family, when exposed on the surface of infected cells or released extracellularly, can act as danger signals (DAMPs) that stimulate innate immune receptors and promote inflammation, potentially contributing to the immunopathology of malaria (Multhoff 2007).

5.4   PfHsp70-x: A Biomarker of Severe Malaria

The World Health Organisation defines clinical manifestations of SMA as marked by the following major complications: cerebral malaria, pulmonary oedema, acute renal failure, severe anaemia, and/or bleeding, acidosis, and hypoglycaemia (WHO 2000). *P. falciparum* is more virulent than other species, causing human malaria due to its ability to infect RBCs of all ages as well as its capability to bind to human endothelial cells, leading to complications such as cerebral malaria. On the other hand, other less virulent species, such as *P. ovale* and *P. vivax*, prefer binding to younger RBC (Sisquella et al. 2017). The export of nearly a tenth of the parasite proteome is linked to the virulence of *P. falciparum* (Maier et al. 2008). Since *P. falciparum* is the only species causing human malaria that exports its Hsp70-x protein homologue to the iRBC, the role of this chaperone in mediating pathogenesis warrants further attention.

The manifestation of fever episodes in malaria is characterised by the up-regulation of several Hsps of the parasite, indicating their significance in the cytoprotection of the parasite against stress (Oakley et al. 2011; Zininga et al. 2016). Previous studies have shown that the host Hsp70 is recruited from the cytosol to the membrane of the iRBC (Banumathy et al. 2002). Surface-exposed Hsp70 is known to act as a danger signal and is thus strategically positioned to activate NK cells (Multhoff 2002). It is possible that the export of PfHsp70-x to the iRBC, while potentially augmenting pathogenesis, may facilitate host immune response. This is supported by the fact that antibodies to parasite Hsp70, Hsp60 and Hsp90 have been reported to circulate in high levels in malaria patients (Zhang et al. 2001a, 2001b). We previously established that PfHsp70-x exhibits chaperone activity and that it binds human Hop (Hsp70-Hsp90 organising protein) in vitro (Mabate et al. 2018). Human Hop, Hsp90, Hip, and Hsp70 are among the chaperones reported to occur in the parasite iRBC (Banumathy et al. 2002). Thus, PfHsp70-x may conduct independent chaperone activity or may operate in concert with human chaperones localised within the iRBC. It remains unclear why the parasite needs to deploy its own Hsp70 to the iRBC, considering that hHsp70 is also present in these cells. Perhaps PfHsp70-x is required to fold parasite proteins exported to the RBC, while hHsp70 shepherds the folding of proteins of human origin. Indeed, there is evidence that Hsp70s of *Plasmodium* are tailored to support the folding of proteins of the parasite (Stephens et al. 2011; Przyborski et al. 2015; Lebepe et al. 2020; Makumire et al. 2021; Makhoba et al. 2016).

Furthermore, there is evidence that PfHsp70-x colocalises and complexes with *P. falciparum* J domain proteins (JDPs)/Hsp40s that serve as co-chaperones of Hsp70. It has been reported to be associated with two JDPs: PfHsp40 (PFA0660w and PFE0055c) and PfEMP1 in specific cellular structures named J dots (Külzer et al. 2010, 2012; Grover et al. 2013; Njunge et al. 2013). Since JDPs are responsible for scanning Hsp70 substrates as well as stimulating the ATPase activity of Hsp70, the association of PfHsp70-x with these co-chaperones within the iRBC suggests that PfHsp70-x facilitates the folding of exported parasite proteins that are directed to it by these two co-chaperones (Dutta et al. 2021a). Furthermore, the two JDPs are reported to stimulate the basal ATPase activity of PfHsp70-x in vitro (Diehl et al. 2021; Dutta et al. 2021b). The association of PfHsp70-x with proteins implicated in knob formation, such as PFEMP1, further highlights a role for this chaperone in augmenting parasite pathogenicity (Charnaud et al. 2017; Day et al. 2019).

PfHsp70-x is trafficked to the cytosol of the iRBC through the PTEX pathway (Fig. 5.3; de Koning-Ward et al. 2009; El Bakkouri et al. 2010; Zhang et al. 2017). Its translocation is mediated by a seven-residue peptide (SNNAEES) that is positioned after the N-terminal secretory signal (Rhiel et al. 2016). The PTEX facilitates the translocation of parasite proteins across the PVM en route to the RBC. Furthermore, PfHsp70-x is thought to associate with PfHsp101, another chaperone that is a key component of the PTEX system (Fig. 5.3; de Koning-Ward et al. 2009; El Bakkouri et al. 2010; Zhang et al. 2017; Ho et al. 2018). Given its association with some molecules that are crucial for the development of pathogenic processes of the parasite, it is surprising that PfHsp70-x is not essential (Külzer et al. 2012). This notion is based on gene function studies conducted on parasites maintained in vitro, which may mask its importance during the development of clinical malaria.

5.5 Granzyme B as a Biomarker of Severe Malaria

One of the mechanisms by which the host immune system eliminates iRBC is by releasing a serine protease, granzyme B (GrB), which is secreted by the NK cells and CD8+ cytolytic T cells (Masson et al. 1986; Garcia-Sanz et al. 1988; Böttger et al. 2012; Kaminski et al. 2019). GrB has a single chain and single domain that folds into double six-stranded β-barrels connected by a segment of three trans-domain (Estebanez-Perpina et al. 2000; Rotonda et al. 2001). GrB possesses a perforin-dependent pro-apoptotic function, which is crucial for its role in targeting tumours and infected cells (Cullen and Martin 2008; Chowdhury and Lieberman 2008). Perforin introduces pores that granzyme B uses to enter cells (Tschopp et al. 2006). Notably, granzyme B exhibits unique substrate selectivity as it prefers substrates with an aspartic acid residue in the P1 position of the recognition sequence (Poe et al. 1991). GrB is also reported to translocate to the nucleus, where it degrades lamin B and PARP1 to induce cell death (Jans et al. 1996; Zhang et al. 2001a, b; Waterhouse et al. 2005; Casciola-Rosen et al. 2007; Chaitanya and Babu 2009).

The population of T lymphocytes that harbour the γδ T-cell receptor (TCR) is known to massively expand in the peripheral blood (Roussilhon et al. 1990; Ho et al. 1994). In addition, it was previously demonstrated that the population of γδ T cells was significantly expanded upon treatment of PMBCs with cell-free parasite iRBC (Hernández-Castañeda et al. 2020). In addition, the γδ T cells were shown to inhibit parasite growth at the blood stage, through cell-to-cell contact with the iRBC (Hernández-Castañeda et al. 2020). Furthermore, the cytolytic response of the γδ T cells was associated with upregulation of IFN-γ (Hernández-Castañeda et al. 2020). It was further suggested that the γδ T cells release GrB, which enters the bound iRBCs. It has been reported that CD8+ T cells upregulate the

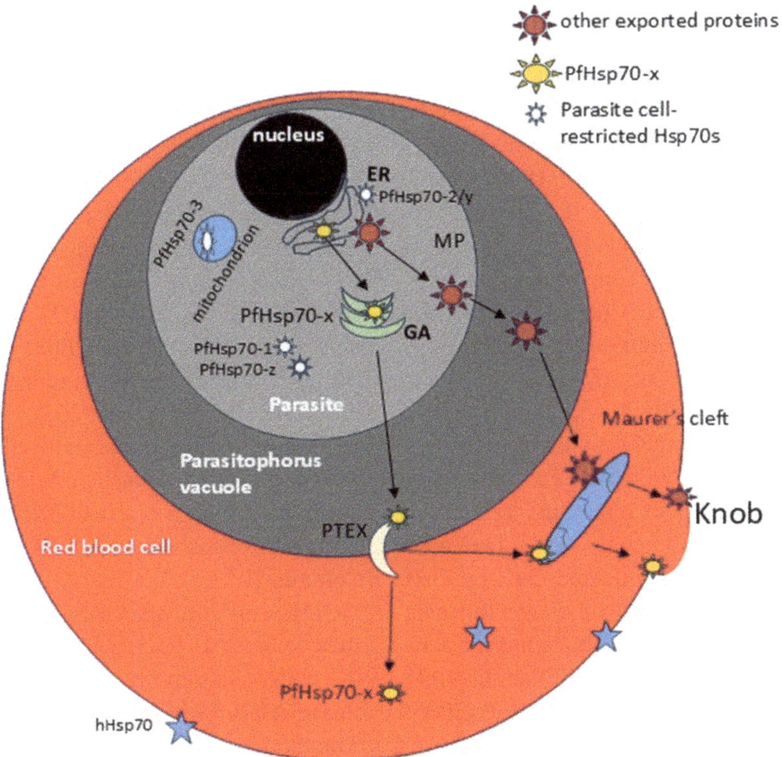

Fig. 5.3 PfHsp70-x export. (PfHsp70-x is trafficked to the infected erythrocyte cytosol through the vacuolar translocon, PTEX. Its translocation is mediated by a seven-residue peptide (SNNAEES) that is positioned after the N-terminal secretory signal. PfHsp70-x is directed to the ER by its signal sequence. At the ER, the signal sequence is then cleaved off. PfHsp70-x is transported from the ER to the Golgi apparatus and eventually reaches the PV. The PTEX component, located in the PVM, traffics it further to the infected erythrocyte (Külzer et al. 2012; Rhiel et al. 2016). Hsp70 localises to the iRBC membrane during the development of the parasite at this stage (Banumathy et al. 2002))

production of GrB and perforin during *Plasmodium vivax* malaria (Burel et al. 2016). The perforin secreted by CD8+ T cells is known to cause endothelial disruptions, leading to fatal oedema (Huggins et al. 2017). It is thought that the cytolytic protein, granulysin (GNLY), produced by cytotoxic granules augments the antiplasmodial activity of $\gamma\delta$ T cells (Farouk et al. 2004; Costa et al. 2011). In this respect, it is thought that GrB is taken up by the iRBC through the mediatory role of membrane-disrupting proteins, including GNLY (Farouk et al. 2004; Costa et al. 2011; Hernández-Castañeda et al. 2020). Further evidence for the GrB-mediated antiplasmodial activity is supported by the observed expansion of the population of CD8[+] T cells, which in turn leads to the upregulated GrB production during *P. vivax* infections (Kapelski et al. 2015). GrB has been found to degrade several proteins of the malaria parasite, and thus, mediates the death of the parasite-infected cells through the lymphocyte-mediated pathway (Hernández-Castañeda et al. 2020).

Of the granzyme family of molecules, which includes A, K, M and H, B is the most dominant in tumour/infected cells, and hence it is the one that most effectively induces cell lysis (Trapani and Sutton 2003). However, GrA also occurs in notably high levels in T cells of patients with infectious diseases (Garcia-Sanz et al. 1990; Lieberman 2003). Furthermore, GrB is the most efficient amongst the members of this group of molecules. This attribute is due to its capability to bind a variety of substrates, such as caspase-3, a key executioner caspase of apoptosis, that is activated through a mitochondrial pathway involving mitochondrial outer membrane permeabilisation (Porter and Jänicke 1999; Cullen et al. 2010). While RBCs lack the nucleus and mitochondria, both of which are involved in classical apoptosis, they can undergo a programmed cell death process (eryptosis) defined by cell shrinkage, membrane blebbing, and phosphatidylserine exposure (Lang et al. 2006). Based on mouse models, it was demonstrated that infected cells lacking GrB undergo a slower death rate compared to wild-type cells, showing the importance of GrB in mediating cell death (Heusel et al. 1994; Pardo et al. 2004). The rapid and efficient apoptosis of infected cells by GrB is dependent on its ability to activate intrinsic proteases (caspase 3 and 7) responsible for activating cell death by degrading cellular substrate proteins (Martin et al. 1996; Adrain et al. 2005; Cullen et al. 2007).

The level of GrB is known to be significantly increased in patients with SMA (Jiang et al. 2011; Kaminski et al. 2019; Dooley et al. 2023). In addition, IFN-γ and GrB, both of which upregulate Ephrin-A2 (EphA2) are crucial for the development of murine experimental cerebral malaria (ECM) (Haque et al. 2011; Swanson et al. 2016; Sorensen et al. 2018). EphA2 is implicated in the disruption of the blood-brain barrier in cerebral malaria (Darling et al. 2011). It has further been observed that mice lacking the CD8 cytotoxic effector molecules perforin or GrB are resistant to the ECM (Nitcheu et al. 2003; Haque et al. 2011). The levels of both GrA and GrB in circulation were found to significantly correlate with malaria severity in African children (Hermsen et al. 2003; Kaminski et al. 2019). It has further been reported that GrB inhibits parasite growth within the nanomolar range, and the antiplasmodial effect was enhanced by eight-fold when GrB was delivered by attachment to a merozoite surface protein (Kapelski et al. 2015).

5.6 The Role of the TKD Motif of Hsp70 in the Activation and Recruitment of Granzyme B

hHsp70 (HSPA1A) uses a 14-mer TKD (TKDNNLLGRFELSG) peptide located in its C-terminal substrate binding domain to stimulate NK cells, leading to the release of GrB (Multhoff et al. 2001; Gross et al. 2003). It has previously been reported that extracellular Hsp70 binds to TREM-1, leading to the secretion of TNF-α and INF-α (Sharapova et al. 2021). TREM-1 is a member of the immunoglobulin superfamily and is a receptor that is displayed on monocytes and neutrophils, which is pro-inflammatory by triggering the release of cytokines and chemokines (Colonna 2023).

CD4⁺ cells are then stimulated by the cytokines to release IL-2, which then activates the NK cells (Sharapova et al. 2021). NK cells release GrB, which kills the infected cells, through one of two mechanisms: perforin-dependent uptake, a process that involves perforin creating pores that facilitate GrB uptake, or through extracellular Hsp70 binding, leading to GrB release followed by its entry into the infected cells (Gross et al. 2003).

In addition, surface-exposed Hsp70 is known to sensitize tumour cells to death by NK cell-mediated killing through perforin-independent GrB uptake (Gross et al. 2003). By orchestrating GrB release by NK cells, extracellular Hsp70 mediates apoptosis in tumour cells (Gross et al. 2003). The role of the TKD peptide of Hsp70 in mediating the stimulation of NK, coupled with the release of GrB, has been demonstrated (Yazdi et al. 2024). In addition, a scrambled peptide (NGLTLKNDFSRLEG) composed of residues constituting the TKD peptide in random order was unable to activate the NK cells, demonstrating that the arrangement of the residues is important for the stimulation of NK cells (Yazdi et al. 2024). All these findings suggest a specificity of the TKD peptides of Hsp70 in activating NK cells. Because of its ability to activate NK cells, the TKD peptide presents a promising complementary antimalarial therapy (Zininga et al. 2021).

The targeted destruction of cancer cells by the Hsp70-GrB pathway draws parallels with the *P. falciparum* iRBC. It has been proposed that Hsp70 present in the malaria parasite iRBCs induces NK cells, leading to the release of GrB (Fig. 5.4; Böttger et al. 2012). However, it is not clear if this role is played by either hHsp70 or PfHsp70-x or both. Since the iRBC lacks a nucleus and hence does not have protein synthesis capability, it is more likely that this role lies with the parasite protein whose expression is inducible (Külzer et al. 2012).

hHsp70 and PfHsp70-x co-exist in the iRBC, and their TKD peptides share 80% identity (Ramatsui et al. 2023). There are 3 amino acid substitutions in the TKD peptide of PfHsp70-x relative to that of hHsp70. Notably, PfHsp70-x exhibits higher affinity for GrB than hHsp70 (Ramatsui et al. 2023). In addition, the TKD peptide of PfHsp70-x exhibits much higher affinity for GrB than the TKD peptide of hHsp70 (Ramatsui et al. 2023). In addition, GrB possesses antiplasmodial activity (Böttger et al. 2012; Ramatsui et al. 2023). Since PfHsp70-x exhibits higher affinity for GrB than the hHsp70, the parasite chaperone may play a more effective role in modulating NK cells, thus this might lead to the selective targeting of the iRBC. It has been reported that the expression of the *hHSP70* gene is suppressed during the development of SMA in children (Kempaiah et al. 2016; Belh et al. 2024). Given the association of PfHsp70-x and GrB with SMA, the GrB-PfHsp70-x pathway may represent a possible weakness to exploit in the quest to develop alternative antimalarial therapies. While the expression of the *hHSP70* gene was reportedly suppressed during the development of SMA in children, the study by Behl et al. (2024), proposed that the protein was RBC

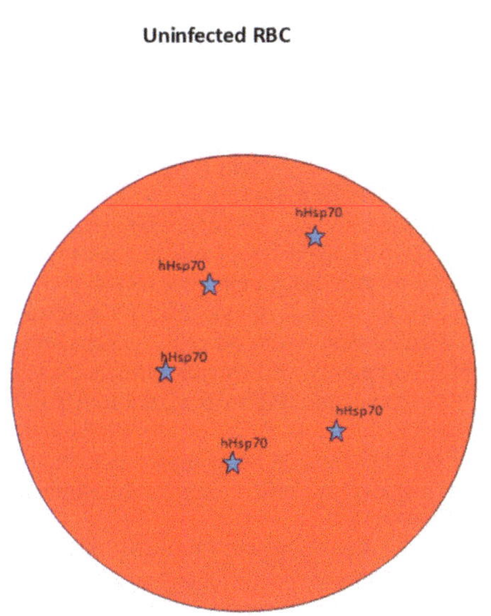
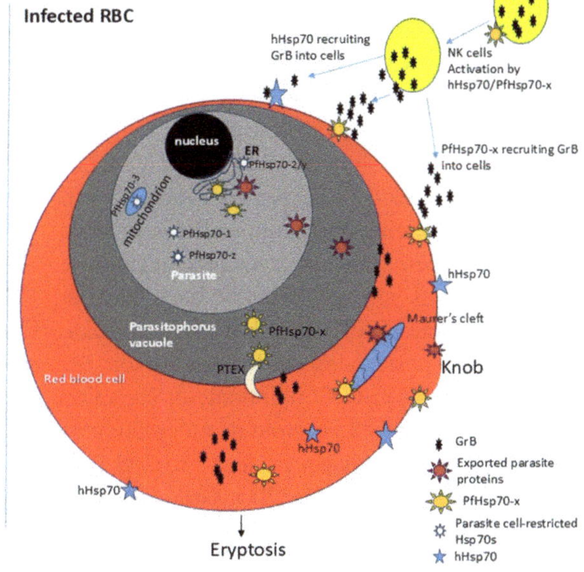

Fig. 5.4 A model for the stimulation of NK cells by extracellular Hsp70 released by parasite-infected red blood cells. (The red blood cell contains haemoglobin and several other proteins. Notably, the uninfected RBC harbours hHsp70, while parasite-infected RBCs harbour both hHsp70 and PfHsp70-x. It has been proposed that Hsp70 present in the iRBC moves to the membrane and occurs on the surface (Banumathy et al. 2002; Behl et al. 2024) and possibly within the extracellular space, thus positioning it to stimulate NK cells, leading to GrB release. PfHsp70-x may play a more important role in this regard because of its inducible production during the progression of clinical malaria. The released GrB subsequently targets the iRBC, leading to their selective destruction)

surface-exposed and appeared to facilitate the invasion of cells by the parasite. Notably, several studies exploring the role and localisation of Hsp70 at the clinical stage of malaria rely on immune assays. It is thus possible that some of the studies are unable to distinguish between hHsp70 and Hsp70 of the parasite due to possible antibody cross-reactivity. Nonetheless, there is growing evidence that several parasite proteins are released into the extracellular space with the *P. falciparum* ER resident chaperone, PfHsp70–2 (PF3D7_0917900; Shonhai et al. 2007; Cortés et al. 2020), amongst them (Fraering et al. 2024).

5.7   Conclusions and Prospects

It has been previously proposed that Hsp70 present in malaria parasite iRBCs modulates NK cells to release GrB (Böttger et al. 2012). We speculate that the export of PfHsp70-x to the iRBCs may constitute a Trojan horse for the selective targeting of the iRBCs through PfHsp70-x-mediated GrB uptake. In support of this, we previously demonstrated that through its TKD motif, PfHsp70-x directly binds human GrB in vitro. However, while PfHsp70-x is known to be exported to the iRBCs, its prospect to stimulate NK cells as well as recruit GrB will require that it be positioned on the outer leaflet of the iRBC membrane. It is therefore important to establish the precise positioning of this protein. Cancer cells secrete Hsp70 as a danger signalling biomarker. It is also conceivable that parasite iRBCs may release PfHsp70-x into the plasma. The possible positioning of PfHsp70-x on the outer leaflet of the iRBC membrane or secretion into the plasma would promote its modulation of NK cells as well as enhance its access to GrB. The exploitation of the PfHsp70-x-GrB interface could usher in a novel way to treat malaria. Encouragingly, GrB has been shown to exhibit antiplasmodial activity (Böttger et al. 2012; Ramatsui et al. 2023). In addition, GrB functionalized with nanoparticles or fused to a peptide to enhance targeted cell delivery demonstrated improved antiplasmodial activity (Kapelski et al. 2015; Shevtsov et al. 2019). In this regard, GrB

fusion proteins combined with existing antimalarial drugs offer a promising solution for the treatment of malaria.

Conflict of Interest   Author THD declares that he has no conflict of interest. Author AS declares that he has no conflict of interest.

Ethical Approval   This chapter does not contain any studies with human participants performed by any of the authors.

Funding   This study was funded by the International Center for Genetic Engineering and Biotechnology (ICGEB) for funding (grant HDI/CRP/012) awarded to AS. We further thank the Council for Scientific and Industrial Research (CSIR) of South Africa for providing a student bursary for THD.

References

Adrain C, Murphy BM, Martin SJ (2005) Molecular ordering of the caspase activation cascade initiated by the cytotoxic T lymphocyte/natural killer (CTL/NK) protease granzyme B. J Biol Chem 280:4663–4673

Albakova Z, Armeev GA, Kanevskiy LM et al (2020) Hsp70 multi-functionality in cancer. Cells 9:587

Arama C, Troye-Blomberg M (2014) The path of malaria vaccine development: challenges and perspectives. Int J Med 275:456–466

Banumathy G, Singh V, Tatu U (2002) Host chaperones are recruited in membrane-bound complexes by *Plasmodium falciparum*. J Biol Chem 277:3902–3912

Batinovic S, McHugh E, Chisholm S et al (2017) An exported protein-interacting complex involved in the trafficking of virulence determinants in *Plasmodium*-infected erythrocytes. Nat Commun 8:16044

Behl A, Kumar V, Bisht A et al (2019) Cholesterol bound *Plasmodium falciparum* co-chaperone 'PFA0660w' complexes with major virulence factor 'PfEMP1' via chaperone 'PfHsp70-x'. Sci Rep 9:1–7

Behl M, Maurya A, Saini P et al (2024) Targeting PfProhibitin 2-Hu-Hsp70A1A complex as a unique approach towards malaria vaccine development. iScience 27:109918

Boddey JA, Moritz RL, Simpson RJ et al (2009) Role of the *Plasmodium* export element in trafficking parasite proteins to the infected erythrocyte. Traffic 10:285–299

Botha M, Chiang AN, Needham PG et al (2011) *Plasmodium falciparum* encodes a single cytosolic type I Hsp40 that functionally interacts with Hsp70 and is upregulated by heat shock. Cell Stress Chaperones 16:389–401

Böttger E, Multhoff G, Kun JF et al (2012) *Plasmodium falciparum*-infected erythrocytes induce granzyme B by NK cells through expression of host-Hsp70. PLoS One 7:e33774

Buchberger A, Schröder H, Hesterkamp T et al (1996) Substrate shuttling between the DnaK and GroEL systems indicates a chaperone network promoting protein folding. J Mol Biol 261:328–333

Bukau B, Walker GC (1989) Delta dnaK mutants of Escherichia coli have defects in chromosome segregation and plasmid maintenance at normal growth temperatures. J Bacteriol 171:6030–6038

Bukau B, Weissman J, Horwich AL (2006) Molecular chaperones and protein quality control. Cell 125:443–451

Bullen HE, Charnaud SC, Kalanon M et al (2012) Biosynthesis, localization, and macromolecular arrangement of the *Plasmodium falciparum* translocon of exported proteins (PTEX). J Biol Chem 287:7871–7884

Burel JG, Apte SH, McCarthy JS et al (2016) *Plasmodium vivax* but not *Plasmodium falciparum* blood-stage infection in humans is associated with the expansion of a CD8+ T cell population with cytotoxic potential. PLoS Negl Trop Dis 10:e0005031

Casciola-Rosen L, Garcia-Calvo M, Bull HG et al (2007) Mouse and human granzyme B have distinct tetrapeptide specificities and abilities to recruit the bid pathway. J Biol Chem 282:4545–4552

Chaitanya GV, Babu PP (2009) Differential PARP cleavage: an indication of heterogeneous forms of cell death and involvement of multiple proteases in the infarct of focal cerebral ischemia in rat. Cell Mol Neurobiol 29:563–573

Chakafana G, Zininga T, Shonhai A (2019a) Comparative structure-function features of Hsp70s of *Plasmodium falciparum* and human origins. Biophys Rev 11:1–12

Chakafana G, Zininga T, Shonhai A (2019b) The link that binds: the linker of Hsp70 as a helm of the protein's function. Biomolecules 9:543

Chakafana G, Middlemiss CJ, Zininga T et al (2024) Swapping the linkers of canonical Hsp70 and Hsp110 chaperones compromises both self-association and client selection. Heliyon 10:e29690

Charnaud SC, Kumarasingha R, Bullen H et al (2017) Knockdown of the translocon protein EXP2 in *Plasmodium* falciparum reduces growth and protein export. PLoS One 13:e0204785

Chiang AN, Valderramos JC, Balachandran R et al (2009) Select pyrimidinones inhibit the propagation of the malarial parasite, *Plasmodium falciparum*. Bioorgan Med Chem 17:1527–1533

Chotivanich K, Udomsangpetch R, McGready R (2002) Central role of the spleen in malaria parasite clearance. J Infect Dis 185:1538–1541

Chowdhury D, Lieberman J (2008) Death by a thousand cuts: granzyme pathways of programmed cell death. Annu Rev Immunol 26:389–420

Cobb DW, Florentin A, Fierro MA et al (2017) The exported chaperone PfHsp70x is dispensable for the *Plasmodium falciparum* intraerythrocytic life cycle. mSphere 2:e00363–17

Colonna M (2023) The biology of TREM receptors. Nat Rev Immunol 23:580–594

Cortés GT, Wiser MF, Gómez-Alegría CJ (2020) Identification of *Plasmodium falciparum* HSP70-2 as a resident of the Plasmodium export compartment. Heliyon 6:e04037

Costa G, Loizon S, Guenot M et al (2011) Control of *Plasmodium falciparum* erythrocytic cycle: γδ T cells target the red blood cell-invasive merozoites. Blood 118:6952–6962

Cullen SP, Martin SJ (2008) Mechanisms of granule-dependent killing. Cell Death Differ 15:251–262

Cullen SP, Adrain C, Luthi AU et al (2007) Human and murine granzyme B exhibit divergent substrate preferences. J Cell Biol 176:435–444

Cullen SP, Brunet M, Martin SJ (2010) Granzymes in cancer and immunity. Cell Death Differ 17:616–623

Damelang T, Aitken EH et al (2021) Antibody mediated activation of natural killer cells in malaria exposed pregnant women. Sci Rep 11:4130

Darling TK, Mimche PN, Bray C et al (2011) EphA2 contributes to disruption of the blood-brain barrier in cerebral malaria. PLoS Pathog 16:e1008261

Day J, Passecker A, Beck HP et al (2019) The *Plasmodium falciparum* Hsp70-x chaperone assists the heat stress response of the malaria parasite. FASEB J 33:14611–14624

de Koning-Ward TF, Gilson PR, Boddey JA et al (2009) A newly discovered protein export machine in malaria parasites. Nature 459:945–949

Diamant S, Peres Ben-Zvi A, Bukau B et al (2000) Size-dependent disaggregation of stable protein aggregates by the DnaK chaperone machinery. J Biol Chem 275:21107–21113

Dick JK, Hart GT (2022) Natural killer cell antibody-dependent cellular cytotoxicity (ADCC) activity against *plasmodium falciparum*-infected red blood cells. Methods Mol Biol 2470:641–657

Diehl M, Roling L, Rohland L et al (2021) Co-chaperone involvement in knob biogenesis implicates host-derived chaperones in malaria virulence. PLoS Pathog 17:e1009969

Dongola TH, Chakafana G, Middlemiss C et al (2024) Insertion of GGMP repeat residues of *Plasmodium falciparum* Hsp70-1 in the lid of DnaK adversely impacts client recognition. Int J Biol Macromol 255:128070

Dooley NL, Chabikwa TG, Pava Z et al (2023) Single cell transcriptomics shows that malaria promotes unique regulatory responses across multiple immune cell subsets. Nat Commun 14:7387

Dups JN, Pepper M, Cockburn IA (2014) Antibody and B cell responses to *Plasmodium* sporozoites. Front Microbiol 5:625

Dutta T, Singh H, Gestwicki JE et al (2021a) Exported plasmodial J domain protein, PFE0055c, and PfHsp70-x form a specific co-chaperone-chaperone partnership. Cell Stress Chaperones 26:355–366

Dutta T, Pesce ER, Maier AG et al (2021b) Role of the J domain protein family in the survival and pathogenesis of *Plasmodium falciparum*. In: Shonhai A, Picard

D, Blatch GL (eds) Heat shock proteins of malaria, Advances in experimental medicine and biology 1340. Springer, Cham

El Bakkouri M, Pow A, Mulichak A et al (2010) The Clp chaperones and proteases of the human malaria parasite *Plasmodium falciparum*. J Mol Biol 404:456–457

Elmallah MIY, Cordonnier M, Vautrot V et al (2020) Membrane-anchored heat-shock protein 70 (Hsp70) in cancer. Cancer Lett 469:134–141

Estebanez-Perpina E, Fuentes-Prior P, Belorgey D et al (2000) Crystal structure of the caspase activator human granzyme B, a proteinase highly specific for an Asp-P1 residue. Biol Chem 381:1203–1214

Fakhari DE, Saidi LG, Wahlster L (2013) Molecular chaperones and protein folding as therapeutic targets in Parkinson's disease and other synucleinopathies. Acta Neuropathol Commun 1:1–79. https://doi.org/10.1186/2051-5960-1-79

Farouk SE, Mincheva-Nilsson L, Krensky A et al (2004) Gamma delta T cells inhibit *in vitro* growth of the asexual blood stages of *Plasmodium falciparum* by a granule exocytosis-dependent cytotoxic pathway that requires granulysin. Eur J Immunol 34:2248–2256

Flaherty KM, DeLuca-Flaherty C, McKay DB et al (1990) Three-dimensional structure of the ATPase fragment of a 70 K heat-shock cognate protein. Nature 346:623–628

Fraering J, Salnot V, Gautier EF et al (2024) Infected erythrocytes and plasma proteomics reveal a specific protein signature of severe malaria. EMBO Mol Med 16:319–333

Garcia-Sanz JA, Velotti F, MacDonald HR et al (1988) Appearance of granule-associated molecules during activation of cytolytic T-lymphocyte precursors by defined stimuli. Immunology 64:129–134

Garcia-Sanz JA, MacDonald HR, Jenne DE, Nabholz M et al (1990) Cell specificity of granzyme gene expression. J Immunol 145:3111–3118

Gehrmann M, Liebisch G, Schmitz G et al (2008) Tumor-specific Hsp70 plasma membrane localization is enabled by the glycosphingolipid Gb3. PLoS One 3:e1925

Gitau GW, Mandal P, Blatch GL et al (2012) Characterization of the *Plasmodium falciparum* Hsp70-Hsp90 organizing protein (PfHop). Cell Stress Chaperones 17:191–202

Gross C, Koelch W, DeMaio A et al (2003) Cell surface-bound heat shock protein 70 (Hsp70) mediates perforin-independent apoptosis by specific binding and uptake of granzyme B. J Biol Chem 278:41173–41181

Grover M, Chaubey S, Ranade S et al (2013) Identification of an exported heat shock protein 70 in *Plasmodium falciparum*. Parasite 20:1123–1132

Haque A, Best SE, Unosson K et al (2011) Granzyme B expression by CD8+ T cells is required for the development of experimental cerebral malaria. J Immunol 186:6148–6156

Hermsen CC, Konijnenberg Y, Mulder L et al (2003) Circulating concentrations of soluble granzyme A and B increase during natural and experimental Plasmodium falciparum infections. Clin Exp Immunol 132:467–472

Hernández-Castañeda MA, Happ K, Cattalani F et al (2020) γδ T cells kill *plasmodium falciparum* in a Granzyme- and Granulysin-dependent mechanism during the late blood stage. J Immunol 204:1798–1809

Heusel JW, Wesselschmidt RL, Shresta S et al (1994) Cytotoxic lymphocytes require granzyme B for the rapid induction of DNA fragmentation and apoptosis in allogeneic target cells. Cell 76:977–987

Hiller NL, Bhattacharjee S, van Ooij C et al (2004) host-targeting signal in virulence proteins reveals a secretome in malarial infection. Science 306:1934–1937

Ho M, Tongtawe P, Kriangkum J et al (1994) Polyclonal expansion of peripheral gamma delta T cells in human *Plasmodium falciparum* malaria. Infect Immun 62:855–862

Ho CM, Beck JR, Lai M et al (2018) Malaria parasite translocon structure and mechanism of effector export. Nature 561:70–75

Huggins MA, Johnson HL, Jin F et al (2017) Perforin expression by CD8 T cells is sufficient to cause fatal brain edema during experimental cerebral malaria. Infect Immun 85:e00985-16

Jäättelä M, Wissing D, Kokholm K et al (1998) Hsp70 exerts its anti-apoptotic function downstream of caspase-3-like proteases. EMBO J 17:6124–6134

Jani D, Nagarkatti R, Beatty W et al (2008) HDP-a novel heme detoxification protein from the malaria parasite. PLoS Pathog 4:e1000053

Jans DA, Jans P, Briggs LJ et al (1996) Nuclear transport of granzyme B (fragmentin-2). Dependence of perforin in vivo and cytosolic factors *in vitro*. J Biol Chem 271:30781–30789

Jiang W, Chai NR, Maric D et al (2011) Unexpected role for Granzyme K in CD56bright NK cell-mediated Immunoregulation of multiple sclerosis. J Immunol 187:781–790

Jonsdottir TK, Gabriela M, Crabb BS (2021) Defining the essential exportome of the malaria parasite. Trends Parasitol 37:664–675

Kafuye-Mlwilo MY, Mukherjee P, Chauhan VS (2012) Kinetics of humoral and memory B cell response induced by the *Plasmodium falciparum* 19-kilodalton merozoite surface protein 1 in mice. Infect Immun 80:633–642

Kaminski LC, Riehn M, Abel A et al (2019) Cytotoxic T cell-derived granzyme B is increased in severe *Plasmodium falciparum* Malaria. Front Immunol 10:2917. https://doi.org/10.3389/fimmu.2019.02917

Kapelski S, de Almeida M, Fischer R et al (2015) Antimalarial activity of granzyme B and its targeted delivery by a granzyme B-single-chain Fv fusion protein. Antimicrob Agents Chemother 59:669–672

Kempaiah P, Dokladny K, Karim Z et al (2016) Reduced Hsp70 and glutamine in pediatric severe malaria anemia: role of hemozoin in suppressing Hsp70 and NF-κB activation. Mol Med 22:570–584

Kim YE, Hipp MS, Bracher A et al (2013) Molecular chaperone functions in protein folding and proteostasis. Annu Rev Biochem 82:323–355

Kudyba HM, Cobb DW, Fierro MA et al (2019) The endoplasmic reticulum chaperone PfGRP170 is essential for asexual development and is linked to stress response in malaria parasites. Cell Microbiol 21:e13042

Külzer S, Rug M, Brinkmann K et al (2010) Parasite-encoded Hsp40 proteins define novel mobile structures in the cytosol of the P. *falciparum*-infected erythrocyte. Cell Microbiol 12:1398–1420

Külzer S, Charnaud S, Dagan T et al (2012) *Plasmodium falciparum*-encoded exported Hsp70/Hsp40 chaperone/co-chaperone complexes within the host erythrocyte. Cell Microbiol 14:1784–1795

Kumar N, Koski G, Harada M et al (1991) Induction and localization of *Plasmodium falciparum* stress proteins related to the heat shock protein 70 family. Mol Biochem Parasitol 48:47–58

Lang F, Lang KS, Lang PA et al (2006) Mechanisms and significance of eryptosis. Antioxid Redox Signal 8:1183–1192

Lebepe CM, Matambanadzo PR, Makhoba et al (2020) Comparative characterization of *Plasmodium falciparum* Hsp70-1 relative to *E. coli* DnaK reveals the functional specificity of the parasite chaperone. Biomolecules 10:856. https://doi.org/10.3390/biom10060856

Lee WC, Russell B, Rénia L (2019) Sticking for a cause: the *Falciparum* Malaria parasites cytoadherence paradigm. Front Immunol 10:1444

Lieberman J (2003) The ABCs of granule-mediated cytotoxicity: new weapons in the arsenal. Nat Rev Immunol 3:361–370

Liehl P, Meireles P, Albuquerque IS et al (2015) Innate immunity induced by *Plasmodium* liver infection inhibits malaria reinfections. Infect Immun 83:1172–1180

Mabate B, Zininga T, Ramatsui L et al (2018) Structural and biochemical characterization of *Plasmodium falciparum* Hsp70-x reveals functional versatility of its C-terminal EEVN motif. Proteins 86:1189–1201

Maier AG, Rug M, O'Neill MT (2008) Exported proteins required for virulence and rigidity of *Plasmodium falciparum*-infected human erythrocytes. Cell 134:48–61

Makhoba XH, Burger A, Coertzen D et al (2016) Use of a chimeric Hsp70 to enhance the quality of recombinant *Plasmodium falciparum* S-Adenosylmethionine decarboxylase protein produced in *Escherichia coli*. PLoS One 11:e0152626

Makumire S, Dongola TH, Chakafana G et al (2021) Mutation of GGMP repeat segments of *Plasmodium falciparum* Hsp70-1 compromises chaperone function and Hop co-chaperone binding. Int J Mol Sci 22:2226. https://doi.org/10.3390/ijms22042226

Marti M, Baum J, Rug M et al (2005) Signal-mediated export of proteins from the malaria parasite to the host erythrocyte. J Cell Biol 171:587–592

Martin SJ, Amarante-Mendes GP, Shi L et al (1996) The cytotoxic cell protease granzyme B initiates apoptosis in a cell-free system by proteolytic processing and activation of the ICE/CED-3 family protease, CPP32, via a novel two-step mechanism. EMBO J 15:2407–2416

Masson D, Nabholz C, Estrade J et al (1986) Granules of cytolytic T-lymphocytes contain two serine esterases. EMBO J 5:1595–1600

Mayer MP, Bukau B (2005) Hsp70 chaperones: cellular functions and molecular mechanism. Cell Mol Life Sci 62:670–684

Menard D, Dondorp A (2017). Antimalarial drug resistance: a threat to malaria elimination. Cold Spring Harb Perspect Med 7(7):a025619. https://doi.org/10.1101/cshperspect.a025619

Meslin B, Barnadas C, Boni V et al (2007) Features of apoptosis in *Plasmodium falciparum* erythrocytic stage through a putative role of PfMCA1 metacaspase-like protein. J Infect Dis 195:1852–1859

Multhoff G (2002) Hyperthermia classic commentary: activation of natural killer (NK) cells by heat shock protein 70, Gabriele Multhoff. Int J Hyperth 18:576–585

Multhoff G (2007) Heat shock protein 70 (Hsp70): membrane location, export and immunological relevance. Methods 43:229–237

Multhoff G, Pfister K, Gehrmann M et al (2001) A 14-mer Hsp70 peptide stimulates natural killer (NK) cell activity. Cell Stress Chaperones 6:337–344

Muralidharan V, Oksman A, Pal P et al (2012) *Plasmodium falciparum* heat shock protein 110 stabilizes the asparagine repeat-rich parasite proteome during malarial fevers. Nat Commun 3:1310. https://doi.org/10.1038/ncomms2306

Nagata Y, Ono S, Matsuo M et al (2002) Differential presentation of a soluble exogenous tumor antigen, NY-ESO-1, by distinct human dendritic cell populations. Proc Natl Acad Sci USA 99:10629–10634

Nasr A, Allam G, Hamid O et al (2014) IFN-gamma and TNF associated with severe *falciparum malaria* infection in Saudi pregnant women. Malar J 13:314. https://doi.org/10.1186/1475-2875-13-314

Nitcheu J, Bonduelle O, Combadiere C et al (2003) Combadiere Perforin-dependent brain-infiltrating cytotoxic CD8 + T lymphocytes mediate experimental cerebral malaria pathogenesis. J Immunol 170:2221–2228

Njunge JM, Ludewig MH, Boshoff A et al (2013) Hsp70s and J proteins of *Plasmodium* parasites infecting rodents and primates: structure, function, clinical relevance, and drug targets. Curr Pharm Des 19:387–403

Njunge JM, Mandal P, Przyborski JM et al (2015) PFB0595w is a *Plasmodium falciparum* J protein that co-localizes with PfHsp70-1 and can stimulate its *in vitro* ATP hydrolysis activity. Int J Biochem Cell Biol 62:47–53

Nollen EA, Salomons FA, Brunsting JF et al (2001) Dynamic changes in the localization of thermally unfolded nuclear proteins associated with chaperone-dependent protection. Proc Natl Acad Sci USA 98:12038–12043

Oakley MS, Gerald N, McCutchan TF, Aravind L, Kumar S (2011) Clinical and molecular aspects of malaria fever. Trends Parasitol 27(10):442–449. https://doi.org/10.1016/j.pt.2011.06.004

Pardo J, Bosque A, Brehm RA et al (2004) Apoptotic pathways are selectively activated by granzyme A and/or granzyme B in CTL-mediated target cell lysis. J Cell Biol 167:457–468

Pavithra SR, Banumathy G, Joy O, Singh V, Tatu U (2004) Recurrent fever promotes *Plasmodium falciparum* development in human erythrocytes. J Biol Chem 279:46692–46699

Poe M, Blake JT, Boulton DA et al (1991) Human cytotoxic lymphocyte granzyme B. Its purification from granules and the characterization of substrate and inhibitor specificity. J Biol Chem 266:98–103

Porter A, Jänicke R (1999) Emerging roles of caspase-3 in apoptosis. Cell Death Differ 6:99–104

Prado M, Eickel N, De Niz M et al (2015) Long-term live imaging reveals cytosolic immune responses of host hepatocytes against *Plasmodium* infection and parasite escape mechanisms. Autophagy 11:1561–1579

Przyborski JM, Diehl M, Blatch GL (2015) *Plasmodial* HSP70s are functionally adapted to the malaria parasite life cycle. Front Mol Biosci 2:34

Rahimi BA, Thakkinstian A, White NJ et al (2014) Severe vivax malaria: a systematic review and meta-analysis of clinical studies since 1900. M alar J 13:481

Ramatsui L, Dongola TH, Zininga T et al (2023) Human granzyme B binds *Plasmodium falciparum* Hsp70-x and mediates antiplasmodial activity in vitro. Cell Stress Chaperones 28:321–331

Rhiel V, Bittl A, Tribensky SC (2016) Trafficking of the exported *P. falciparum* chaperone PfHsp70x. Sci Rep 6:36174

Risco-Castillo V, Topcu S, Marinach C et al (2015) Malaria sporozoites traverse host cells within transient vacuoles. Cell Host Microbe 18:593–603

Rotonda J, Garcia-Calvo M, Bull HG et al (2001) The three-dimensional structure of human granzyme B compared to caspase-3, key mediators of cell death with cleavage specificity for aspartic acid. Chem Biol 8:357–368

Roussilhon C, Agrapart M, Ballet JJ (1990) T lymphocytes bearing the gamma delta T cell receptor in patients with acute *Plasmodium falciparum* malaria. J Infect Dis 162:283–285

Ryu SW, Stewart R, Pectol DC et al (2020) Proteome-wide identification of HSP70/HSC70 chaperone clients in human cells. PLoS Biol 18:e3000606

Saibil H (2013) Chaperone machines for protein folding, unfolding and disaggregation. Nat Rev Mol Cell Biol 14:630–642

Schmidt CQ, Kennedy AT, Tham WH (2015) More than just immune evasion: hijacking complement by Plasmodium falciparum. Mol Immunol 67:71–84

Sharapova TN, Romanova EA, Ivanova OK et al (2021) Hsp70 interacts with the TREM-1 receptor expressed on monocytes and thereby stimulates generation of cytotoxic lymphocytes active against MHC-negative tumor cells. Int J Mol Sci 22:6889

Shevtsov M, Stangl S, Nikolaev B et al (2019) Granzyme B functionalized nanoparticles targeting membrane Hsp70-positive tumors for multimodal cancer theranostics. Small 15:e1900205

Shonhai A, Boshoff A, Blatch GL (2005) Molecular characterization of the chaperone properties of *plasmodium falciparum* heat shock protein 70. Mol Gen Genet 274:70–78

Shonhai A, Boshoff A, Blatch GL (2007) The structural and functional diversity of Hsp70 proteins from *Plasmodium falciparum*. Protein Sci 16:1803–1818

Shonhai A (2021) The Role of Hsp70s in the Development and Pathogenicity of Plasmodium falciparum. Adv Exp Med Biol 1340:75–95. https://doi.org/10.1007/978-3-030-78397-6_3

Sisquella X, Nebl T, Thompson JK et al (2017) *Plasmodium falciparum* ligand binding to erythrocytes induce alterations in deformability essential for invasion. elife 6:e21083

Sorensen EW, Lian J, Ozga AJ et al (2018) CXCL10 stabilizes T cell-brain endothelial cell adhesion leading to the induction of cerebral malaria. JCI Insight 3:e98911

Srivastava P (2002) Roles of heat-shock proteins in innate and adaptive immunity. Nat Rev Immunol 2:185–194

Srivastava IK, Schmidt M, Grall M, Certa U, Garcia AM, Perrin LH (1992) Identification and purification of glucose phosphate isomerase of *Plasmodium falciparum*. Mol Biochem Parasitol 54:153–164

Stephens LL, Shonhai A, Blatch GL (2011) Co-expression of the *Plasmodium falciparum* molecular chaperone, PfHsp70, improves the heterologous production of the antimalarial drug target GTP cyclohydrolase I, PfGCHI. Protein Expr Purif 77:159–165

Swanson PA, Hart GT, Russo MV (2016) CD8+ T cells induce fatal brainstem pathology during cerebral malaria via luminal antigen-specific engagement of brain vasculature. PLoS Pathog 12:e1006022

Trampuz A, Jereb M, Muzlovic I et al (2003) Clinical review: severe malaria. Crit Care 27:315–323

Trapani JA, Sutton VR (2003) Granzyme B: pro-apoptotic, antiviral and antitumor functions. Curr Opin Immunol 15:533–543

Tschopp CM, Spiegl N, Didichenko S et al (2006) Granzyme B, a novel mediator of allergic inflammation: its induction and release in blood basophils and human asthma. Blood 108:2290–2299

Vostakolaei MA, Hatami-Baroogh L, Babaei G et al (2021) Hsp70 in cancer: a double agent in the battle between survival and death. J Cell Physiol 236:3420–3444

Wang TF, Chang J, Wang C (1993) Identification of the peptide binding domain of hsc70. 18-Kilodalton fragment located immediately after ATPase domain is sufficient for high affinity binding. J Biol Chem 268:26049–26051

Waterhouse NJ, Sedelies KA, Browne KA et al (2005) A central role for Bid in granzyme B-induced apoptosis. J Biol Chem 280:4476–4482

WHO (2021) https://www.who.int/publications/i/item/9789240015791. Accessed Nov 2022

WHO (2022) https://www.who.int/news/item/08-12-2022. Accessed January 2024

WHO (2024) World Health Organization world malaria report. https://www.who.int/teams/global-malaria-programme/reports/world-malaria-report-2024

William T, Menon J, Rajahram G et al (2011) Severe *Plasmodium knowlesi* malaria in a tertiary care hospital, Sabah, Malaysia. Emerg Infect Dis 17:1248–1255

World Health Organization (2000) Severe falciparum malaria. Trans R Soc Trop Med Hyg 94(suppl 1):S1–S90

Yano T, Kurata S (2011) Intracellular recognition of pathogens and autophagy as an innate immune host defence. J Biochem 150:143–149

Yazdi M, Hasanzadeh Kafshgari M, Khademi Moghadam F et al (2024) Crosstalk between NK cell receptors and tumor membrane Hsp70-Derived peptide: a combined computational and experimental study. Adv Sci (Weinh) 11(14):e2305998. https://doi.org/10.1002/advs.202305998

Zhang D, Beresford PJ, Greenberg AH et al (2001a) Granzymes A and B directly cleave lamins and disrupt the nuclear lamina during granule-mediated cytolysis. Proc Natl Acad Sci USA 98:5746–5751

Zhang M, Hisaeda H, Kano S (2001b) Antibodies specific for heat shock proteins in human and murine malaria. Microbes Infect 3:363–367

Zhang Q, Ma C, Oberli A et al (2017) Proteomic analysis of exported chaperone/co-chaperone complexes of *P. falciparum* reveals an array of complex protein–protein interactions. Sci Rep 7:42188

Zheng H, Tan Z, Xu W (2014) Immune evasion strategies of pre-erythrocytic malaria parasites. Mediat Inflamm 2014:362605

Zininga T, Achilonu I, Hoppe H et al (2016) *Plasmodium falciparum* Hsp70-z, an Hsp110 homologue, exhibits independent chaperone activity and interacts with Hsp70-1 in a nucleotide-dependent fashion. Cell Stress Chaperones 21:499–513

Zininga T, Böttger E, Multhoff G (2021) Role of heat shock proteins in immune modulation in malaria. In: Shonhai A, Picard D, Blatch GL (eds) Heat shock proteins of malaria, Advances in experimental medicine and biology 1340. Springer, Cham

Zorzi E, Bonvini P (2011) Inducible hsp70 in the regulation of cancer cell survival: analysis of chaperone induction, expression and activity. Cancers (Basel) 3(4):3921–3656. https://doi.org/10.3390/cancers3043921

Prevalence of Drug-Resistant *Mycobacterium tuberculosis* in Rural Communities of Vhembe District, South Africa

M. S. Mashilo, N. T. Banda, V. Mavumengwana, M. C. Rikhotso, J. P. Kabue Ngandu, C. J. Kinnear, N. Potgieter, and A. N. Traoré

Abstract

Drug-resistant tuberculosis (DR-TB) is still a concern for public health and health security, making TB the world's largest cause of death from a single infectious agent. This Chapter was aimed at evaluating the prevalence of DR-MTB in rural communities of South Africa. A cross-sectional study in rural Vhembe District, South Africa, recruited 175 active TB patients. Data on lifestyle behaviour, socioeconomic, and environmental characteristics were collected via a questionnaire, and samples (blood and sputum) were collected. The U-rapid blood test confirmed HIV status. DNA was extracted from sputum and tested using multiplex real-time PCR assays with Anyplex and Allplex kits to detect Mycobacterium tuberculosis/nontuberculous mycobacteria, as well as DR-TB strains, respectively. The participants' ages ranged from 18 to 82 years (mean = 44 ± 13.0), with the majority being males (57.1%, 100/175) and unemployed (61.1%, 107/175). The co-infections detected include NTM and HIV. The prevalence of DR-MTB was 2.3% (4/175). A 20.6% (36/175) success rate was achieved in treatment outcomes. DR-TB remains low and is effectively managed, with favourable treatment outcomes reflecting effective clinical and public health interventions. However, the high prevalence of emerging NTM and HIV infections may compromise these gains, posing a potential threat to sustained TB control.

M. S. Mashilo (✉) · N. T. Banda · M. C. Rikhotso · J. P. K. Ngandu · N. Potgieter · A. N. Traoré
Department of Biochemistry and Microbiology, Faculty of Sciences, Engineering & Agriculture, University of Venda, Thohoyandou, South Africa

V. Mavumengwana
DSI-NRF Centre of Excellence for Biomedical Tuberculosis Research, South African Medical Research Council Centre for Tuberculosis Research, Division of Molecular Biology and Human Genetics, Faculty of Medicine and Health Sciences, Stellenbosch University, Cape Town, South Africa

C. J. Kinnear
DSI-NRF Centre of Excellence for Biomedical Tuberculosis Research, South African Medical Research Council Centre for Tuberculosis Research, Division of Molecular Biology and Human Genetics, Faculty of Medicine and Health Sciences, Stellenbosch University, Cape Town, South Africa

South African Medical Research Council Genomics Platform, Cape Town, South Africa

Keywords

Drug-resistant *Mycobacterium tuberculosis* · HIV · Nontuberculous mycobacteria · Risk factors · Rural communities · Treatment outcome

© The Author(s) 2026
A. Shonhai et al. (eds.), *Advances in Biochemistry and Molecular Biology to meet Africa´s Needs*, Advances in Experimental Medicine and Biology 1507, https://doi.org/10.1007/978-3-032-24254-9_6

6.1 Introduction

Tuberculosis (TB), a communicable disease, remains a public health crisis (He et al. 2023) and a health security threat. It is the leading infectious disease cause of mortality (WHO 2022) and is among the top ten causes of death globally (Freschi et al. 2021). In 2021, the global TB incidence was approximately 10.6 million, with about 1.6 million reported deaths (Bagcchi 2023). The *Human immunodeficiency virus* (HIV) has a significant impact on the TB symptoms and progression, thus leading to increased transmission and amplification of drug-resistant tuberculosis (DR-MTB) (Khan et al. 2019). In 2020, TB and HIV/AIDS were reported to be the leading causes of mortality in the Vhembe district of South Africa for females and males aged 25–64 (VDMP 2020). The etiological agent of the disease is *Mycobacterium tuberculosis* (MTB), which forms part of the Mycobacterium tuberculosis complex (MTBC). The MTBC species comprises *M. tuberculosis, M. africanum, M. bovis, M. microti, M. canetti,* and *M. pinnipedii* (Richter et al. 2004). These species can cause TB in both humans and animals (Clarke et al. 2022); however, MTB is the major etiological agent of human TB (Lazzarini et al. 2007).

MTB is not only restricted to attacking the lungs; thus, it can also affect other body parts, such as the spine, kidneys, and brain (Zaman 2010). It spreads through airborne droplets when an individual with TB coughs, sneezes, sings, or speaks. Consequently, individuals nearby may inhale MTB-containing droplets and become infected (CDC 2022). In addition, MTB is an opportunistic pathogen that may take advantage of an individual's impaired immune system due to chronic illnesses or poor lifestyle choices such as substance abuse (WHO 2023). Hence, TB is a multifactorial disease in which the environment interacts with host-related factors (Lienhardt et al. 2005), such as host genetic susceptibility to TB and response to therapy (Lauschke and Ingelman-Sundberg 2019; Azad et al. 2012).

TB is preventable and curable; however, the emergence of DR-MTB is a major obstacle to controlling TB. According to the CDC (2022), DR-MTB comprises multiple drug-resistant MTB (MDR-MTB), Pre-Extensively Drug-resistant (pre-XDR-MTB), and extensively drug-resistant TB (XDR-MTB). The MDR-MTB bacterium is resistant to at least rifampicin (RIF) and isoniazid (INH), whereas the pre-XDR-MTB bacterium is resistant to INH, RIF, plus any fluoroquinolones (FQ) and second-line injectables, and a bacterium is resistant to INH, RIF, plus any FQ. The resistance to multiple first-line anti-TB drugs, except the INH and RIF, is defined as poly-resistance. In addition, mono-resistance to INH includes MTB strains that are resistant to INH, but sensitive to RIF, as isoniazid-resistant TB (Hr-TB) (WHO 2018) and RIF. Therefore, this chapter was aimed at evaluating the prevalence of drug-resistant MTB in rural communities of South Africa.

6.2 Materials and Methods

6.2.1 Study Area

The study was conducted in the Vhembe District, Limpopo Province, South Africa (Fig. 6.1). This district is situated in the North-eastern part of SA and shares borders with Zimbabwe, Mozambique, and Botswana (Fig. 6.1a). It is comprised of 4 local municipalities (Fig. 6.1b) with an estimated population size of 1.4 million people (VDMP 2020).

6.2.2 Inclusion and Exclusion Criteria

Eligible patients recruited in this study had a laboratory-confirmed diagnosis of TB and were on treatment at the time of sampling. A total of 175 patients with active pulmonary TB aged 18 and above were enrolled, and all participants

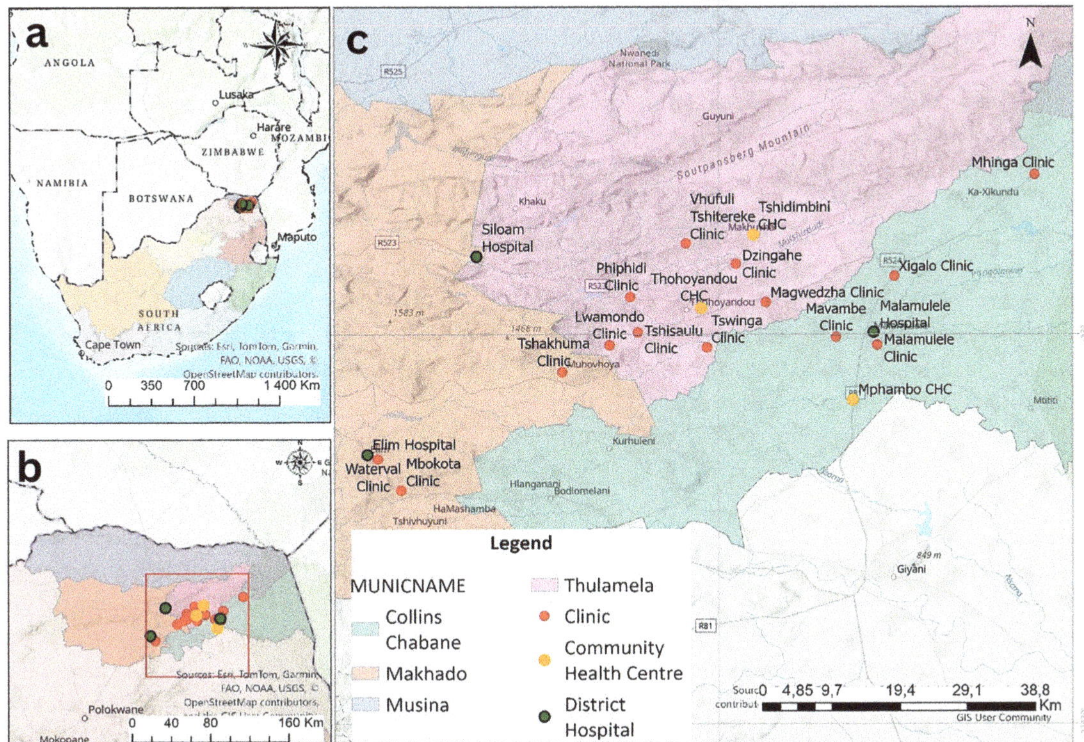

Fig. 6.1 Map illustrating the geographical context of the study area. (**a**) South Africa and its neighbouring countries (**b**) Limpopo province, highlighting the Vhembe District (**c**), shows the distribution of healthcare facilities included in the study. (Created by UNIVEN GIS Oct 2025, projection WGS84, Source: STATSA2011)

included in the study provided a signed informed consent. Patients with any of the following were excluded: (i) those who were critically ill, and (ii) prisoners.

6.2.3 Ethical Consideration

Approval was secured from the University of Venda Research Ethics Committee (SMNS/20/MBY/13/2104). Provincial ethical clearance was obtained from the Limpopo Provincial Department of Health (DoH, LP_2021-11-001, LP_2023-05-013, and LP_2024-10-032). Permission was granted to conduct the study in the selected facilities from the DoH Vhembe District. The aim of the study was shared with all participants, and patient confidentiality was

ensured using study codes assigned by the researcher.

6.2.4 Patients and Specimen Collection

The study assumed a cross-sectional design conducted from 2022 to 2025, with participants enrolled from 23 healthcare facilities in the Vhembe district. A total of 175 active patients undergoing TB treatment were enrolled. Blood and the early morning sputum specimens were collected from each patient and transported to the Microbiology unit TB lab, University of Venda (Univen) for processing. Each collection bottle contained 5 mL of a 4% sodium hydroxide (NaOH) solution (Sigma-Aldrich, St. Louis,

Missouri, United States) for decontamination. Interviews were conducted using a detailed structured questionnaire consisting of patients' demographic data and other patient-related information, included variables such as date of birth, age, gender, educational status, religion, vaccination status against TB, contacts with a history of TB, the record of chronic illnesses, residential features, type of environment they spend their time, and country of birth. Samples were collected when patients came for their TB treatment, specifically on the dates scheduled for medication collection ("return dates"). Upon arrival at the laboratory, all samples were recorded in a logbook for the project.

6.2.5   Sputum Pre-treatment

Upon arrival, the collected samples were incubated at room temperature for 15 min. A total of 1.5 mL of the mixture was transferred to sterile tubes and centrifuged (1-14K SIGMA; Osterod; Germany) at 15,000× g for 5 min. The supernatant was carefully discarded from each tube using a sterile pipette. About 1 mL of 1X PBS (Merck, Germany) solution was added to the tube and mixed well by vortexing for 10 s. The tubes were centrifuged at 15,000× g for 5 min, and the supernatant was carefully discarded. This step was repeated twice. The pellet was used for DNA extraction.

6.2.6   DNA Extraction

All processed specimens were subjected to DNA extraction using an Allplex™ kit (Seegene, Seoul, Korea) according to the manufacturer's instructions. The pellet was resuspended with 1 mL of sterile water, and a volume of 10 µL of MTB/DRe internal control (IC) was added and centrifuged at 15,000× g for 5 min. The supernatant was carefully discarded from each tube using a sterile pipette. A volume of 100 µL of DNA Extraction Solution was added to the pellet and vortexed for 30 s. The tubes were cap-locked and

placed on the heat block to boil for 20 min, followed by centrifugation at 15,000× g for 5 min. A total of 5 µL of the supernatant was stored as a PCR template for multiplex polymerase chain (mPCR). The DNA samples were stored at −20 °C until further use.

6.2.7   Multiplex Real-Time PCR Using Anyplex MTB/ MTN and Allplex MTB/MDR/XDRe Detection Kit

Anyplex™ MTB/NTM uses Seegene's patented DPO™ technology to concurrently identify MTB and NTM by targeting MTB-specific genes (MPB6464, IS6110) and a pan-mycobacterial 16S rRNA gene for NTM. It offers elevated sensitivity and specificity for distinguishing MTB from NTM in clinical specimens, mitigates carryover contamination using a UDG system, and facilitates automated data interpretation (Luukinen et al. 2020). The Allplex™ MTB/ MDR/XDR kit enhances detection of MTB and identifies resistance mutations to first- and second-line tuberculosis medications, focusing on genes linked to INH (*katG, inhA*), RIF (*rpoB*), FQ (*gyrA, gyrB*), aminoglycosides (*rrs, eis*), and macrolides (*erm*) in NTM species. It facilitates the swift detection of MDR and XDR-MTB strains (mutation analysis) using real-time PCR (Seegene, nd). These assays both include an internal control to enable effective detection of MTB and NTM to enhance the diagnosis and treatment of TB patients (Seegene Inc. 2017; Seegene n.d.).

The Anyplex™ MTB/MTN kit (Cat. No. TB7200X, Seegene Korea, Seoul) was used as described in the manufacturer's instructions to analyse the DNA extracted from sputum samples for the detection of targeted sequences of *M. tuberculosis*. Subsequently, Allplex™ MTB/ MDR/XDRe (Cat. No. TB10173Y, Seegene Korea, Seoul) was performed to qualitatively test for the detection of multiple MTB strains and their resistance to first-line anti-tuberculosis drugs (RIF and INH) and second-line drugs (FQ and injectable drugs) from sputum specimens from symptomatic patients. A DNA template of 5 μL was added to the 15 μL Anyplex ™ MTB/NTM master mix, resulting in a final volume of 20 μL. Similar volumes were used for the Allplex™ MTB/MDR/XDRe reaction. The DNA amplifications were achieved using a CFX96 Real-time PCR system (Bio-Rad, California, United States) in two separate reactions.

6.2.8 HIV Test

HIV test was performed on all blood specimens using a U-Test HIV/AIDS rapid test (Humor Diagnostica, Hermanstad, SA) following the manufacturer's instructions (internal control, considered as a negative control, was included in the cassette, represented by the 1st line, and if the 2nd line appeared on the cassette, the test was considered positive).

6.2.9 Data Analysis

Data were described as means and percentages/ frequencies. Descriptive statistics were used to determine the prevalence using the Statistical Package for Social Sciences (SPSS) version 24 software (IBM Corporation, Armonk, USA). A p-value of ‹0.05 was considered significant.

6.3 Results

6.3.1 Characteristics of the Study Population

The majority of the recruited patients in the study were males (57.1%, 100/175) compared to females (42.9%, 75/175). Among the *Mycobacterium tuberculosis* (MTB) patients enrolled in the study, the majority were middle-aged individuals (35–59 years) (61.1%, 107/175),

followed by the youth age group (18–34 years) (26.3%, 46/175), as shown in Table 6.1. Most participants were unemployed (65.1%, 114/175), a category that also included students. In terms of educational attainment, a large proportion (74.3%, 130/175) had only basic education (Grades 1–7 and Grades 8–12), while a smaller fraction (8%) reported having no formal education. The study population primarily consisted of South African native citizens (90.3%, 158/175), with a minority comprising migrants (9.7%, 17/175).

About 69.1% (121/175) of the population were exposed to dust, and only 29.1% (51/175) wore/used face masks in dusty areas or public places regularly. A large proportion, 58.9% (103/175), of the participants relied on public transport, whereas 8.6% (15/175) used personal cars as their mode of transport daily. Most participants, 52% (91/175), visited the healthcare facilities after a month of exhibiting signs and symptoms, and 12.6% (22/175) of them consulted traditional healer(s) prior to visiting healthcare facilities (Table 6.1). Of these, only 8.6% (15/175) were administered herbal remedies. A proportion of 60% (105/175) of participants were vaccinated with the BCG vaccine. Most participants disclosed their TB status to their family members, 95.4% (167/175), contrary to disclosing to their friends, colleagues, and or community members, 48% (84/175).

The current findings comprised more new cases, 78.3% (137/175), compared to recurring MTB cases, 21.7% (38/175), and 29.7% (52/175) of the participants had a family history of TB. Fifty-six percent (98/175) of the respondents answered yes for living with HIV (as per the surveys on chronic illness and laboratory confirmed HIV status), and among those, 25.1% (44/175) were NTM + HIV co-infected. In addition, more chronic illnesses such as diabetes, hypertension, asthma, ulcers, nerve damage, and heart failure were recorded in the survey (data not shown). Table 6.1 shows the characteristics of the study population.

Table 6.1 Description of the characteristics of MTB study population

Variable	Category	Frequency	Percent
Age	18–34	46	26,3
	35–59	107	61,1
	60–82	22	12,6
Gender	Female	75	42,9
	Male	100	57,1
Occupation status	Student	7	4
	Unemployed	107	61,1
	employed	35	20
	Self-employed	26	14,9
Educational level	No education	14	8
	Grade1–7	32	18,3
	Grade 8–12	98	56
	Tertiary	31	17,7
Religion	African	30	17,1
	Christian	137	78,3
	Other	5	2,9
	Atheist	3	1,7
Place of birth	Native	158	90,3
	Migrant	17	9,7
Type of house	Missing data	1	0,6
	Traditional house	7	4
	Modern house	131	74,9
	RDP house	25	14,3
	Rental	6	3,4
	Squatter camp	5	2,9
Are you exposed to dust daily?	Missing data	5	2,9
	Yes	81	46,3
	No	49	28
	Sometimes	40	22,9
Wearing of mask in public or dusty areas	Missing data	2	1,1
	Yes	51	29,1
	No	26	14,9
	Sometimes	96	54,9
Spending most of the time in an open space	Missing data	8	4,6
	Yes	146	83,4
	No	14	8
	Sometimes	7	4
Mode of transport	Walking	56	32
	Public	103	58,9
	Own car	15	8,6
	Other	1	0,6
Number of family members	Missing data	1	0,6
	[0–4]	90	51,4
	[5–9]	83	47,4
	10/above	1	0,6
Number of bedrooms in the house	Missing data	3	1,7
	1	15	8,6
	2	42	24
	3	56	32
	4	35	20
	≥ 5	24	13,7

(continued)

Table 6.1 (continued)

Variable	Category	Frequency	Percent
Are you sharing a room?	Missing data	1	0,6
	Yes	85	48,6
	No	81	46,3
	Sometimes	8	4,6
Family history of TB	Missing data	1	0,6
	Yes	52	29,7
	No	121	69,1
	Do not know	1	0,6
Have you ever consulted a traditional healer?	Yes	22	12.6
	No	153	87,4
Did the healer give you a herbal remedy?	Yes	15	8,6
	No	146	83,4
Healthcare consultation period	Do not recall	7	4
	1–3 weeks	74	42,3
	$\geq$1 month	91	52
	Diagnosed during a routine check	3	1,7
Vaccinated (BCG vaccine)	Yes	105	60
	No	66	37,7
	Do not know	4	2,3
Latent Tuberculosis Infection (LTBI) skin test?	Yes	6	3,4
	No	169	96,6
Disclosing to family members	Yes	167	95,4
	No	8	4,6
How long did you take to disclose to family members?	1–7 days	159	90,9
	2 weeks	4	2,3
	Months	5	2,9
	Do not recall	7	4
Does your family seem to be supportive?	Yes	158	90,3
	No	11	3,4
	Sometimes	6	3,4
Disclosing to colleagues, community members or friends after testing positive for TB	Yes	84	48
	No	91	52
Do they seem to be supportive?	Yes	76	43,4
	No	95	54,3
	Sometimes	4	2,3
History of TB recurrent patients	Yes	38	21,7
	No	137	78,3
Do you have any chronic illnesses?	Yes	79	87,5
	No	96	87,5
HIV status	Yes	90	51,4
	No	64	36,6
	Not laboratory confirmed	21	12
Total		175	100

Table 6.2 Detection rate using Anyplex MTB/ MTN and Allplex MTB/MDR/XDRe

	Detection rate	Frequency	Percent	*P-value*
MTB/MTN (Anyplex)	NTM only	76	43,4	0.000
	MTB only	37	21,1	
	MTB + NTM	26	14,9	
	None detected/treated	36	20,6	
Resistance (Allplex)	None	171	97.7	0.000
	INH-R	1	0,6	
	FQ-R	1	0,6	
	RIF-R	1	0,6	
	RIF-R, INH-R, FQ-R	1	0,6	
Total		175	100	

6.3.2 MTB/NTM Detection and Prevalence of Drug-Resistant TB

Of the 175 patients enrolled in the study undergoing treatment for MTB, 36% (63/175) of their sputum samples tested positive for MTB, 43.4% (76/175) for NTM only, and no detection was recorded for 20.6% (36/175) of the participants (Table 6.2). The observed variation in detection rates was highly significant ($p < 0.001$).

6.3.3 MDR/XDRe-MTB Detection

The overall prevalence of DR-MTB detection was 2.3% (4/175). Among these, one case each exhibited resistance to RIF, INH and FQ, while another case had a combination of resistance to all 3 drugs (Table 6.2). Thirty-three percent (59/175) of the samples were identified as sensitive Allplex.

6.3.4 Distribution of Various Co-infections in the Recruited Patients

The most prevalent co-infection seen was NTM + HIV (25.1%, 44/175), followed by MTB + HIV (11.4%, including 1/175 XDR-MTB + HIV) and 8% (14/175) MTB + HIV + NTM. In addition, the lowest co-infection rate of MTB + NTM, which accounted

for 6.9% (12/175), included 3/175 DR-MTB (Fig. 6.2). Furthermore, drug resistance included 1.1% (2/175) MDR-MTB + NTM and 0.6% (1/35) XDR-MTB + NTM co-infections. Meanwhile, 0.6% (1/175) of specimens tested positive for XDR-MTB + HIV co-infection.

6.4 Discussion

Drug-resistant MTB continues to be a global health menace worldwide, including in the Vhembe district. The current study identified the prevalence of MTB, NTM, and DR-MTB, as well as the factors associated with them. A high prevalence of MTB was observed in the age group 35–59 (middle-aged). This indicated that the middle-aged, active population/economically productive class is the most susceptible to TB. Thus, this could be a result of weakened immune systems since most TB patients tested positive for MTB + HIV and or MTB + NTM. Hence, the observed high co-infection rate. Ma et al. (2024) linked diminished counts of lymphocyte subsets to impaired TB-specific immunity.

Despite the overall TB burden, male patients constituted a more significant proportion of the study population (57.1%) and exhibited a higher rate of MTB positivity, hinting at possible biological or social factors that might predispose men to TB infection. In addition, literature has demonstrated that men have a higher risk of TB than women. Hence, behavioural factors (like

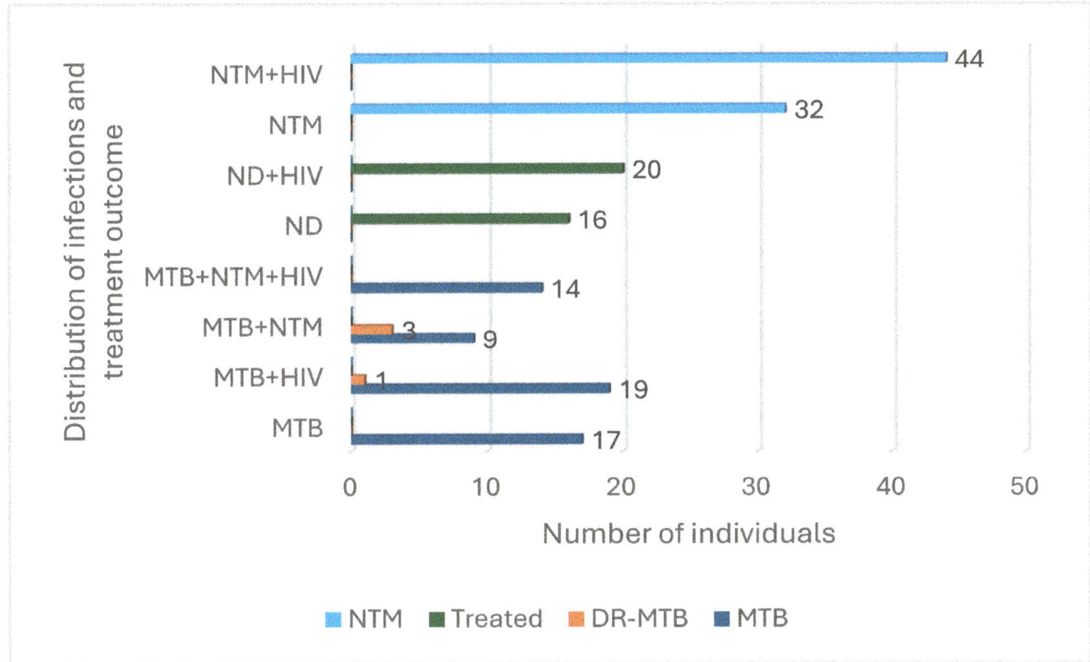

Fig. 6.2 Distribution of MTB, NTM, successful treatment outcome (treated patients), and various co-infections in the recruited patients. MTB *Mycobacterium tuberculo-sis*, NTM *Nontuberculous mycobacteria*, DR-MTB Drug--Resistant *Mycobacterium tuberculosis*, HIV *Human immunodeficiency virus*, ND None Detected

alcohol intake and smoking), social influences (such as healthcare access and high-risk occupational exposures), and biological variation (such as immune modulation and hormonal differences) are attributable to this observation (Rickman et al. 2025; Mlondo et al. 2022; Marçôa et al. 2018; Nhamoyebonde and Leslie 2014). DR-MTB cases showed an equal distribution of MDR-MTB and XDR-MTB among both genders.

The occurrence of MDR-MTB in younger adults (18–34) versus XDR-MTB in middle-aged individuals (35–59) could point to evolving dynamics in drug resistance that vary by age group. Dalton et al. (2012) reported that compared to men, FQ-resistance was more common in females, which increased their likelihood of developing XDR-TB. Contrarily, the current study has shown that males were resistant to FQ, thereby increasing their likelihood of developing XDR-TB, while females were mono-resistant to INH and RIF, which increased their risk of developing MDR-TB. However, more investigations with a larger population size are needed to further understand this trend. In addition, Bu et al. (2023) demonstrated the occurrence of DR-MTB in the age group above 35, contrary to the high prevalence of DR-MTB in the 15–34 age group (Lee et al. 2020).

Despite expectations for a progression from multi-drug resistance to extensive drug resistance, the study unexpectedly found equal prevalence (1.1%) of both MDR-MTB and XDR-MTB, suggesting a potential plateau or simultaneous emergence of severe resistance patterns. In this chapter, none of the DR-TB individuals were TB recurrent patients, which might be a result of acquiring first-hand DR-MTB, also known as primary drug resistance (CDC 2022; An et al. 2020; Song et al. 2019). Conversely, the prevalence of DR-TB has been reported in previously treated TB cases or TB recurring patients (Tiberi et al. 2022; Lohiya et al. 2020). A study conducted in Iraq has reported a high prevalence of DR-TB in previously treated cases relative to newly diagnosed cases (Mohammed et al. 2022). They also further reported higher primary drug resistance in females compared to males in their study. Tuberculosis is an opportunistic infection (Fenner et al. 2013) that is influenced by various factors that affect its progression.

The high unemployment rate in this study is a factor because it contributes to further financial challenges for affected individuals and their families, thus indicating inadequate dietary habits and living circumstances (Ciobanu et al. 2024; Barter et al. 2012; Marais et al. 2009). Furthermore, this poverty status often results in reduced access to essential social services such as healthcare and transportation (Nyathi et al. 2024; Ngumbela et al. 2020; Centre for Social Development in Africa 2019). This narrative also affects the Vhembe district communities as well (Health Systems Trust 2015). Predominantly, more than 75% of Vhembe's population resides in rural areas (Health Systems Trust 2015; Moetlo et al. 2011).

Public transport was identified as a factor for DS-MTB and DR-TB in the current study. This could result from poor ventilation, close contact, and extended travel duration that promote airborne spread of MTB during transport. Nevertheless, the immune system of the exposed individuals should be considered. Deol et al. (2022) concluded that public transportation is a viable target for preventing the spread of airborne infections.

One of the study's primary predisposing factors was delayed healthcare consultations, of which 12.6% (22/175) were attributable to visits to traditional healers, thereby resulting in delayed diagnosis, treatment initiation, and potential transmission of the infection across households and communities. Claessens et al. (2001) conducted a study on traditional healers and pulmonary tuberculosis in Malawi, suggesting that there is a likelihood of continuous visitation by active TB patients to traditional healers compared to healthcare facilities, particularly in rural areas. They further recognised a need for healthcare workers to educate or encourage traditional healers to refer patients with TB symptoms to healthcare facilities for screening. Approximately 52% (91/175) of the study patients, including the

DR-TB individuals, only went to consult at the healthcare facilities after ≥1 month of TB symptoms manifestation. Similar trends were observed in the study conducted by Yasobant et al. (2023), where the most common delay in the TB care cascade was the delay of the first official consultation following the manifestation of TB symptoms. In addition, a study in KwaZulu-Natal reported that approximately 40% of individuals delayed getting TB treatment due to reasons such as lengthy lines at the healthcare facilities, not feeling sick, and long distance (Chiposi et al. 2021).

Fifty-two percent of the enrolled patients did not disclose their TB status to either their friends, community members, or colleagues, including both the MDR-TB/ XDR-TB participants. This might be due to the stigma and others opting to seek healthcare services outside their communities. Thus, further increasing the exposure of TB within the population, facilitating the transmission of infection, and increasing the possibility of TB and DR-TB prevalence. Thereby posing a potential risk to them. The patients highlighted the discomfort of disclosing and avoiding the stigma around TB. Several studies have reported that most TB patients encounter stigmatisation daily (Lee et al. 2017; Chang and Cataldo 2014; Courtwright and Turner 2010). Lee et al. (2017) reported that non-disclosure by active TB patients of their TB status has a strong link with depression.

The prevalence of recurrent active TB was found to be higher (13.1%, 23/175) in males compared to females (8.6%, 15/175). This disagrees with the study conducted by Korhonen et al. (2017) in Finland. They reported that females were at a higher risk of recurrence. The reason seems to be linked to the relapse observed in low-incidence countries. South Africa is classified as a high TB incidence country (WHO 2013), and the Vhembe District is a very rural region with limited resources compared to urban areas.

The NTM species have emerged to be significant in human lung diseases. In contrast to MTB aerosols, NTM aerosols derived from various environmental niches are responsible for spreading NTM illnesses (Gopalaswamy et al. 2020; Lopez-Luis et al. 2020). Studies have reported that patients with suspected MDR-TB had MTB + NTM and or NTM only (Huang et al. 2022; Shahraki et al. 2015). These studies further highlighted the possibility of misdiagnosing MTB + NTM as MDR-TB. The current study revealed 58.3% (102/175) MTB + NTM co-infections and NTM-only cases, suggesting that nontuberculous mycobacteria may be under-recognised in TB-like disease presentations. Seventy-five percent of MDR-MTB + NTM (2/4) and XDR-MTB + NTM (1/4) cases were identified. This aligns with the findings reported by Gao et al. (2020), wherein they identified various NTM species in DR-TB patients. This study did not monitor the effect of NTMs in co-existence with MTB; therefore, it was not possible to confirm the evolution of their co-existence. Sarro et al. (2018) reported that patients with TB and NTM co-infection may receive inappropriate treatment due to inaccurate results from Xpert™ MTB/RIF and sputum smear microscopy (SSM). Consequently, this leads to the development of DR-TB. NTM and MTB infections are clinically indistinguishable, and diagnostic tools such as SSM (Gopalaswamy et al. 2020; Sarro et al. 2018) and Xpert™ MTB/RIF are not exempt from this dilemma (Sarro et al. 2018; Pang et al. 2017). Hence, the current study highlights the need to diagnose these infections and their drug resistance status concurrently. This is imperative to ensure accurate treatment choices (Sarro et al. 2018).

Nearly half of the samples 43.4% (76/175) tested positive for NTM only, suggesting that: (i) the TB treatment administered to patients may have been effective in eliminating the MTB and the possibility that studied patients harboured the MTB + NTM co-infection; (ii) in this study population, the main diagnostic tools for MTB detection are Xpert MTB/RIF and Sputum Smear Microscopy (SSM) and several studies have reported the misdiagnosis of NTM as MTB which poses a significant challenge to effective control strategies (Opperman et al. 2024; Pang et al. 2017; Shahraki et al. 2015; Maiga et al. 2012). The findings from this study, therefore,

suggest a possible misdiagnosis of NTM as MTB by GeneXpert MTB/RIF or SSM, as previously documented in multiple studies. In addition to this, the observed occurrence of NTM in 43.4% of the population suggests that TB treatment may alter the lung or tissue environment, creating conditions that favour NTM colonisation or proliferation, thereby increasing their detectability.

The high incidence of NTM co-infections might influence the observed drug resistance profiles, potentially mask underlying resistance trends or contribute to an atypical resistance development pathway. This study observed 25% (1/4) of XDR-MTB + HIV. Similar findings have been previously reported on the association between DR-MTB and HIV/AIDS (Okonkwo et al. 2017; Putong et al. 2002). Lotz et al. (2023) observed a high prevalence of DR-TB+ HIV co-infections. This might be due to the weakened immune system, particularly in HIV patients with a low number of CD4+ T cells (CDC 2024; Qi et al. 2023). A study by Chingonzoh et al. (2018) highlighted that HIV co-infection, amongst others, was a predisposing factor for DR-TB deaths in patients not receiving antiretroviral therapy (ART). Interestingly, the low incidence of HIV co-infection observed in this study, with only one case of XDR-MTB + HIV, contrasts with commonly reported patterns in TB studies. This finding may reflect changing trends in HIV infection or regional and demographic variation in the study population, warranting further investigation.

The current study utilised the Anyplex™ MTB/NTM and Allplex™ MTB/MDR/XDR assays for the detection of MTB/NTM and determination of drug resistance mutations, respectively. These assays are economical, sensitive, and specific for identifying MTB and determining drug-resistant mutations, making them suitable for routine application in resource-limited settings with a high TB disease burden (Chumpa et al. 2020; Luukinen et al. 2020). The Anyplex™ MTB/NTM assay has demonstrated high specificity and sensitivity for MTB screening, with values of approximately 98% and 87.5%, respectively (Sawatpanich et al. 2022; Kim et al. 2020; Lim et al. 2014). Additionally, Luukinen et al.

(2020) reported 100% specificity for both Anyplex™ MTB/NTM and Allplex™ MTB/MDR/XDR assays. In this study, the use of these two assays enabled a successful differentiation of MTB from NTM and the accurate detection of drug-resistant profiles. These findings further support the reliability and diagnostic value of these two platforms, even in situations with varying bacterial loads, particularly in high-burden, resource-constrained environments, where prompt and precise diagnosis is crucial.

The low prevalence of resistance to first-line and second-line drugs (each at 1.1%) might indicate that resistance mechanisms are emerging concurrently, calling for integrated treatment strategies. The identical resistance rates for first-line (RIF, INH) and second-line (FQ-R, RIF + INH + FQ-R) drugs imply that resistance mechanisms may be developing concurrently rather than sequentially, which is an unexpected deviation from typical resistance progression models. Chumpa et al. (2020) reported that individuals with INH-R cases on the standardised treatment exhibited a significantly elevated rate of acquired drug resistance, 8% compared to drug susceptible TB (0.3%).

The findings showed a 20.6% rate of successful treatment outcomes, indicating the treatment's efficacy. However, this chapter failed to record the treatment duration at the point of sample collection per participant. In this study, over half (52.8%, 19/36) of the treated patients were males, contrasting national patterns indicating that females achieve superior TB treatment outcomes (TB Accountability Consortium 2024).

6.5 Conclusion

This chapter's molecular analysis of patients undergoing TB treatment identified predominant NTM and common cases of NTM + HIV co-infection, highlighting the diagnostic challenges of mycobacterial infections in high-burden settings. These NTM cases were also identified in participants living without HIV. Despite their clinical similarities, TB and NTM pulmonary disease differ greatly in their antibiotic therapy, and the comparatively low rate of DR-TB (2.3%) contrasts with the high frequency of NTM detection, indicating that many patients might be subjected to unnecessary or incorrect anti-TB treatment. These findings highlight the necessity to integrate precise molecular diagnostics into conventional TB screening programs to effectively distinguish MTB from NTM, enhance patient care, and mitigate medication resistance associated with misclassification. In addition, various factors, including delayed healthcare consultation, unemployment, lack of disclosure of active TB patient status, and public transport reliance, were revealed. Furthermore, enhancing diagnostic tools at the point of care is crucial for improving global TB control and promoting judicious antibiotic utilisation in areas where TB and HIV converge. There is a need for innovative lifestyle behaviour, social and environmental protection policies to help mitigate the transmission and burden of DR-TB in the rural communities of Vhembe district, SA.

Future efforts should focus on characterising the NTM species within the research area to distinguish virulent from avirulent strains and enhance diagnostic precision. Rapid molecular tools that can differentiate MTB from NTM are crucial for tailored treatment, and age-specific treatment for MDR and XDR TB should be investigated with integrated guidelines for managing MTB + NTM co-infections established to optimise outcomes. Public health policies and strengthening surveillance, with regular monitoring of resistance patterns and co-infection rates, will be essential for effectively adapting national TB control strategies.

Acknowledgments We would like to thank all the medical staff and respective healthcare facilities for their assistance during the study, and the TB active patients for their willingness to participate in the study.

Conflict of Interest Author MSM declares that she has no conflict of interest. Author NTB declares that she has no conflict of interest. Author VM declares that he has no conflict of interest. Author MCR declares that he has no conflict of interest. Author JPKN declares that he has no conflict of interest. Author CJK declares that he has no conflict of interest. Author NP declares that she has no conflict of interest. Author ANT declares that she has no conflict of interest.

Ethical Approval Ethical approval for the study involving human participants was secured from the University of Venda Research Ethics Committee (SMNS/20/MBY/13/2104). Provincial ethical clearance was obtained from the Limpopo Provincial Department of Health (LP_2021–11-001, LP_2023-05-013, and LP_2024-10-032). Permission was granted to conduct the study in the selected facilities from the DoH Vhembe District. The aim of the study was shared with all participants, and patient confidentiality was ensured using study codes assigned by the researcher.

Funding The authors greatly acknowledge the South African Medical Research Council, Research Capacity Development (SAMRC, RCD), for financial support.

References

An Q, Song W, Liu J et al (2020) Primary drug-resistance pattern and trend in elderly tuberculosis patients in Shandong, China, from 2004 to 2019. Infect Drug Resist Volume 13:4133–4145

Azad AK, Sadee W, Schlesinger LS (2012) Innate immune gene polymorphisms in tuberculosis. Infect Immun 80(10):3343–3359

Bagcchi S (2023) WHO's global tuberculosis report 2022. Lancet Microbe 4(1):20

Barter DM, Agboola SO, Murray MB, Bärnighausen T (2012) Tuberculosis and poverty: the contribution of patient costs in sub-Saharan Africa–a systematic review. BMC Public Health 12:1–21

Bu Q, Qiang R, Fang L et al (2023) Global trends in the incidence rates of MDR and XDR tuberculosis: findings from the global burden of disease study 2019. Front Pharmacol 14:1156249

Centers for Disease Control and Prevention (CDC) (2022) How TB spreads. https://www.cdc.gov/tb/topic/basics/howtbspreads.htm. Accessed 26 Sept 2023

Centers for Disease Control and Prevention (CDC) (2024) TB risk and people with HIV. https://www.cdc.gov/tb/riskfactors/hiv.html#:~:text=People%20with%20HIV%20are%20more,body%20to%20fight%20TB%20germs. Accessed 10 Sept 2024

Centre for Social Development in Africa (2019) Poverty, inequality and social exclusion in South Africa: a systematic assessment of how key policies, strategies and flagship programmes address poverty, inequality and equity issues. Report submitted to National Development Agency. University of Johannesburg. Available at: https://nda.org.za/assets/resources/CF824421-4FA0-41EE-AB69-. Accessed 2 Aug 2025

Chang SH, Cataldo JK (2014) A systematic review of global cultural variations in knowledge, attitudes and health responses to tuberculosis stigma. Int J Tuberc Lung Dis 18(2):168–173

Chingonzoh R, Manesen MR, Madlavu MJ (2018) Risk factors for mortality among adults registered on the routine drug resistant tuberculosis reporting database in the Eastern Cape Province, South Africa, 2011 to 2013. PLoS One 13(8):0202469

Chiposi L, Cele LP, Mokgatle M (2021) Prevalence of delay in seeking tuberculosis care and the health care seeking behaviour profile of tuberculous patients in a rural district of KwaZulu Natal, South Africa Pan African Med J 39(1)

Chumpa N, Kawkitinarong K, Rotcheewaphan S (2020) Evaluation of Anyplex™ II MTB/MDR kit's performance to rapidly detect isoniazid and rifampicin resistant *Mycobacterium tuberculosis* from various clinical specimens. Mol Biol Rep 47(4):2501–2508

Ciobanu A, Plesca V, Doltu S et al (2024) TB and poverty: the effect of rifampicin-resistant TB on household income. IJTLD OPEN 1(4):181–188

Claessens NJM, Gausi FF, Meijnen S et al (2001) Traditional healers and pulmonary tuberculosis in Malawi. Malawi Med J 13(4):7–8

Clarke C, Kerr TJ, Warren RM et al (2022) identification and characterisation of nontuberculous mycobacteria in African buffaloes (Syncerus caffer), South Africa. Microorganisms 10(9):1861

Courtwright A, Turner AN (2010) Tuberculosis and stigmatization: pathways and interventions. Public Health Rep 125:34–42

Dalton T, Cegielski P, Akksilp S et al (2012) Prevalence of and risk factors for resistance to second-line drugs in people with multidrug-resistant tuberculosis in eight countries: a prospective cohort study. Lancet 380(9851):1406–1417

Deol AK, Shaikh N, Middelkoop K et al (2022) Importance of ventilation and occupancy to Mycobacterium tuberculosis transmission rates in congregate settings. BMC Public Health 22(1):1772

Fenner L, Reid SE, Fox MP et al (2013) Tuberculosis and the risk of opportunistic infections and cancers in HIV-infected patients starting ART in Southern Africa. Trop Med Int Health 18(2):194–198

Freschi L, Vargas R Jr, Husain A et al (2021) Population structure, biogeography and transmissibility of *Mycobacterium tuberculosis*. Nat Commun 12(1):6099

Gao J, Pei Y, Yan X et al (2020) Emergence of nontuberculous mycobacteria infections during bedaquiline-containing regimens in multidrug-resistant tuberculosis patients. Int J Infect Dis 100:196–198

Gopalaswamy R, Shanmugam S, Mondal R, Subbian S (2020) Of tuberculosis and non-tuberculous mycobacterial infections–a comparative analysis of epidemiology, diagnosis and treatment. J Biomed Sci 27:1–17

He W, Tan Y, Song Z et al (2023) Endogenous relapse and exogenous reinfection in recurrent pulmonary tuberculosis: a retrospective study revealed by whole genome sequencing. Front Microbiol 14:1115295

Health Systems Trust (2015) Disease profile for Vhembe Health District Limpopo. Available at: https://hst.org.

za/publications/HST%20Publications/Disease%20pr. Accessed 2 Aug 2025

Huang M, Tan Y, Zhang X et al (2022) Effect of mixed infections with *Mycobacterium tuberculosis* and non-tuberculous mycobacteria on diagnosis of multidrug-resistant tuberculosis: a retrospective multicentre study in China. Infect Drug Resist:157–166

Khan PY, Yates TA, Osman M et al (2019) Transmission of drug-resistant tuberculosis in HIV-endemic settings. Lancet Infect Dis 19(3):77–88

Kim J, Choi Q, Kim JW (2020) Comparison of the Genedia MTB/NTM detection kit and Anyplex plus MTB/NTM detection kit for detection of mycobacterium tuberculosis complex and nontuberculous mycobacteria in clinical specimens. J Clin Lab Anal 34(1):23021

Korhonen V, Soini H, Vasankari T (2017) Recurrent tuberculosis in Finland 1995–2013: a clinical and epidemiological cohort study. BMC Infect Dis 17:1–7

Lauschke VM, Ingelman-Sundberg M (2019) Prediction of drug response and adverse drug reactions: from twin studies to next generation sequencing. Eur J Pharm Sci 130:65–77

Lazzarini LCO, Huard RC, Boechat NL et al (2007) Discovery of a novel Mycobacterium tuberculosis lineage that is a major cause of tuberculosis in Rio de Janeiro, Brazil. J Clin Microbiol 45(12):3891–3902

Lee LY, Tung HH, Chen SC, Fu CH (2017) Perceived stigma and depression in initially diagnosed pulmonary tuberculosis patients. J Clin Nurs 26(23–24):4813–4821

Lee EG, Min J, Kang JY et al (2020) Age-stratified anti-tuberculosis drug resistance profiles in South Korea: a multicenter retrospective study. BMC Infect Dis 20:1–10

Lienhardt C, Fielding K, Sillah JS et al (2005) Investigation of the risk factors for tuberculosis: a case–control study in three countries in West Africa. Int J Epidemiol 34(4):914–923

Lim J, Kim J, Kim JW et al (2014) Multicenter evaluation of Seegene Anyplex TB PCR for the detection of Mycobacterium tuberculosis in respiratory specimens. J Microbiol Biotechnol 24(7):1004–1007

Lohiya A, Abdulkader RS, Rath RS et al (2020) Prevalence and patterns of drug resistant pulmonary tuberculosis in India—a systematic review and meta-analysis. J Global Antimicrob Resist 22:308–316

Lopez-Luis BA, Sifuentes-Osornio J, Pérez-Gutiérrez MT et al (2020) Nontuberculous mycobacterial infection in a tertiary care center in Mexico, 2001-2017. Braz J Infect Dis 24(3):213–220

Lotz JK, Porter JD, Conradie HH et al (2023) Treating drug-resistant tuberculosis in an era of shorter regimens: insights from rural South Africa. S Afr Med J 113(11):1491–1500

Luukinen BV, Vuento R, Hirvonen JJ (2020) Evaluation of a semi-automated Seegene PCR workflow with MTB, MDR, and NTM detection for rapid screening of tuberculosis in a low-prevalence setting. APMIS 128(5):406–413

Ma WW, Wang LC, Zhao DA et al (2024) Analysis of T-lymphocyte subsets and risk factors in children with tuberculosis. Tuberculosis 146:102496

Maiga M, Siddiqui S, Diallo S et al (2012) Failure to recognize nontuberculous mycobacteria leads to misdiagnosis of chronic pulmonary tuberculosis. PLoS One 7(5):36902

Marais BJ, Hesseling AC, Cotton MF (2009) Eur Respir J 33(4):943–944

Marçôa R, Ribeiro AI, Zão I, Duarte R (2018) Tuberculosis and gender–factors influencing the risk of tuberculosis among men and women by age group. Pulmonology 24(3):199–202

Mlondo MN, Melesse SF, Mwambi HG (2022) Risk factors associated with tuberculosis among men; a study of South Africa. Open Public Health J 15(1)

Moetlo GJ, Pengpid S, Peltzer K (2011) An evaluation of the implementation of integrated community home-based care services in Vhembe District, South Africa. Indian J Palliat Care 17(2):137

Mohammed KA, Khudhair GS, Al-Rabeai DB (2022) Prevalence and drug resistance pattern of isolated from tuberculosis patients in Basra, Iraq. Polish J Microbiol 71(2):205–215

Ngumbela XG, Khalema EN, Nzimakwe TI (2020) Local worlds: vulnerability and food insecurity in the eastern cape province of South Africa. Jàmbá J Disast Risk Stud 12(1):1–9

Nhamoyebonde S, Leslie A (2014) Biological differences between the sexes and susceptibility to tuberculosis. J Infect Dis:100–106

Nyathi L, Balogun T, De Lange J et al (2024) Social issues affecting social cohesion in low-resource communities in South Africa. Afr J Govern Dev 13(2):135–162

Okonkwo RC, Onwunzo MC, Chukwuka CP et al (2017) The use of the GeneXpert mycobacterium tuberculosis/Rifampicin (MTB/Rif) assay in detection of multi-drug resistant tuberculosis (MDRTB) in Nnamdi Azikiwe University Teaching Hospital, Nnewi, Nigeria. J HIV Retrovirus 3:1

Opperman CJ, Ntshangase Z, Gumede M et al (2024) Misdiagnosis of non-tuberculous mycobacteria as tuberculous by the GeneXpert MTB/RIF ultra: fact or fiction? South Afr J Infect Dis 39(1):576

Pang Y, Lu J, Su B et al (2017) Misdiagnosis of tuberculosis associated with some species of nontuberculous mycobacteria by GeneXpert MTB/RIF assay. Infection 45:677–681

Putong NM, Pitisuttithum P, Supanaranond W et al (2002) Mycobacterium tuberculosis infection among HIV/AIDS patients in Thailand: clinical manifestations and outcomes. Blood 46:17–10

Qi CC, Xu LR, Zhao CJ (2023) Prevalence and risk factors of tuberculosis among people living with HIV/AIDS in China: a systematic review and meta-analysis. BMC Infect Dis 23(1):584

Richter E, Weizenegger M, Fahr AM, Rüsch-Gerdes S (2004) Usefulness of the GenoType MTBC assay for differentiating species of the Mycobacterium tuberculosis complex in cultures obtained from clinical specimens. J Clin Microbiol 42(9):4303–4306

Rickman HM, Phiri MD, Feasey HR et al (2025) Sex differences in the risk of Mycobacterium tuberculosis infection: a systematic review and meta-analysis of population-based immunoreactivity surveys. Lancet Public Health 10(7):588–598

Sarro YD, Kone B, Diarra B et al (2018) Simultaneous diagnosis of tuberculous and non-tuberculous mycobacterial diseases: time for a better patient management. Clin Microbiol Infect Dis 3(3)

Sawatpanich A, Petsong S, Tumwasorn S, Rotcheewaphan S (2022) Diagnostic performance of the Anyplex MTB/NTM real-time PCR in detection of Mycobacterium tuberculosis complex and nontuberculous mycobacteria from pulmonary and extrapulmonary specimens. Heliyon 8(12):e11935

Seegene (n.d.) Allplex™ MTB/MDR/XDRe detection. Available at: https://seegene.com/assays/allplex_mtb_mdr_xdre_detection. Accessed 2 Nov 2025

Seegene Inc. (2017) Anyplex™ MTB/NTM real-time detection brochure. Seegene Inc, Seoul.

Shahraki AH, Heidarieh P, Bostanabad SZ et al (2015) "Multidrug-resistant tuberculosis" may be nontuberculous mycobacteria. Eur J Intern Med 26(4):279–284

Song WM, Li YF, Ma XB et al (2019) Primary drug resistance of *Mycobacterium tuberculosis* in Shandong, China, 2004–2018. Respir Res 20:1–12

TB Accountability Consortium (2024) The State of TB in South Africa: what about TB Mortality? March 2024. TB Accountability Consortium, Pretoria. Available at: https://tbac.org.za/wp-content/uploads/2024/03/state-of-TB-22-04-24digi.pdf. Accessed 14 Nov 2025

Tiberi S, Utjesanovic N, Galvin J et al (2022) Drug resistant TB–latest developments in epidemiology, diagnostics and management. Int J Infect Dis 124:20–25

Vhembe District Municipality Profile (VDMP) (2020). https://www.cogta.gov.za/ddm/wp-content/uploads/2020/11/vhembe-october-2020.pdf. Accessed 28 June 2023

World Health Organization (WHO) (2013) Global tuberculosis report 2013. https://books.google.co.za/books?hl=en&lr=&id=1rQXDAAAQBAJ&oi=fnd&pg=PP1&dq=World+Health+Organization.+Global+tuberculosis+report+2017.+2017&ots=la_aSt1v3V&sig=LjwfTOWmy3obyTHI-SCLjU1dLOQ&redir_esc=y#v=onepage&q=World%20Health%20Organization.%20Global%20tuberculosis%20report%202017.%202017&f=false. Accessed 26 Sept 2023

World Health Organization (WHO) (2018) WHO treatment guidelines for isoniazid-resistant tuberculosis: supplement to the WHO treatment guidelines for drug-resistant tuberculosis. World Health Organization, Geneva. Available at: http://www.who.int/tb/publications/2018/WHO_guidelines_isoniazid_resistant_TB/en/. Accessed 1 Nov 2025

World Health Organization (WHO) (2022). https://www.iom.int/sites/g/files/tmzbdl486/files/documents/2023-03/Global-TB-Report-2022.pdf. Accessed 26 Sept 2023

World Health Organization (WHO) (2023) Global Tuberculosis report. https://cdn.who.int/media/docs/default-source/hq-tuberculosis/global-tuberculosis-report-2023/global-tb-report-2023_factsheet.pdf?sfvrsn=f0dfc8a4_4&download=true. Accessed 28 Aug 2024

Yasobant S, Shah H, Bhavsar P et al (2023) Why and where?—delay in tuberculosis care cascade: a cross-sectional assessment in two Indian states, Jharkhand, Gujarat. Front Public Health 11:1015024

Zaman K (2010) Tuberculosis: a global health problem. J Health Popul Nutr 28(2):111

Neuroinflammatory and Neurodegenerative Roles of HIV-1 Tat: A Review of Recent Evidence

Monray. E. Williams, Tshengedzeni Muvenda, Vhuthuhawe Mugwena, and Simo S. Zulu

Abstract

HIV-1 remains a major global health concern, exerting detrimental effects not only on the immune system but also on the central nervous system (CNS). Soon after HIV-1 infection, the virus can breach the blood–brain barrier (BBB), initiating processes that result in neuronal damage and cell death. These effects manifest clinically as a spectrum of cognitive, behavioural, and motor impairments collectively termed HIV-associated neurocognitive disorders (HAND). Among the viral proteins implicated in CNS pathology, the Trans-Activator of Transcription (Tat) protein stands out due to its pronounced neurotoxic and neurotropic characteristics. Tat is secreted by infected cells and can accumulate in the CNS, persisting even in individuals undergoing effective antiretroviral therapy (ART), thereby contributing to ongoing neurocognitive complications. Although Tat has been extensively studied since the 1980s, current literature examining its effects within the context of modern treatment regimens remains sparse. This chapter offers a comprehensive and updated review of the role of HIV-1 Tat in neuropathogenesis, drawing on research from the past 25 years to highlight its contribution to HAND and to outline potential avenues for therapeutic advancement.

Keywords

Tat · Neurotropism · Neuroinflammation · Apoptosis · Mitochondrial dysfunction

7.1 Introduction

HIV-1 remains a significant global health concern, not only for its impact on the immune system but also for its effects on the central nervous system (CNS). Shortly after infection, HIV-1 is capable of entering the CNS after crossing the blood-brain barrier (BBB), where it exerts both direct and indirect effects that contribute to neuronal injury and cell death (González-Scarano and Martín-García 2005). These pathogenic mechanisms culminate in a range of cognitive, behavioural, and motor impairments collectively known as HIV-associated neurocognitive disorders (HAND) (Antinori et al. 2007).

M. E. Williams (✉) · T. Muvenda · V. Mugwena
Biomedical and Molecular Metabolism Research (BioMMet), North-West University, Potchefstroom, South Africa
e-mail: Monray.Williams@nwu.ac.za

S. S. Zulu
Department of Human Biology and Integrated Pathology, Faculty of Health Sciences, Nelson Mandela University, Gqeberha, South Africa

A. Shonhai et al. (eds.), *Advances in Biochemistry and Molecular Biology to meet Africa´s Needs*, Advances in Experimental Medicine and Biology 1507,
https://doi.org/10.1007/978-3-032-24254-9_7

In the context of HIV-1 neuropathogenesis, several viral proteins, including Trans-Activator of Transcription (Tat), Viral protein R (Vpr), Negative Regulatory Factor (Nef), Regulator of Virion Expression (Rev), and Glycoprotein (gp)120, have been implicated in CNS dysfunction (Nath 2002). Among these, the Tat protein stands out as one of the most extensively studied. Tat is a potent neurotoxic protein with strong neurotropic properties (Bagashev and Sawaya 2013). Importantly, Tat is actively secreted by infected cells and can be detected in the CNS even in individuals receiving suppressive antiretroviral therapy (ART) (Marino et al. 2020b). This unique ability to act independently of active viral replication makes Tat a key factor in the persistence of HAND in the modern ART era.

The role of Tat in HIV-1-associated neuropathogenesis has been under investigation since the early 1980s. Despite the foundational research conducted over the past few decades, recent updates specifically focusing on the ongoing impact of Tat under contemporary treatment regimens have been limited. As such, this chapter provides an up-to-date overview of HIV-1 Tat-mediated neuropathogenesis, with a focus on findings from the past 25 years. By synthesizing recent insights, this chapter aims to clarify the continuing role of Tat in HAND and identify future directions for therapeutic interventions.

7.2 HIV-1 and the Development of HAND

HIV-1 is capable of invading the CNS early during infection, primarily through a mechanism known as the "Trojan horse" hypothesis (González-Scarano and Martín-García 2005) (Fig. 7.1). In this process, HIV-infected monocytes cross the BBB and subsequently differentiate into macrophages within the CNS (González-Scarano and Martín-García 2005). Another possibility is that free HIV can cross BBB through the paracellular mechanism (McRae 2016). These infected macrophages and free virus contributes to viral persistence and can transmit the virus to other resident immune cells,

including microglia and perivascular macrophages, amplifying infection within the brain (González-Scarano and Martín-García 2005). Once HIV-1 has entered the CNS, it contributes to neurodegeneration through indirect and direct mechanisms, despite the effective plasma viral control (Fig. 7.1).

Indirectly, HIV-1 viral particles and proteins, including but not limited to gp120, Tat, and Vpr, can activate resident immune cells in the brain, including macrophages, microglia, and astrocytes (González-Scarano and Martín-García 2005) (Fig. 7.1). This activation leads to chronic production and release of pro-inflammatory cytokines and chemokines, resulting in sustained neuroinflammation and neuronal injury (Williams et al. 2021; Williams et al. 2020a; Williams and Naudé 2024). In the modern ART era, it is believed that this indirect, chronic neuroinflammation is responsible for the persistent development of HAND (Harezlak et al. 2011). Moreover, the presence of viral proteins can disrupt the integrity of the BBB, allowing increased transmigration of peripheral immune cells into the CNS, thereby exacerbating the inflammatory milieu, compounding neuronal stress and injury (McRae 2016). Both chronic inflammation and viral protein activity contribute to the progressive breakdown of the BBB and the advancement of neurological damage (Fig. 7.1).

Direct mechanisms of HIV-1 protein neurotoxicity are supported by various fundamental studies (Fig. 7.1), which demonstrate that the direct application of Tat (Rao et al. 2013), Vpr (Jones et al. 2007), gp120 (Haughey and Mattson 2002), Rev (Mabrouk et al. 1991) and Nef (Yarandi et al. 2021) on neuronal cultures results in reduced cell viability. Among these viral proteins, the Tat protein has been extensively studied for its ability to trigger excitotoxicity, oxidative stress, and apoptosis in neurons. Such continuous neuronal injury contributes to the development of HAND.

Tat-induced mechanisms persist in the CNS despite viral control in the periphery because of the limited ability of cART to effectively cross the BBB (Osborne et al. 2020).

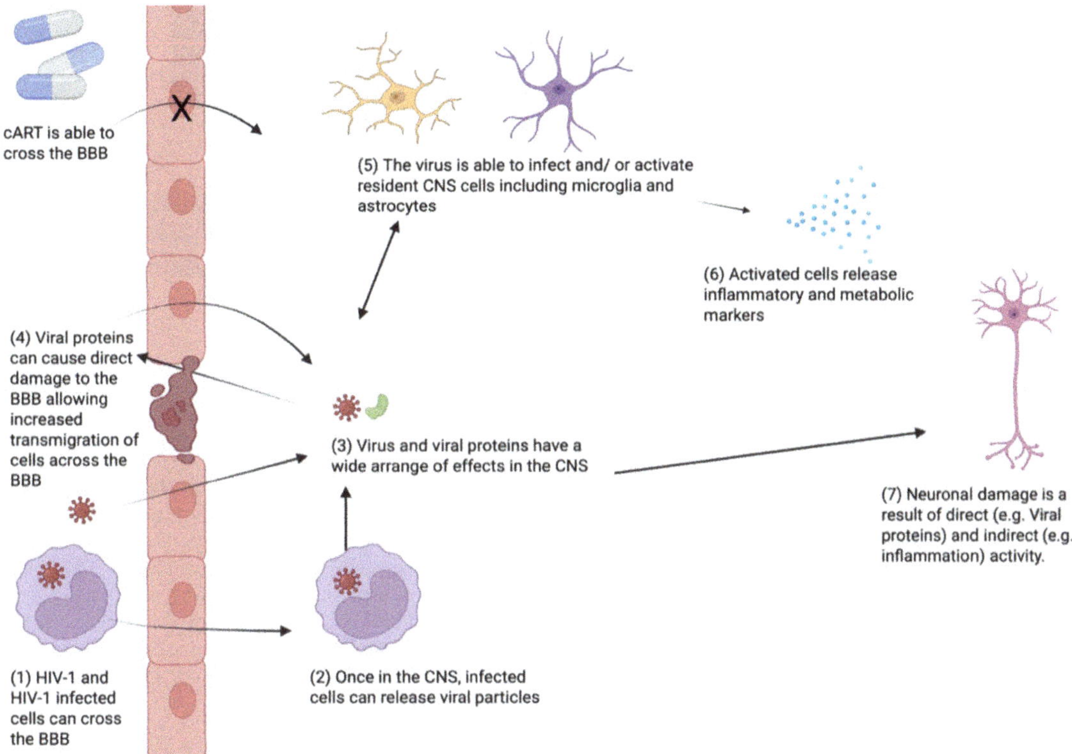

Fig. 7.1 The neuropathogenesis of HIV in the modern antiretroviral therapy (cART) era. (1) HIV-infected cells can cross the (BBB through a mechanism known as the "Trojan horse." Free HIV can also cross the BBB through the paracellular mechanism. (2) Once HIV is in the central nervous system (CNS), these infected cells release active viral particles, a key event in the development of HIV-associated neuropathology. (3) These viral particles contribute to neurotoxic events within the CNS. (4) They can directly damage the BBB, resulting in increased transmigration of immune cells into the CNS. (5) The virus and its particles infect and/or activate glial cells including astrocytes and microglia. (6) These activated glial cells release a range of inflammatory and metabolic mediators. (7) These mediators contribute to neuronal damage through several mechanisms. In addition, viral proteins themselves can cause direct damage to the BBB. Collectively, these processes contribute to the clinical presentation of HIV-associated neurocognitive disorders (HAND). One of the reasons for the persistent development of HAND is the limited ability of cART to cross the BBB (X). (All visual elements were created in BioRender. Mason, S. (2026). https://BioRender.com/sn3h1kg)

HAND is classified using a criterion developed in 2007 by a working group convened by the U.S. National Institute of Mental Health (Antinori et al. 2007). Asymptomatic neurocognitive impairment (ANI) is characterized by cognitive performance that falls at least one standard deviation below the mean of demographically adjusted normative scores in at least two cognitive domains (Rumbaugh and Tyor 2015). These domains may include attention-information processing, language, abstraction-executive function, complex perceptual motor skills, memory (including learning and recall), simple motor skills, or sensory-perceptual abilities (Rumbaugh and Tyor 2015). The criteria also specify that at least five cognitive domains must be examined to make this diagnosis. Mild neurocognitive disorder (MND) may be underdiagnosed if cognitive symptoms and signs are not carefully assessed. It results in mild interference with daily functioning, such as inefficiency at work, difficulties managing household responsibilities, or problems in social situations (Rumbaugh and Tyor 2015). This contrasts with HIV-associated dementia (HAD), which significantly impairs activities of daily living (ADLs) (Rumbaugh and

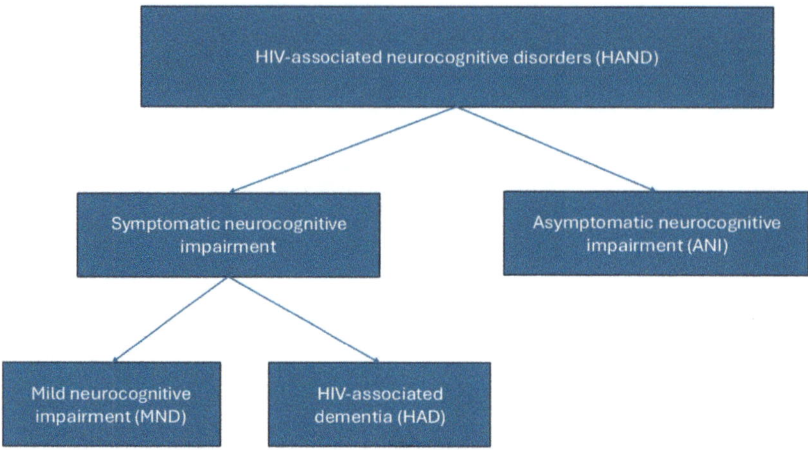

Fig. 7.2 Classification of HIV-associated neurocognitive disorders (HAND) and its subgroups. HAND is broadly categorized into three clinical subtypes based on severity: asymptomatic neurocognitive impairment (ANI), mild neurocognitive disorder (MND), and HIV-associated dementia (HAD). ANI is characterized by measurable cognitive impairment on neuropsychological testing without observable impact on daily functioning. MND involves mild-to-moderate cognitive deficits that begin to interfere with everyday activities, such as work performance or social functioning. HAD represents the most severe form, with marked cognitive decline and significant impairment in daily living. This classification helps guide clinical assessment, monitoring, and management of neurocognitive impairment in people living with HIV, particularly in the context of effective ART, where overt dementia has become less common, but milder forms remain prevalent

Tyor 2015). A diagnosis of MND requires neuropsychological testing that shows abnormalities in at least two cognitive domains (Fig. 7.2). Finally, HAD is diagnosed based on the presence of moderate-to-severe acquired cognitive impairment, evidenced by performance at least two standard deviations below the demographically corrected normative means in at least two different cognitive areas (Rumbaugh and Tyor 2015). In addition, there must be marked interference with ADLs due to cognitive deficits. The impairment should not be attributable to delirium or adequately explained by other comorbid conditions (Fig. 7.2). The introduction of ART has significantly reduced the incidence of the most severe forms of HAND, such as HAD. However, milder manifestations, including ANI and MND, remain common, even in individuals with long-term viral suppression (Alford and Vera 2018).

This persistence of HAND suggests that latent viral reservoirs, ongoing low-grade inflammation, and the continued presence of viral proteins within the CNS contribute to neuropathology despite effective systemic viral control. Notably, several viral proteins have been implicated in the development of HAND, with particular focus on Tat due to its dual role in immune activation and direct neurotoxicity.

7.3 HIV-1 Tat Protein: Structure and Function

The HIV-1 Tat protein comprises 86 to 101 amino acids (depending on specific viral isolate), with a variable weight of 9–11 kDa, and is encoded by two exons (Gotora et al. 2023; Gotora et al. 2024) (Fig. 7.3). The first exon, spanning amino acids 1–72, encodes the majority of the functional domains of Tat, while the second exon encodes the C-terminal region, which contains additional functional motifs that contribute to the extracellular activities of Tat (Li et al. 2012). Tat is structurally organized into six functionally distinct regions, each associated with specific molecular roles in viral replication and host interaction (Williams et al. 2020b) (Fig. 7.3). Region 1 (residues 1–21), known as the proline-rich domain, contains a conserved tryptophan residue at position 20 that may facilitate protein-protein interac-

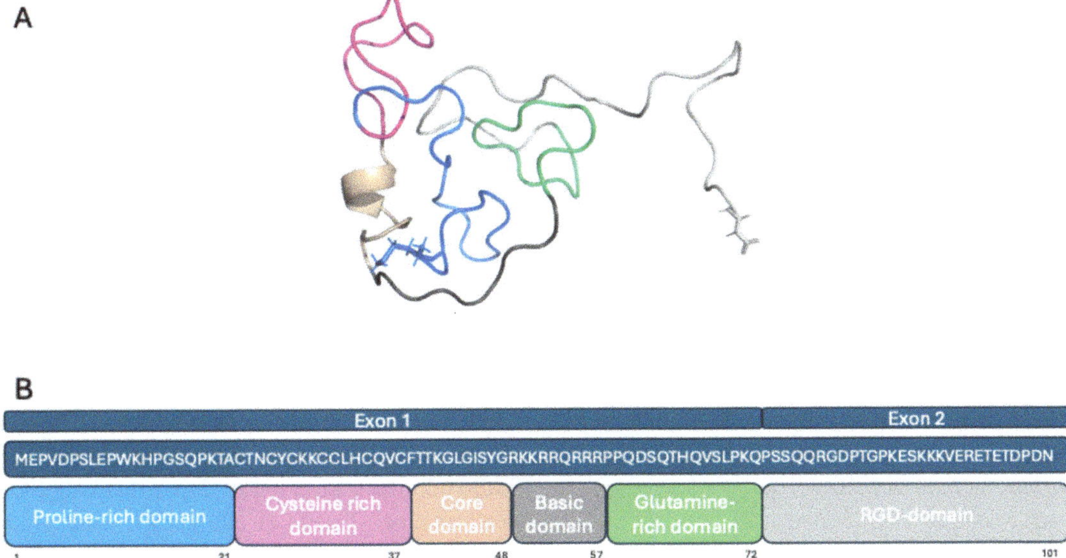

Fig. 7.3 Structural and functional representation of the HIV-1 Tat protein. (**a**) Three-dimensional (3D) structure of the HIV-1 Tat protein, generated using MODELLER. The structural model illustrates the spatial conformation of key functional regions. (**b**) Linear amino acid sequence of the HIV-1 Tat protein (group M, subtype B, isolate JRCSF), annotated with functional domains and exonic regions. Exon 1 and Exon 2 are indicated, along with the six major functional domains: the proline-rich region, cysteine-rich domain, core region, basic domain, glutamine-rich region, and RGD motif. The colour scheme used in panel **a** corresponds to the coloured regions in panel **b**, allowing for direct structural-to-functional domain mapping

tions (Fig. 7.3) (Spector et al. 2019). Region 2 (residues 22–37), known as the cysteine-rich domain, includes seven highly conserved cysteine residues essential for maintaining the structural integrity of the Tat protein via disulfide bond formation. This domain is one of the most extensively studied, particularly in HIV-1 subtype B, due to its critical role in transactivation and redox regulation (Spector et al. 2019). Region 3 (residues 38–48), referred to as the core domain, features conserved motifs such as phenylalanine at position 38 (F38) and the LGISYG sequence. This region contributes to the structural stability of Tat and its interactions with cellular cofactors (Campbell and Loret 2009). Region 4 (residues 49–57), often called the basic or arginine-rich domain, contains the conserved sequence RKKRRQRRRPP. This region is vital for binding to the Trans-Activation Response (TAR) RNA element and facilitates recruitment of positive transcription elongation factor b (P-TEFb) (Gotora et al. 2023; Gotora et al. 2024; Campbell

and Loret 2009), thereby enhancing transcriptional elongation by RNA polymerase II. Notably, this domain also functions as a cell-penetrating peptide (CPP), allowing Tat to cross cellular membranes (Ruiz et al. 2019). Region 5 (residues 58–72) is a glutamine-rich segment and represents the most variable region among HIV-1 subtypes. Region 6, encoded by the second exon, includes the C-terminal domain that contains an RGD motif, which may mediate interactions with integrins and other cellular proteins (Clark et al. 2017).

Tat plays a central role in HIV-1 gene expression through its interaction with the TAR element of nascent viral transcripts. This binding facilitates the recruitment of P-TEFb, promoting phosphorylation of the RNA polymerase II complex and enabling the transition from transcriptional pausing to productive elongation (Karn 2011). Recent evidence suggests that amino acid variations in the Tat protein may influence its interaction with the TAR element, highlighting the critical role of the basic domain in this function (Gotora et al. 2023; Gotora et al. 2024; Williams and Cloete 2022a; Ronsard et al. 2017). In addition to its nuclear functions, Tat is actively secreted by infected cells and can be internalized by uninfected bystander cells, including neurons and glial cells, via its CPP domain (Ruiz et al. 2019). This cell-penetrating capacity enables Tat to exert pathological effects on non-infected cells, significantly contributing to HIV-associated neuropathogenesis. Numerous studies have underscored the neurotropic and neurotoxic properties of Tat, including its ability to induce oxidative stress, neuroinflammation, and excitotoxicity, mechanisms that play key roles in the development of HAND.

7.4 Tat and Clinical Presentations of HAND

In human studies, the HIV-1 Tat protein has been directly associated with the clinical development of HAND. One study reported that the presence of Tat in the cerebrospinal fluid (CSF) was significantly associated with poorer performance in the cognitive domains of psychomotor speed ($p = 0.04$) and information processing ($p = 0.02$) (Henderson et al. 2019). An early study conducted in 2000 aimed to determine the role of Tat in mediating HIV encephalitis (HIVE) (Hudson et al. 2000). Tat mRNA and protein were measured in tissue extracts from nine participants with HIVE and seven patients without HIVE. Despite extended autopsy intervals and significant tissue degradation in this study, Tat mRNA was detected in 4 out of 9 participants with HIVE, but in none of the seven patients without dementia (Hudson et al. 2000). Another study examining anti-Tat antibodies in the CSF of a well-characterized cohort of 52 people living with HIV (PLWH) and 13 HIV-negative controls found that anti-Tat antibody levels were higher in individuals without neurocognitive impairment compared to those with HAND (Bachani et al. 2013). These findings suggest a potential protective role of anti-Tat immune responses. However, the number of studies that have directly measured

Tat levels in human blood or CSF concerning neurocognitive performance remains limited.

Similar findings have been echoed in in vivo models. A study investigating the effects of Tat on glutamatergic circuitry reported that recombinant Tat86 (86 amino acid variant) or saline was injected into the medial prefrontal cortex (mPFC) of male Sprague–Dawley rats to examine potential mechanisms underlying cognitive impairment (Duffy et al. 2024) (Table 7.1). The animals were evaluated using behavioural tasks dependent on mPFC function, including novel object recognition (NOR), spatial object recognition (SOR), and temporal order (TO) tasks at 1 and 2 weeks postoperatively. The study found that Tat86 administration led to impairments in SOR performance (Duffy et al. 2024). One study hypothesized that HIV-1 Tat impairs cognitive behaviour and selectively injures specific hippocampal interneuron subtypes (Marks et al. 2016). In this study, male transgenic mice with inducible expression of HIV-1 Tat, along with non-expressing control mice, were assessed for cognitive performance and examined for alterations in hippocampal CA1 interneuron subpopulations using specific cellular markers (Marks et al. 2016). Tat expression was found to impair spatial memory in the Barnes maze and reduce mnemonic performance in the novel object recognition task (Marks et al. 2016). Several other studies, as reported in a recent review, show that the presence of Tat in the brains of in vivo models significantly contributes to poorer neurocognitive performance (Sil et al. 2021; Harricharan et al. 2015) (Table 7.1). Therefore, evidence suggests that the presence of Tat in the brain influences clinical and behavioural outcomes. As highlighted above, Tat can also influence BBB function, allowing increased transmigration of cells into the CNS.

7.5 Tat and the Blood-Brain Barrier (BBB)

HIV and its associated viral particles exacerbate damage to the BBB, a selective interface that regulates substance exchange between the blood-

Table 7.1 Key studies reporting the relationship between Tat and neuropathogenic mechanisms

Reference	Tat model	Tat association
Clinical HAND		
Henderson et al. (2019)	Human, CSF	Psychomotor speed and information processing
Hudson et al. (2000)	Human, brain tissue	HIVE
Duffy et al. (2024)	Animal, rat	Spatial object recognition
Marks et al. (2016)	Animal, Tat transgenic mouse	Spatial memory and mnemonic performance
Direct neuropathogenesis		
Yadav-Samudrala et al. (2025)	Animal, Tat transgenic mouse	Dendritic simplification, axonal disruption, synaptic loss, and dysregulation of synaptic plasticity genes
Schier et al. (2017)	Animal, Tat transgenic mouse	Synaptodendritic injury
Hahn et al. (2015)	Animal, Tat transgenic mouse	Dendritic spine deficits, cellular spine deficit and cognitive impairment
Krogh et al. (2014)	Animal, cell culture	Mitochondrial dysfunction
Hahn et al. (2015)	Cell culture	Neurotrophic support
Bergonzini et al. (2004)	Cell culture	Neurotrophic support
Liu et al. (2017)	Animal model, cell culture	Demyelination
Zou et al. (2015)	Cell cultures	Demyelination
Hauser et al. (2009)	Animal model, cell culture	Demyelination
Neuroinflammation		
Yadav-Samudrala et al. (2025)	Animal, Tat Transgenic mouse model	Neuronal function Object recognition memory. Recognition memory Cognitive impairment Cognitive deficits in HAND.
Deshetty et al. (2024)	Human cell line, Cell culture	Mitochondrial dysfunction
BBB damage		
Andras et al. (2003)	Animal, cell culture	BBB dysfunction
Banks et al. (2005)	Animal, brain tissue and blood	BBB dysfunction
	Animal, brain tissue and blood	BBB dysfunction
Leibrand et al. (2017)	Animal model, Transgenic mouse model	BBB function Neuroinflammation
Fan and He (2016)	Cell culture	BBB function
Leibrand et al. (2019)	Animal model, Mouse	BBB function
Niu et al. (2021)	Animal model, Mouse Human cell line, Cell culture	BBB damage BBB damage
Subtype variation and amino acid variation		
Ruiz et al. (2019)	In vitro study model	Neuroinflammatory responses
Williams and Cloete (2022b)	Molecular docking	Tat protein bind/ Tat structure

stream and the CNS (Kadry et al. 2020). The BBB primarily consists of tightly connected endothelial cells lining brain capillaries, supported by pericytes within the basement membrane, and enveloped by astrocyte end-feet (Kadry et al. 2020). This structure maintains the protected CNS environment.

During HIV infection, the BBB initially acts as a barrier to viral entry but eventually becomes a conduit for the virus to access the brain, leading to neurological complications (Strazza et al. 2011). Tat disrupts tight junction proteins, induces oxidative stress, and upregulates matrix metalloproteinases (MMPs), leading to increased BBB permeability. This disruption promotes and facilitates the transmigration of HIV-infected monocytes into the CNS, increasing the viral load in the CNS (Said and Venketaraman 2025). The role of the Tat protein in cells of the BBB has been extensively reviewed in several recent studies. Sun and colleagues examined how HIV-1 Tat affects brain microvascular endothelial cells and the expression of tight junction proteins (Sun et al. 2023). Additionally, Atluri and colleagues provided a comprehensive review of the impact of HIV-1 on the BBB, including a dedicated section outlining the mechanisms of Tat-induced BBB disruption (Atluri et al. 2015). Tat causes damage to the BBB through several cellular mechanisms, as highlighted below.

Tat promotes apoptosis in brain microvascular endothelial cells (BMVECs). This effect is mediated, in part, by endoplasmic reticulum (ER) stress (Fig. 7.4). One mechanism by which Tat does this is through the accumulation of misfolded or unfolded proteins within the ER, thereby disrupting proteostasis and activating the unfolded protein response (UPR) (Fan and He 2016; Ma et al. 2016). Tat has also been shown to

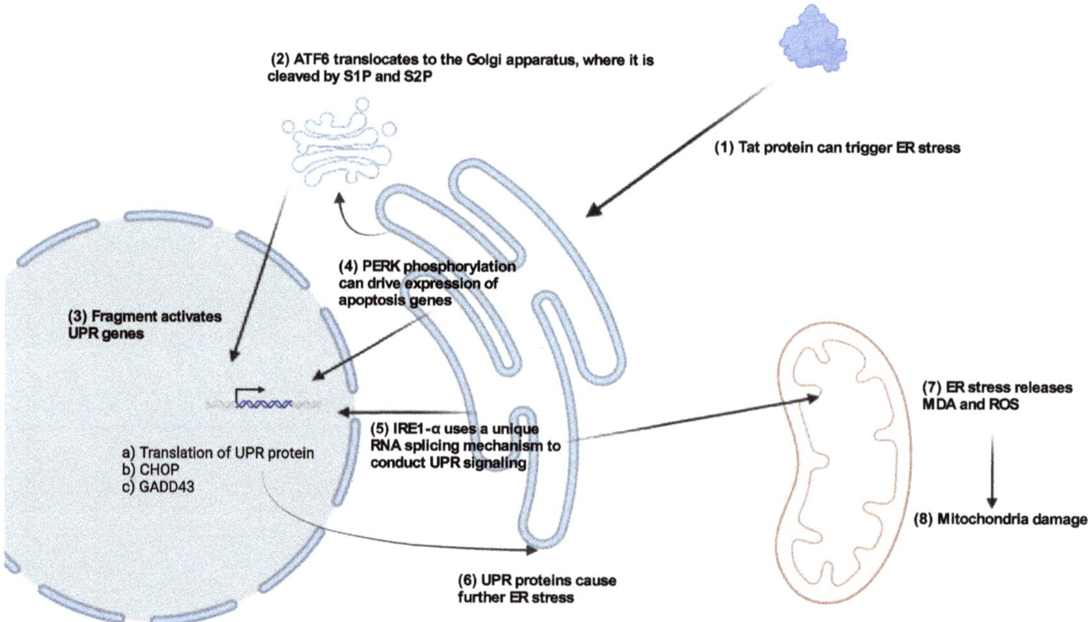

Fig. 7.4 Tat-mediated endothelial cell damage. (1) HIV-1 Tat protein initiates endoplasmic reticulum (ER) stress in brain microvascular endothelial cells (BMVECs), triggering the unfolded protein response (UPR). (2) In response, the membrane sensor ATF6 translocates to the Golgi apparatus, where it is cleaved by site-1 and site-2 proteases (S1P and S2P), (3) releasing a transcriptionally active fragment that enters the nucleus to activate UPR target genes. (4) Concurrently, PERK is activated and phosphorylates eIF2α, promoting the translation of ATF4, which induces the expression of pro-apoptotic genes such as CHOP and GADD34. (5) IRE1α, another ER membrane sensor, becomes activated and splices XBP1 mRNA, generating a transcription factor that upregulates UPR genes and can further contribute to apoptotic signalling. (6) Sustained UPR activation exacerbates ER stress, (7) leading to increased production of reactive oxygen species (ROS) and malondialdehyde (MDA), a marker of lipid peroxidation. (8) These oxidative stress responses result in mitochondrial dysfunction and ultimately endothelial death. (All visual elements were created in BioRender. Mason, S. (2026) https://BioRender.com/05ouudz. The figure was adapted from Sun et al. (2023))

trigger ER loss of calcium via the activation of the ryanodine receptor (RyR) (Norman et al. 2008), which then initiates protein misfolding and promotes ER stress (Hu 2016). Recent studies have shown that Tat expression can trigger UPR in BMVECs (Fan and He 2016) (Fig. 7.4).

The UPR is a cellular mechanism activated in response to ER stress and involves three primary sensors, ATF6, PERK, and IRE1α. Upon ER stress, (1) ATF6 translocate to the Golgi apparatus, where it is cleaved by site-1 and site-2 proteases (S1P and S2P), releasing a fragment that enters the nucleus to activate UPR target genes (Shen and Prywes 2004) (Fig. 7.4). (2) PERK phosphorylates eukaryotic initiation factor 2α (eIF2α), resulting in the translation of activating transcription factor 4 (ATF4), which induces genes such as CHOP and GADD34, linked to apoptosis and feedback regulation (Dai et al. 2019) (Fig. 7.4). (3) IRE1α, once activated, splices XBP1 mRNA, producing a transcription factor that not only upregulates UPR genes but can also activate apoptotic pathways (Prasad et al. 2020). Collectively, these pathways aim to restore ER homeostasis or, if unresolved, initiate apoptosis (Fig. 7.4). These cellular components also present potential targeted for therapeutic intervention.

Tat also contributes to BBB dysfunction through indirect mechanisms. Studies have demonstrated that Tat exposure in BMVECs increases the expression of NADPH oxidase and the production of ROS. This process involves the Transient receptor potential melastatin 2 (TRPM2) pathway which leads to the downregulation of antioxidant enzymes such as catalase (CAT), glutathione peroxidase (GSH-PX), and superoxide dismutase (SOD), while simultaneously upregulating ROS levels, including malondialdehyde (MDA), a marker of lipid peroxidation (Mishra et al. 2008) (Fig. 7.4).

Mitochondrial dysfunction is another hallmark of Tat-mediated BBB damage. Tat-treated BMVECs exhibit significant mitochondrial oxidative DNA damage. This damage is closely associated with the increased stability of hypoxia-inducible factor 1-α (HIF-1α), which in turn suppresses the expression of mitochondrial transcription factor A (TFAM) (Kim et al. 2019; Sharma et al. 2022). The Tat-induced stabilization of HIF-1α is thought to be mediated primarily through the activation of protein kinase C (PKC) and extracellular signal-regulated kinases (ERK1/2) (Kim et al. 2019; Sharma et al. 2022) (Fig. 7.4). Furthermore, Tat promotes the release of pro-inflammatory cytokines from BBB-associated cells (e.g., BMVECs and astrocytes), including interleukin (IL)-6, IL-1β, and tumour necrosis factor-alpha (TNF-α), thereby amplifying the local inflammatory environment (Toborek et al. 2003; Acheampong et al. 2002; Thangaraj et al. 2018). These findings collectively underscore the multifaceted role of HIV-1 Tat in compromising BBB integrity, promoting neuroinflammation, and contributing to the neuropathogenesis observed in NeuroHIV.

7.6 Tat and Neuroinflammation: Indirect Neuropathogenesis

When HIV is present in the CNS, it can infect/activate resident immune cells within the brain. Infected microglia and astrocytes produce viral proteins that both trigger and contribute to neuroinflammation, ultimately leading to HIV-1-associated neuronal damage (González-Scarano and Martín-García 2005) (Fig. 7.1).

Several clinical studies in humans have shown that, in the modern ART era, peripheral inflammation (Williams et al. 2020a), CSF inflammation (Williams et al. 2021), and brain tissue inflammation (Williams and Naudé 2024) are associated with neurocognitive impairment in PLWH, as highlighted in several recent systematic reviews. Similarly, both review studies (Thaney et al. 2018; Vigorito et al. 2015) and fundamental research (Bai et al. 2023; Borrajo et al. 2021) have demonstrated that animal models on HAND indicate that inflammation is a key driver of neuronal damage and neurocognitive impairment.

The HIV-1 Tat protein is one of the key drivers of neuroinflammation and exhibits both cytotoxic and pro-inflammatory effects under pathological conditions in both in vivo and in vitro models

(Minghetti et al. 2004). Once Tat is present in the CNS, it can activate immune cells, resulting in the chronic release of inflammatory markers (Fig. 7.1). Its pro-inflammatory activity includes the induction of cytokines such as tumour necrosis factor-alpha (TNF-α), IL-6, IL-1, including C-C motif chemokine ligand 2 (CCL2), which contributes to neuronal loss and dysregulated signalling pathways. Muvenda and colleagues reported that from several Tat inflammation-mediated studies on primary cells as well as cell lines, Tat amplifies neuroinflammatory response by upregulating pro-inflammatory cytokines such as CCL2, IL-6, IL-8, and TNF-α. These pro-inflammatory cytokines were identified as having the most consistent findings across all studies investigating Tat-exposed cell cultures and primary cell models and their upregulation is regulated through signalling pathways including ERK1/2 MAPK, PI3K, p38 MAPK, and NF-kB (Muvenda et al. 2024). The dysregulation of these markers is especially notorious for contributing to neurocognitive impairment.

Current evidence indicates that HIV does not directly infect neurons. However, neuronal injury primarily occurs through indirect mechanisms, such as the release of cytokines. Conversely, viral proteins can directly contribute to neuronal injury through several mechanisms, ultimately leading to neurocognitive dysfunction.

7.7 Tat-Induced Neurotoxicity: Direct Neuropathogenesis

Tat can be taken up via various mechanisms within neurons including via receptor-mediated endocytosis or passive diffusion, due to its basic domain (CPP), which allows cell penetration. Once inside the neuron, Tat localizes in both the cytoplasm and nucleus, where it interferes with normal cellular processes by initiating a cascade of pathological events (Haughey et al. 2001). One of the most prominent outcomes of Tat exposure is the induction of neuronal apoptosis. Tat-initiated neuronal cell death takes place through both intrinsic and extrinsic pathways, which are

closely linked to mitochondrial dysfunction and excessive intracellular calcium accumulation (Buscemi et al. 2007) (Fig. 7.5).

Following interaction with neuronal surface receptors, particularly the low-density lipoprotein receptor-related protein (LRP), internalization of Tat occurs, promoting the formation of a multiprotein complex that includes PSD95, NMDA receptors (NMDAR), tyrosine kinase Pyk2, and neuronal nitric oxide synthase (nNOS). This complex facilitates a pathogenic calcium influx through NMDARs, initiating a series of intracellular disruptions (Haughey et al. 2001; Eugenin et al. 2007). Tat can directly injure neurons through multiple complex mechanisms that are yet to be fully understood. Therefore, this section discusses recent evidence on key pathways, including mitochondrial dysfunction, oxidative stress, glutamate excitotoxicity, and impairment of neurotrophic support.

7.7.1 Mitochondrial Dysfunction

Mitochondrial dysfunction refers to the impaired ability of mitochondria to perform their essential functions, such as producing ATP (energy), regulating cellular metabolism, and controlling cell death (Nicolson 2014). When mitochondria fail to operate properly, it leads to increased production of reactive oxygen species (ROS), leading to oxidative stress and disruption of energy generation. Mitochondrial dysfunction is associated with a range of diseases, including neurodegenerative disorders, metabolic, and cardiovascular diseases (Guo et al. 2013). In Tat neuropathogenesis, mitochondrial function is disrupted by an overload of calcium entry (Krogh et al. 2014) (Fig. 7.5). Calcium accumulation induces mitochondrial membrane depolarization, which subsequently opens the mitochondrial permeability transition pore and initiates the production of ROS (Norman et al. 2007). Cytochrome c is then released into the cytosol due to oxidative stress combined with calcium overload. Caspase-3, a key executioner enzyme in initiating the DNA fragmentation characteristic of apoptosis, is acti-

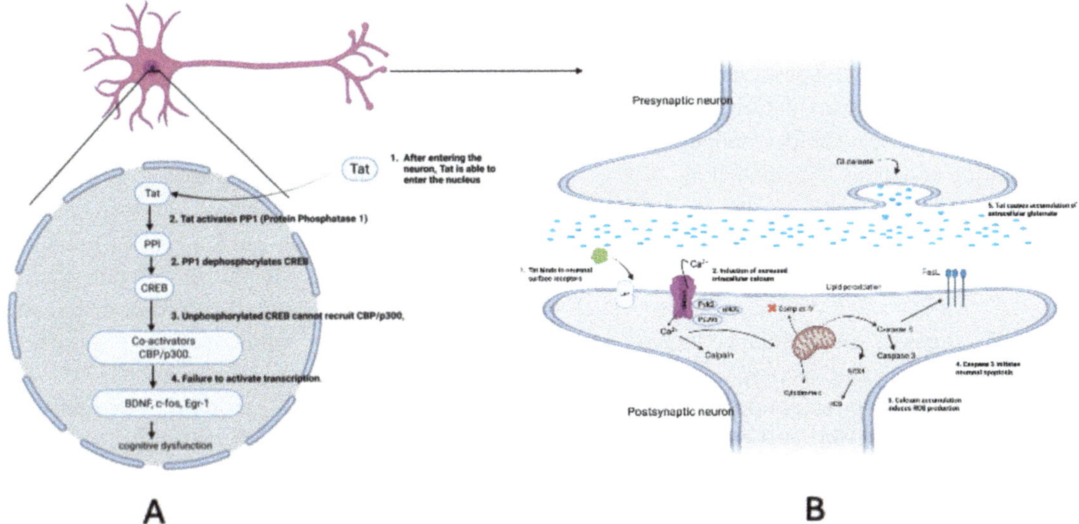

Fig. 7.5 Direct neuropathogenesis of Tat. The Tat protein exerts neurotoxic effects through multiple interconnected mechanisms in the nucleus (**a**) or cytosol (**b**). (**a**) Tat enters the neuron and localizes to the nucleus, where it activates Protein Phosphatase 1 (PP1). In both the cytosol and nucleus, Tat disrupts signalling pathways that normally phosphorylate CREB. Activated PP1 dephosphorylates CREB at Ser133, rendering it transcriptionally inactive. As a result, CREB fails to promote the expression of key neuroplasticity genes such as BDNF, c-fos, and Egr-1, ultimately impairing synaptic function and contributing to cognitive dysfunction. (**b**) (1) Tat binds to specific neuronal surface receptors, such as the low-density lipoprotein receptor-related protein (LRP), initiating a cascade of intracellular events. (2) This receptor engagement facilitates an abnormal influx of calcium ions (Ca^{2+}) into the neuron. (3) Elevated intracellular calcium levels disrupt mitochondrial function by causing depolarization of the mitochondrial membrane. This depolarization impairs normal mitochondrial activity and promotes the excessive generation of reactive oxygen species (ROS). (4) Accumulated ROS, in turn, activate the intrinsic apoptotic pathway, primarily through the activation of caspases, ultimately leading to programmed neuronal cell death (apoptosis). In addition to its direct effects on neurons, Tat also interacts with glial cells, although this aspect is not depicted in the figure. Specifically, Tat binds to glial receptors responsible for maintaining extracellular glutamate homeostasis, particularly through glutamate uptake transporters. Tat-induced glial dysfunction impairs the clearance of extracellular glutamate, resulting in its pathological accumulation in the synaptic cleft. (5) The elevated levels of glutamate contribute to excitotoxic stress on neurons by overactivating glutamate receptors, such as NMDA receptors, which further exacerbates calcium influx and neuronal damage. Together, these pathways illustrate the multifaceted role of Tat in promoting oxidative stress, excitotoxicity, and apoptosis, all of which contribute to HIV-associated neurodegeneration. (All visual elements were created in BioRender. Mason, S. (2026) https://BioRender.com/j218suu and Mason, S. (2026) https://BioRender.com/9szel8z)

vated (Singh et al. 2004). Calcium also activates phospholipase A2, which degrades membrane phospholipids, releasing arachidonic acid, further contributing to inflammation and oxidative stress (Caro and Cederbaum 2003). Caspase-mediated cleavage of gasdermin E has been implicated in neuronal pyroptosis, a highly inflammatory form of programmed cell death, adding a new dimension to the role of Tat in HAND pathogenesis (Fernandes et al. 2024). Additionally, Tat sensitizes neurons and sur-rounding glial cells to extrinsic death signals by increasing the expression of caspase-8, enhancing susceptibility to Fas ligand (FasL) and TRAIL-induced apoptosis (Buscemi et al. 2007) (Fig. 7.5). This amplifies cell death in both infected and uninfected cells. As apoptosis unfolds, Tat further disrupts mitochondrial function by inhibiting complex IV (cytochrome c oxidase) of the electron transport chain, leading to mitochondrial hyperpolarization, a drop in intracellular pH, decreased NADPH levels, and impaired calcium buffering, collectively contributing to energy failure and redox imbalance (Norman et al. 2007). These metabolic disturbances favour lipid peroxidation, protein oxidation, and nucleic acid damage, further exacerbating apoptosis (Norman et al. 2007).

7.7.2 Oxidative Stress

Oxidative stress is an imbalance between the production of reactive oxygen species (ROS) and the body's ability to neutralize or detoxify them using antioxidants (Pizzino et al. 2017). ROS are highly reactive molecules that contain oxygen. They include free radicals like superoxide ($O_2\bullet$) and hydroxyl radicals ($OH\bullet$), as well as non-radical molecules like hydrogen peroxide (H_2O_2) (Pizzino et al. 2017). Tat induces ROS production through NADPH oxidase 4 (NOX4), which is primarily localized in the endoplasmic reticulum (ER), leading to elevated H_2O_2 levels in the ER of HIV-infected cells (Wu et al. 2010). Suppression of NOX4 in Tat-expressing cells results in the reduction of H_2O_2 in the ER (Wu et al. 2010) (Fig. 7.5). A study has shown that Tat can induce ROS production by upregulating spermine oxidase activity, which in turn leads to activation of the NMDA glutamate receptor in neuroblastoma cells (Capone et al. 2013). Another study demonstrated that Tat contributes to neurotoxicity by inducing oxidative stress (Aksenov et al. 2001). In the study by Aksenov and colleagues, Tat injection into the rat striatum led to a marked increase in protein carbonylation—an indicator of protein oxidation—which occurred early and preceded the onset of astrogliosis. Further, Immunohistochemical analysis revealed strong protein carbonyl immunoreactivity surrounding the injection site, particularly within cell bodies, suggesting that oxidative protein damage is a key early event in Tat-mediated neurodegeneration (Aksenov et al. 2001) (Fig. 7.5). The role of Tat exposure in increasing oxidative stress through increased ROS and NO production, including induction of ER stress, has been widely reported (Baker and Petersen 2018) (Fig. 7.5).

7.7.3 Glutamate Excitotoxicity

Glutamate, the primary excitatory neurotransmitter in the CNS, is tightly regulated, as elevated levels can cause excitotoxicity and lead to neuronal death. Glutamate excitotoxicity is another key mechanism through which Tat exerts its neurotoxicity (Potter et al. 2013) (Fig. 7.5). Tat disrupts glutamatergic signalling by potentiating NMDA receptor activity, thus promoting receptor phosphorylation while inhibiting astrocytic glutamate transporters such as GLT-1 and GLAST, primarily through NF-κB and MAPK/ERK pathways in glial cells (Marino et al. 2020a; Gorska and Eugenin 2020) (Fig. 7.5). This leads to elevated extracellular glutamate levels. Surplus glutamate overstimulates extra-synaptic NMDA receptors, which are strongly associated with pro-death signalling. The result is excessive calcium influx, which activates downstream calcium-dependent enzymatic pathways. These enzymes, including calpains, degrade cellular structures by cleaving cytoskeletal proteins and inducing DNA fragmentation (Haughey et al. 2001). Moreover, Tat-induced astrocyte dysfunction severely impedes glutamate reuptake, exacerbating this excitotoxic environment. These glutamate-mediated effects extend beyond cell death and deeply affect synaptic function and plasticity, essential for communication between neurons (Perry et al. 2005) (Fig. 7.5).

7.7.4 Synaptic Damage, Dendritic Simplification and Demyelination

Sustained Tat exposure causes neurons to exhibit significant synaptic damage and dendritic simplification. Tat reduces the expression of critical synaptic proteins such as synaptophysin, PSD95, as well as β3-tubulin (Liu et al. 2018). Loss of these markers indicates synaptic degeneration, dendritic spine loss, and simplified dendritic branching. These structural alterations impair synaptic connectivity and plasticity, crucial for cognitive processes such as learning and memory (McLane et al. 2022) (Fig. 7.5).

Myelin damage is a common pathologic feature that is frequently reported in HIV-infected individuals (Gorry et al. 2003). Post-mortem analysis of white matter in people with HIV reports up-regulation of pro-apoptotic markers, p53 and BAX in oligodendrocytes (OLs) (Jayadev et al. 2007), indicating activation of cell death signalling. Most researchers agree that HIV-1 does not directly infect oligodendrocytes; however, their injury is believed to be mediated through viral proteins shed off from virions released by infected cells (Hauser et al. 2009; Bernardo et al. 1997; Liu et al. 2017). These HIV-1 viral proteins include gp120, Tat, and Nef (Hauser et al. 2009). Among these viral proteins, Tat has consistently been detected in both infected and uninfected oligodendrocytes in the brain of PLWH (Del Valle et al. 2000). Tat activates OL cell death pathways by altering potassium (K^+) and calcium (Ca^{2+}) levels within the cells. Tat exposure on oligodendrocytes was shown to activate the voltage-gated K^+ channels in oligodendrocytes, leading to outward K^+ current and cell death (Liu et al. 2017). Experiments using primary rat oligodendrocyte cultures found that Tat triggered Ca^{2+} influx, activating pro-apoptotic molecules in glycogen synthase kinase 3β (GSK3β) (Zou et al. 2015). Authors further found that immature oligodendrocytes are more susceptible to Tat toxicity due to their elevated basal levels of GSK3β compared to mature OLs, which have large amounts of cytoplasmic CaMKIIβ, and in mature OLs the activity of GSK3β is suppressed by CaMKIIβ (Zou et al. 2015). In addition, Tat-mediated OL damage may occur through an indirect mechanism called the "bystander effect," which is mediated by proinflammatory cytokines such as TNF-α and Interleukin-1. For instance, interleukin-1 release inhibits glutamate uptake by astrocytes leading to excitotoxicity of oligodendrocytes via glutamatergic receptor overactivation particularly AMPA and kainate receptors (Takahashi et al. 2003).

7.7.5 Impairment of Neurotrophic Support

HIV-1 Tat impairs neurotrophic support, which is essential for the maintenance, growth, and survival of neurons in the CNS (Li et al. 2022). One of the primary mechanisms involves the downregulation of key neurotrophic factors such as brain-derived neurotrophic factor (BDNF) and nerve growth factor (NGF) (Bathina and Das 2015; King et al. 2006). Tat reduces their expression at both transcriptional and translational levels, thereby limiting the availability of these critical molecules. Tat disrupts NGF and BDNF receptor signalling by hijacking the transcriptional machinery (specifically p300 and PCAF), leading to decreased neuroprotection and increased neuronal vulnerability, a critical contributor to HIV-associated neurodegeneration (Hahn et al. 2015) (Fig. 7.5). Additionally, Tat alters key survival pathways such as MAPK and PI3K, and upregulates Inhibitor of DNA binding 1 (Id1), an inhibitor of neuronal differentiation, particularly in developing neurons (Bergonzini et al. 2004). These actions contribute to neuronal apoptosis, reduced differentiation, and increased vulnerability of the developing brain to HIV-induced damage. Tat also disrupts the CREB/BDNF signalling axis by entering neurons and localizing to the nucleus, where it activates Protein Phosphatase 1 (PP1). PP1 dephosphorylates CREB at Ser133, thereby preventing CREB from activating the transcription of neuroprotective genes such as BDNF, c-fos, and Egr-1. This indirect suppression of CREB-dependent gene expression further contributes to synaptic dysfunction and memory impairment (Fig. 7.5a). This combined loss of trophic support leads to synaptic instability, dendritic retraction, and ultimately contributes to the neuronal damage and cognitive decline observed in HAND (Fig. 7.5).

7.8 Tat Subtype and Amino Acid Variation

The HIV-1 Tat protein exhibits considerable sequence diversity across viral subtypes, with multiple regions showing notable amino acid variation (Fig. 7.6). These differences are especially prominent in functional domains such as the cysteine-rich region, core domain, and basic domain. Although some individual residues have been linked to changes in transactivation, immune modulation, and cellular uptake, the full impact of subtype-specific sequence variation on Tat function is not yet fully understood. In particular, many polymorphisms beyond the well-characterized motifs have not been thoroughly examined for their roles in neurotoxicity, immune activation, or blood–brain barrier disruption. These variations likely influence Tat-mediated neuropathogenesis in ways that remain to be clarified. Among the various subtypes, the differences between subtype B and subtype C have been studied most extensively (Santerre et al. 2019). These two subtypes differ at several key positions that are known or suspected to affect Tat function and neuroinflammatory potential (Williams et al. 2020b).

Tat from subtype B, predominantly found in North America and Europe, contains a conserved dicysteine motif at positions C30 and C31 (Rao et al. 2013). This motif enhances the ability of Tat to mimic chemokines, promoting monocyte recruitment across the BBB and increasing neuroinflammation (Mishra et al. 2008; Rao et al. 2013). Studies have shown that subtype B Tat induces higher levels of glutamate in neuronal cells, upregulates numerous synaptic plasticity genes, and leads to greater neurotoxicity compared to subtype C Tat (Samikkannu et al. 2014). In contrast, subtype C Tat, prevalent in regions like Southern Africa and India, often exhibits a C31S polymorphism (Rao et al. 2013), disrupting

Fig. 7.6 Sequence alignment of HIV-1 Tat proteins from multiple subtypes, highlighting key amino acid residues linked to differential neuropathogenesis. This alignment compares the Tat protein sequences across major HIV-1 subtypes (A, B, C, D, G, H, J, and K), with functional domains annotated across the top: proline-rich, cysteine-rich, core, arginine-rich, glutamine-rich, and the RGD domain. Residue conservation is indicated below the alignment using standard Clustal notation (*,:, .). Red boxes highlight amino acid positions known to influence Tat-mediated neuropathogenesis, including the dicysteine motif (C30–C31), R57, and Q63/E63. (Figure adopted from Gotora and colleagues (Gotora et al. 2023))

the dicysteine motif. This alteration reduces the chemokine mimicry of Tat, leading to diminished monocyte chemotaxis (Ranga et al. 2004) and a lower incidence of HAND in populations where subtype C predominates. The presence of the dicysteine motif in Tat is crucial for its neurotoxic effects. Subtype B Tat with this motif has been associated with elevated glutamate levels in neuronal cells (Samikkannu et al. 2014), upregulation of synaptic plasticity genes (Samikkannu et al. 2014), increased neurotoxicity and NMDA receptor activation (Li et al. 2008), enhanced monocyte chemotaxis and inflammation (Campbell et al. 2007; Gandhi et al. 2009), and greater disruption of BBB integrity. The R57 residue in the basic domain of Tat influences its cellular uptake and inflammatory potential. Subtype B Tat retains R57, facilitating efficient cellular entry and robust induction of proinflammatory cytokines like IL-6, IL-8, IL-1β, and CXCL1 (Ruiz et al. 2019). Conversely, the S57 variant found in some subtype C Tat proteins exhibits reduced cellular uptake and a diminished inflammatory response (Fig. 7.6).

Understanding the molecular differences between Tat subtypes is vital for developing targeted therapies for HAND. While subtype B Tat exhibits higher neurotoxicity, subtype C Tat effects are comparatively milder, correlating with the lower prevalence of HAND in regions where subtype C is dominant. However, direct comparative studies between these subtypes remain limited, highlighting the need for further research to elucidate the mechanisms underlying Tat-mediated neuropathogenesis.

A full description of key studies and associations of Tat with clinical outcomes is presented in Table 7.1.

7.9 Therapeutic Strategies Targeting Tat

Therapeutic strategies targeting HIV-1 Tat in the CNS aim to mitigate its neurotoxic effects and contribute to the management of HAND. Given that Tat can interact with a diverse array of host proteins across CNS cell types, many of which

contribute to HAND pathology, a better understanding of these interactions can inform therapeutic development. The interactome highlighted below shows how the Tat protein engages with molecular targets that disrupt homeostasis in neurons, astrocytes, microglia, and endothelial cells, several of which are already being explored as therapeutic entry points (Fig. 7.7).

In neurons, Tat also disrupts glutamate homeostasis by inhibiting astrocytic glutamate transporters and potentiating NMDA receptor activity, leading to excitotoxicity and calcium overload (Haughey et al. 2001; Gorska and Eugenin 2020; Marino et al. 2020b). Agents such as memantine (an NMDA receptor antagonist) and calcium channel blockers have shown neuroprotective potential in this context (Wu et al. 2009). Tat-mediated suppression of CREB and SIRT1 impairs mitochondrial health and neurotrophic signalling, suggesting that SIRT1 activators (e.g., resveratrol) or CREB pathway modulators could help restore neuronal function (Kwon et al. 2008) (Fig. 7.7). Furthermore, in line with mitochondrial health, Tat impairs mitochondrial function and promotes oxidative damage; thus, mitochondrial stabilizers and ROS scavengers like CoQ10 or mitoQ may offer neuroprotective benefits (Wani et al. 2011). Another critical pathological mechanism is the suppression of Tat on neurotrophic support, including reduced expression of brain-derived neurotrophic factor (BDNF) and nerve growth factor (NGF). This opens the door for therapies using BDNF mimetics or transcriptional coactivator modulators (e.g., HDAC inhibitors) to restore neuronal resilience (Tao et al. 2015) (Fig. 7.7).

In astrocytes, Tat downregulates EAAT2, leading to glutamate accumulation (Ye et al. 2017). Efforts to enhance EAAT2 expression (e.g., using ceftriaxone) have shown promise in restoring glutamate clearance (Zaitsev et al. 2019). Tat also affects STAT3 signalling, making JAK/STAT inhibitors another possible avenue for modulating astrocyte-driven inflammation (Hashioka et al. 2011) (Fig. 7.7).

In microglia, Tat activates TLR4, NF-κB, and the NLRP3 inflammasome, triggering chronic neuroinflammation (Chivero et al. 2017)

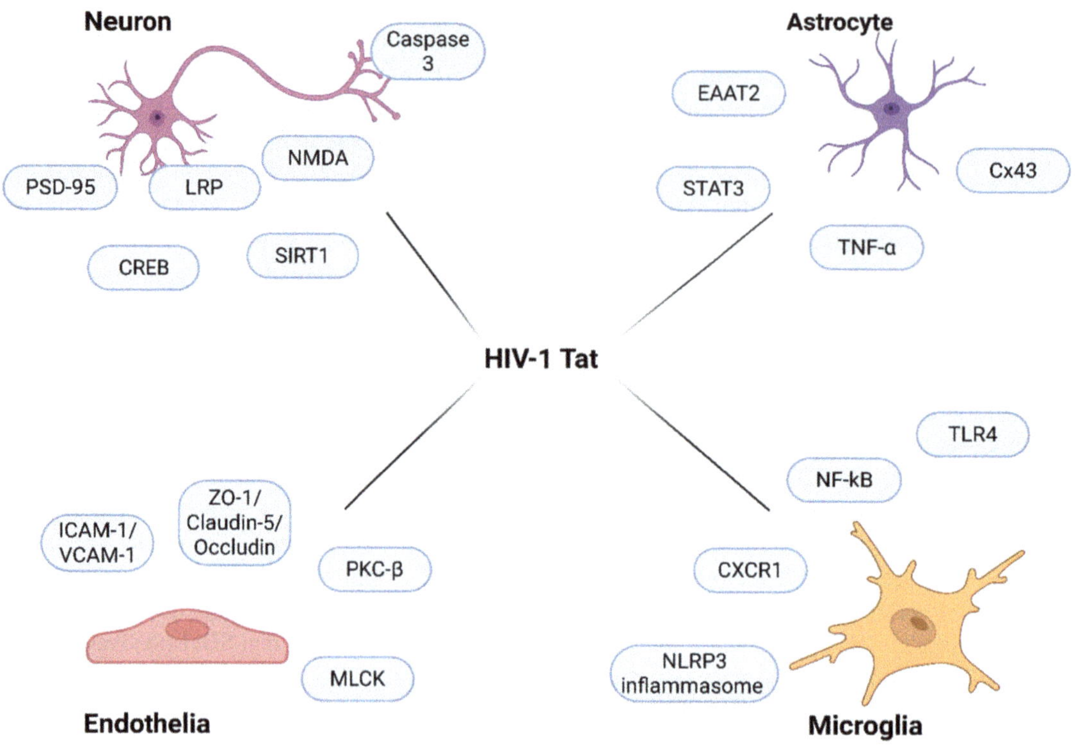

Fig. 7.7 Tat interactome across CNS cell types and its relevance to therapeutic targeting in HAND. This schematic illustrates key protein–protein interactions between HIV-1 Tat and host cellular factors within neurons, astrocytes, microglia, and endothelial cells. In neurons, Tat activates NMDA receptors, suppresses CREB and SIRT1, and disrupts synaptic proteins, contributing to excitotoxicity and mitochondrial dysfunction. In astrocytes, Tat downregulates glutamate transporter EAAT2 and alters STAT3 signalling, promoting impaired glutamate clearance and neuroinflammation. Microglial interactions involve activation of TLR4, NF-κB, and the NLRP3 inflammasome, driving sustained cytokine release. In endothelial cells, Tat impairs tight junction proteins (ZO-1, Claudin-5, Occludin) and upregulates adhesion molecules (ICAM-1, VCAM-1), leading to blood–brain barrier disruption. (All visual elements were created in BioRender. Mason, S. (2026) https://BioRender.com/n64nvtu)

(Fig. 7.7). Targeting these pathways with TLR4 antagonists, NF-κB inhibitors, or NLRP3 blockers (such as MCC950) holds potential to reduce microglial activation and cytokine-mediated neurotoxicity (Ma 2023). Similarly, Tat-driven neuroinflammation, which involves activation of glial cells and chronic cytokine release (e.g., TNF-α, IL-6, CCL2), may be attenuated using antinflammatory agents such as minocycline (Kielian et al. 2007), p38 MAPK inhibitors (Viorel et al. 2024), or IL-1β blockers (Klementiev et al. 2014) (Fig. 7.7).

In endothelial cells, Tat disrupts tight junction proteins (ZO-1, Claudin-5, Occludin) and upregulates ICAM-1/VCAM-1, compromising BBB integrity (Strazza et al. 2011). Therapeutics aimed at preserving junctional proteins, inhibiting MLCK or PKC-β, or blocking leukocyte adhesion may help maintain BBB stability (Li et al. 2021). Tat-induced damage to the BBB, mediated by oxidative stress, ER stress, and upregulation of matrix metalloproteinases, can be therapeutically targeted using antioxidants (e.g., N-acetylcysteine) (Mulè et al. 2024), MMP inhibitors (Sunny et al. 2024), or ER stress modulators like salubrinal (Qie et al. 2017) (Fig. 7.7).

Other strategies have also been investigated. Monoclonal antibodies against Tat have been developed to neutralize its activity. For instance, the anti-Tat antibody Hutat2:Fc has demonstrated neuroprotective effects by inhibiting Tat-induced microglial activation and neuronal damage (Kang et al. 2014). Additionally, broader immunotherapies, such as broadly neutralizing antibodies, CAR-T cell therapy, checkpoint inhibitors, and therapeutic vaccines, are being explored to address CNS HIV reservoirs and associated pathologies (Kapoor and Tan 2020). To enhance drug delivery across the BBB, nanocarriers like hybrid magneto-plasmonic liposomes (MPLs) have been developed. These MPLs can encapsulate antiretroviral drugs and, guided by external magnetic fields, target HIV-infected cells within the CNS, offering a promising strategy to counteract the effects of Tat (Tomitaka et al. 2017). Tat contributes to the maintenance of HIV latency in the CNS. Epigenetic therapies and latency-reversing agents (LRAs) are being investigated to disrupt this latency, allowing for the activation and subsequent elimination of latent HIV reservoirs within the brain (Khoury et al. 2018).

Therapeutic strategies may also benefit from considering Tat subtype-specific differences. Subtype B Tat, with its intact C30–C31 motif and R57 residue, has been shown to be more neurotoxic than subtype C variants (Rao et al. 2013; Ruiz et al. 2019). Future interventions could be subtype-informed, particularly in regions where different Tat variants predominate.

Lastly, combination therapies that simultaneously address Tat-induced inflammation, excitotoxicity, oxidative stress, and latency maintenance, while also improving CNS drug delivery, may offer the most comprehensive approach to preventing or reversing HAND. Clinical trials explicitly targeting Tat-mediated pathways are urgently needed and should be informed by biomarker-guided stratification, such as CSF Tat levels or anti-Tat antibody titers.

7.10 Conclusion

Despite the widespread use of cART, HAND continues to affect a significant proportion of people living with HIV, highlighting the need to understand the persistent factors driving neuropathology. Among the various HIV-1 proteins implicated in CNS dysfunction, Tat remains a critical contributor due to its ability to cross the BBB, induce neuroinflammation, trigger oxidative stress, and disrupt mitochondrial and endothelial function, even in the absence of active viral replication. This chapter has highlighted the multifaceted mechanisms by which Tat mediates neuronal injury, with a particular focus on studies conducted over the last 25 years. Recent findings reinforce the central role of Tat in the pathogenesis of HAND and suggest that targeting Tat-mediated pathways, such as ER stress responses, oxidative damage, and inflammatory signalling, could offer promising therapeutic strategies. Continued research into the molecular effects of Tat in the CNS will be essential for developing targeted interventions that can mitigate long-term neurological complications in the modern ART era.

Conflict of Interest Author MEW declares that he has no conflict of interest. Author TM declares that she has no conflict of interest. Author VM declares that she has no conflict of interest. Author SSZ declares that he has no conflict of interest.

Ethical Approval This chapter does not contain any studies with human participants performed by any of the authors.

References

Acheampong E, Mukhtar M, Parveen Z, Ngoubilly N, Ahmad N, Patel C, Pomerantz RJ (2002) Ethanol strongly potentiates apoptosis induced by HIV-1 proteins in primary human brain microvascular endothelial cells. Virology 304:222–234

Aksenov MY, Hasselrot U, Bansal AK, Guanghan W, Nath A, Anderson C, Mactutus CF, Booze RM (2001) Oxidative damage induced by the injection of HIV-1 Tat protein in the rat striatum. Neurosci Lett 305:5–8

Alford K, Vera JH (2018) Cognitive impairment in people living with HIV in the ART era: A review. Br Med Bull 127:55–68

András IE, Pu H, Deli MA et al. (2003) HIV-1 Tat protein alters tight junction protein expression and distribution in cultured brain endothelial cells. J Neurosci Res 74(2):255–265. https://doi.org/10.1002/jnr.10762

Antinori A, Arendt G, Becker JT, Brew BJ, Byrd DA, Cherner M, Clifford DB, Cinque P, Epstein LG,

Goodkin K, Gisslen M, Grant I, Heaton RK, Joseph J, Marder K, Marra CM, McArthur JC, Nunn M, Price RW, Pulliam L, Robertson KR, Sacktor N, Valcour V, Wojna VE (2007) Updated research nosology for HIV-associated neurocognitive disorders. Neurology 69:1789–1799

Atluri VS, Hidalgo M, Samikkannu T, Kurapati KR, Jayant RD, Sagar V, Nair MP (2015) Effect of human immunodeficiency virus on blood-brain barrier integrity and function: an update. Front Cell Neurosci 9:212

Bachani M, Sacktor N, McArthur JC, Nath A, Rumbaugh J (2013) Detection of anti-tat antibodies in CSF of individuals with HIV-associated neurocognitive disorders. J Neurovirol 19:82–88

Bagashev A, Sawaya BE (2013) Roles and functions of HIV-1 tat protein in the CNS: an overview. Virol J 10:358

Bai R, Song C, Lv S, Chang L, Hua W, Weng W, Wu H, Dai L (2023) Role of microglia in HIV-1 infection. AIDS Res Ther 20:16

Banks WA, Robinson SM, Nath A (2005) Permeability of the blood–brain barrier to HIV-1 Tat. Exp Neurol193(1):218–227. https://doi.org/10.1016/j.expneurol.2004.11.019

Baker DJ, Petersen RC (2018) Cellular senescence in brain aging and neurodegenerative diseases: evidence and perspectives. J Clin Invest 128:1208–1216

Bathina S, Das UN (2015) Brain-derived neurotrophic factor and its clinical implications. Arch Med Sci 11:1164–1178

Bergonzini V, Delbue S, Wang JY, Reiss K, Prisco M, Amini S, Khalili K, Peruzzi F (2004) HIV-Tat promotes cellular proliferation and inhibits NGF-induced differentiation through mechanisms involving Id1 regulation. Oncogene 23:7701–7711

Bernardo A, Agresti C, Levi G (1997) HIV-gp120 affects the functional activity of oligodendrocytes and their susceptibility to complement. J Neurosci Res 50:946–957

Borrajo A, Spuch C, Penedo MA, Olivares JM, Agís-Balboa RC (2021) Important role of microglia in HIV-1 associated neurocognitive disorders and the molecular pathways implicated in its pathogenesis. Ann Med 53:43–69

Buscemi L, Ramonet D, Geiger JD (2007) Human immunodeficiency virus type-1 protein Tat induces tumor necrosis factor-α-mediated neurotoxicity. Neurobiol Dis 26:661–670

Campbell GR, Loret EP (2009) What does the structure-function relationship of the HIV-1 Tat protein teach us about developing an AIDS vaccine? Retrovirology 6:50

Campbell GR, Watkins JD, Singh KK, Loret EP, Spector SA (2007) Human immunodeficiency virus type 1 subtype C Tat fails to induce intracellular calcium flux and induces reduced tumor necrosis factor production from monocytes. J Virol 81:5919–5928

Capone C, Cervelli M, Angelucci E, Colasanti M, Macone A, Mariottini P, Persichini T (2013) A role for spermine oxidase as a mediator of reactive oxygen species production in HIV-Tat-induced neuronal toxicity. Free Radic Biol Med 63:99–107

Caro AA, Cederbaum AI (2003) Role of phospholipase A2 activation and calcium in CYP2E1-dependent toxicity in HepG2 cells. J Biol Chem 278:33866–33877

Chivero ET, Guo ML, Periyasamy P, Liao K, Callen SE, Buch S (2017) HIV-1 tat primes and activates microglial NLRP3 inflammasome-mediated neuroinflammation. J Neurosci 37:3599–3609

Clark E, Nava B, Caputi M (2017) Tat is a multifunctional viral protein that modulates cellular gene expression and functions. Oncotarget 8:27569–27581

Dai H, Shen K, Yang Y, Xingxing S, Luo Y, Jiang Y, Shuai L, Zheng P, Chen Z, Bie P (2019) PUM1 knockdown prevents tumor progression by activating the PERK/eIF2/ATF4 signaling pathway in pancreatic adenocarcinoma cells. Cell Death Dis 10:595

Del Valle L, Croul S, Morgello S, Amini S, Rappaport J, Khalili K (2000) Detection of HIV-1 Tat and JCV capsid protein, VP1, in AIDS brain with progressive multifocal leukoencephalopathy. J Neurovirol 6:221–228

Deshetty UM, Chatterjee N, Buch S, Periyasamy P (2024) HIV-1 tat-mediated human Müller glial cell senescence involves endoplasmic reticulum stress and dysregulated autophagy. Viruses 16:903

Duffy BC, King KM, Nepal B, Nonnemacher MR, Kortagere S (2024) Acute administration of HIV-1 tat protein drives glutamatergic alterations in a rodent model of HIV-associated neurocognitive disorders. Mol Neurobiol 61:8467–8480

Eugenin EA, King JE, Nath A, Calderon TM, Zukin RS, Bennett MVL, Berman JW (2007) HIV-tat induces formation of an LRP–PSD-95–NMDAR–nNOS complex that promotes apoptosis in neurons and astrocytes. Proc Natl Acad Sci 104:3438–3443

Fan Y, He JJ (2016) HIV-1 Tat induces unfolded protein response and endoplasmic reticulum stress in astrocytes and causes neurotoxicity through glial fibrillary acidic protein (GFAP) activation and aggregation. J Biol Chem 291:22819–22829

Fernandes JP, Branton WG, Cohen EA, Koopman G, Kondova I, Gelman BB, Power C (2024) Caspase cleavage of gasdermin E causes neuronal pyroptosis in HIV-associated neurocognitive disorder. Brain 147:717–734

Gandhi N, Saiyed Z, Thangavel S, Rodriguez J, Rao KVK, Nair MPN (2009) Differential effects of HIV type 1 clade B and clade C Tat protein on expression of proinflammatory and antiinflammatory cytokines by primary monocytes. AIDS Res Hum Retrovir 25:691–699

González-Scarano F, Martín-García J (2005) The neuropathogenesis of AIDS. Nat Rev Immunol 5:69–81

Gorry PR, Ong C, Thorpe J, Bannwarth S, Thompson KA, Gatignol A, Vesselingh SL, Purcell DF (2003) Astrocyte infection by HIV-1: mechanisms of restricted virus replication, and role in the pathogenesis of HIV-1-associated dementia. Curr HIV Res 1:463–473

Gorska AM, Eugenin EA (2020) The glutamate system as a crucial regulator of CNS toxicity and survival of HIV reservoirs. Front Cell Infect Microbiol 10:261

Gotora PT, Brown K, Martin DR, van der Sluis R, Cloete R, Williams ME (2024) Impact of subtype C-specific amino acid variants on HIV-1 Tat-TAR interaction: insights from molecular modelling and dynamics. Virol J 21:144

Gotora PT, van der Sluis R, Williams ME (2023) HIV-1 Tat amino acid residues that influence Tat-TAR binding affinity: a scoping review. BMC Infect Dis 23:164

Guo C, Sun L, Chen X, Zhang D (2013) Oxidative stress, mitochondrial damage and neurodegenerative diseases. Neural Regen Res 8:2003–2014

Hahn YK, Podhaizer EM, Farris SP, Miles MF, Hauser KF, Knapp PE (2015) Effects of chronic HIV-1 Tat exposure in the CNS: heightened vulnerability of males versus females to changes in cell numbers, synaptic integrity, and behavior. Brain Struct Funct 220:605–623

Harezlak J, Buchthal S, Taylor M, Schifitto G, Zhong J, Daar E, Alger J, Singer E, Campbell T, Yiannoutsos C, Cohen R, Navia B, the HIV Neuroimaging Consortium (2011) Persistence of HIV-associated cognitive impairment, inflammation, and neuronal injury in era of highly active antiretroviral treatment. AIDS 25:625–633

Harricharan R, Thaver V, Russell VA, Daniels WMU (2015) Tat-induced histopathological alterations mediate hippocampus-associated behavioural impairments in rats. Behav Brain Funct 11:3

Hashioka S, Klegeris A, Qing H, McGeer PL (2011) STAT3 inhibitors attenuate interferon-γ-induced neurotoxicity and inflammatory molecule production by human astrocytes. Neurobiol Dis 41:299–307

Haughey NJ, Mattson MP (2002) Calcium dysregulation and neuronal apoptosis by the HIV-1 proteins Tat and gp120. J Acquir Immune Defic Syndr 31(Suppl 2):S55–S61

Haughey NJ, Nath A, Mattson MP, Slevin JT, Geiger JD (2001) HIV-1 Tat through phosphorylation of NMDA receptors potentiates glutamate excitotoxicity. J Neurochem 78:457–467

Hauser KF, Hahn YK, Adjan VV, Zou S, Buch SK, Nath A, Bruce-Keller AJ, Knapp PE (2009) HIV-1 Tat and morphine have interactive effects on oligodendrocyte survival and morphology. Glia 57:194–206

Henderson LJ, Johnson TP, Smith BR, Reoma LB, Santamaria UA, Bachani M, Demarino C, Barclay RA, Snow J, Sacktor N, McArthur J, Letendre S, Steiner J, Kashanchi F, Nath A (2019) Presence of Tat and transactivation response element in spinal fluid despite antiretroviral therapy. AIDS 33(Suppl 2):S145–Ss57

Hu XT (2016) HIV-1 Tat-mediated calcium dysregulation and neuronal dysfunction in vulnerable brain regions. Curr Drug Targets 17:4–14

Hudson L, Liu J, Nath A, Jones M, Raghavan R, Narayan O, Male D, Everall I (2000) Detection of the human immunodeficiency virus regulatory protein tat in CNS tissues. J Neurovirol 6:145–155

Jayadev S, Yun B, Nguyen H, Yokoo H, Morrison RS, Garden GA (2007) The glial response to CNS HIV infection includes p53 activation and increased expression of p53 target genes. J Neuroimmune Pharmacol 2:359–370

Jones GJ, Barsby NL, Cohen EA, Holden J, Harris K, Dickie P, Jhamandas J, Power C (2007) HIV-1 Vpr causes neuronal apoptosis and in vivo neurodegeneration. J Neurosci 27:3703–3711

Kadry H, Noorani B, Cucullo L (2020) A blood–brain barrier overview on structure, function, impairment, and biomarkers of integrity. Fluids Barriers CNS 17:69

Kang W, Marasco WA, Tong H-I, Byron MM, Chengxiang W, Shi Y, Sun S, Sun Y, Lu Y (2014) Anti-tat Hutat2:Fc mediated protection against tat-induced neurotoxicity and HIV-1 replication in human monocyte-derived macrophages. J Neuroinflammation 11:195

Kapoor A, Tan CS (2020) Immunotherapeutics to treat HIV in the central nervous system. Curr HIV/AIDS Rep 17:499–506

Karn J (2011) The molecular biology of HIV latency: breaking and restoring the Tat-dependent transcriptional circuit. Curr Opin HIV AIDS 6:4–11

Khoury G, Mota TM, Li S, Tumpach C, Lee MY, Jacobson J, Harty L, Anderson JL, Lewin SR, Purcell DFJ (2018) HIV latency reversing agents act through Tat post translational modifications. Retrovirology 15:36

Kielian T, Esen N, Liu S, Phulwani NK, Syed MM, Phillips N, Nishina K, Cheung AL, Schwartzman JD, Ruhe JJ (2007) Minocycline modulates neuroinflammation independently of its antimicrobial activity in staphylococcus aureus-induced brain abscess. Am J Pathol 171:1199–1214

Kim S, Lee M, Choi YK (2019) The role of a neurovascular signaling pathway involving hypoxia-inducible factor and notch in the function of the central nervous system. Biomol Ther 28:45

King JE, Eugenin EA, Buckner CM, Berman JW (2006) HIV tat and neurotoxicity. Microbes Infect 8:1347–1357

Klementiev B, Li S, Korshunova I, Dmytriyeva O, Pankratova S, Walmod PS, Kjær LK, Dahllöf MS, Lundh M, Christensen DP, Mandrup-Poulsen T, Bock E, Berezin V (2014) Anti-inflammatory properties of a novel peptide interleukin 1 receptor antagonist. J Neuroinflammation 11:27

Krogh KA, Wydeven N, Wickman K, Thayer SA (2014) HIV-1 protein Tat produces biphasic changes in NMDA-evoked increases in intracellular Ca2+ concentration via activation of Src kinase and nitric oxide signaling pathways. J Neurochem 130:642–656

Kwon H-S, Brent MM, Getachew R, Jayakumar P, Chen L-F, Schnolzer M, McBurney MW, Marmorstein R, Greene WC, Ott M (2008) Human immunodeficiency virus type 1 tat protein inhibits the SIRT1 deacetylase and induces T cell hyperactivation. Cell Host Microbe 3:158–167

Leibrand CR, Paris JJ, Said Ghandour M, Knapp PE, Kim W-K, Hauser KF, McRae MP (2017) HIV-1 Tat disrupts blood-brain barrier integrity and increases phagocytic perivascular macrophages and microglia in the dorsal striatum of transgenic mice. Neurosci Lett 640:136–143

Leibrand CR, Paris JJ, Jones AM, Masuda QN, Halquist MS, Kim W-K, Knapp PE, Kashuba ADM, Hauser KF, McRae MP (2019) HIV-1 Tat and opioids act independently to limit antiretroviral brain concentrations and reduce blood–brain barrier integrity. J Neurovirol 25:560–577

Li J, Zheng M, Shimoni O, Banks WA, Bush AI, Gamble JR, Shi B (2021) Development of novel therapeutics targeting the blood-brain barrier: from barrier to carrier. Adv Sci 8:e2101090

Li L, Dahiya S, Kortagere S, Aiamkitsumrit B, Cunningham D, Pirrone V, Nonnemacher MR, Wigdahl B (2012) Impact of tat genetic variation on HIV-1 disease. Adv Virol 2012:123605

Li W, Huang Y, Reid R, Steiner J, Malpica-Llanos T, Darden TA, Shankar SK, Mahadevan A, Satishchandra P, Nath A (2008) NMDA receptor activation by HIV-Tat protein is clade dependent. J Neurosci 28:12190–12198

Li Y, Li F, Qin D, Chen H, Wang J, Wang J, Song S, Wang C, Wang Y, Liu S, Gao D, Wang ZH (2022) The role of brain derived neurotrophic factor in central nervous system. Front Aging Neurosci 14:986443

Liu H, Liu J, Xu E, Tu G, Guo M, Liang S, Xiong H (2017) Human immunodeficiency virus protein Tat induces oligodendrocyte injury by enhancing outward K(+) current conducted by K(V)1.3. Neurobiol Dis 97:1–10

Liu Y, Zhou D, Feng J, Liu Z, Yue H, Liu C, Kong X (2018) HIV-1 protein tat 1–72 impairs neuronal dendrites via activation of PP1 and regulation of the CREB/BDNF pathway. Virol Sin 33:261–269

Ma Q (2023) Pharmacological inhibition of the NLRP3 inflammasome: structure, molecular activation, and inhibitor-NLRP3 interaction. Pharmacol Rev 75:487–520

Ma R, Yang L, Niu F, Buch S (2016) HIV tat-mediated induction of human brain microvascular endothelial cell apoptosis involves endoplasmic reticulum stress and mitochondrial dysfunction. Mol Neurobiol 53:132–142

Mabrouk K, Van Rietschoten J, Vives E, Darbon H, Rochat H, Sabatier J-M (1991) Lethal neurotoxicity in mice of the basic domains of HIV and SIV Rev proteins Study of these regions by circular dichroism. FEBS Lett 289:13–17

Marino J, Maubert ME, Mele AR, Spector C, Wigdahl B, Nonnemacher MR (2020b) Functional impact of HIV-1 Tat on cells of the CNS and its role in HAND. Cell Mol Life Sci 77:5079–5099

Marino J, Wigdahl B, Nonnemacher MR (2020a) Extracellular HIV-1 Tat mediates increased glutamate in the CNS leading to onset of senescence and progression of HAND. Front Aging Neurosci 12:168

Marks WD, Paris JJ, Schier CJ, Denton MD, Fitting S, McQuiston AR, Knapp PE, Hauser KF (2016) HIV-1 Tat causes cognitive deficits and selective loss of parvalbumin, somatostatin, and neuronal nitric oxide synthase expressing hippocampal CA1 interneuron subpopulations. J Neurovirol 22:747–762

McLane VD, Lark ARS, Nass SR, Knapp PE, Hauser KF (2022) HIV-1 Tat reduces apical dendritic spine density throughout the trisynaptic pathway in the hippocampus of male transgenic mice. Neurosci Lett 782:136688

McRae M (2016) HIV and viral protein effects on the blood brain barrier. Tissue Barriers 4:e1143543

Minghetti L, Visentin S, Patrizio M, Franchini L, Ajmone-Cat MA, Levi G (2004) Multiple actions of the human immunodeficiency virus type-1 Tat protein on microglial cell functions. Neurochem Res 29:965–978

Mishra M, Vetrivel S, Siddappa NB, Ranga U, Seth P (2008) Clade-specific differences in neurotoxicity of human immunodeficiency virus-1 B and C Tat of human neurons: significance of dicysteine C30C31 motif. Ann Neurol 63:366–376

Mulè S, Ferrari S, Rosso G, Brovero A, Botta M, Congiusta A, Galla R, Molinari C, Uberti F (2024) The combined antioxidant effects of N-acetylcysteine, vitamin D3, and glutathione from the intestinal–neuronal in vitro model. Foods 13:774

Muvenda T, Williams AA, Williams ME (2024) Transactivator of transcription (tat)-induced neuroinflammation as a key pathway in neuronal dysfunction: a scoping review. Mol Neurobiol 61:9320–9346

Nath A (2002) Human immunodeficiency virus (HIV) proteins in neuropathogenesis of HIV dementia. J Infect Dis 186(Suppl 2):S193–S198

Nicolson GL (2014) Mitochondrial dysfunction and chronic disease: treatment with natural supplements. Integr Med 13:35–43

Niu F, Liao K, Hu G, Moidunny S, Roy S, Buch S (2021) HIV tat-mediated induction of monocyte transmigration across the blood–brain barrier: role of chemokine receptor CXCR3. Front Cell Dev Biol 9:724970

Norman JP, Perry SW, Reynolds HM, Kiebala M, De Mesy Bentley KL, Trejo M, Volsky DJ, Maggirwar SB, Dewhurst S, Masliah E, Gelbard HA (2008) HIV-1 Tat activates neuronal ryanodine receptors with rapid induction of the unfolded protein response and mitochondrial hyperpolarization. PLoS One 3:e3731

Norman JP, Perry SW, Kasischke KA, Volsky DJ, Gelbard HA (2007) HIV-1 trans activator of transcription protein elicits mitochondrial hyperpolarization and respiratory deficit, with dysregulation of complex IV and nicotinamide adenine dinucleotide homeostasis in cortical neurons. J Immunol 178:869–876

Osborne O, Peyravian N, Nair M, Daunert S, Toborek M (2020) The paradox of HIV blood-brain barrier penetrance and antiretroviral drug delivery deficiencies. Trends Neurosci 43:695–708

Perry SW, Norman JP, Litzburg A, Zhang D, Dewhurst S, Gelbard HA (2005) HIV-1 transactivator of transcription protein induces mitochondrial hyperpolarization and synaptic stress leading to apoptosis. J Immunol 174:4333–4344

Pizzino G, Irrera N, Cucinotta M, Pallio G, Mannino F, Arcoraci V, Squadrito F, Altavilla D, Bitto A (2017) Oxidative stress: harms and benefits for human health. Oxidative Med Cell Longev 2017:8416763

Potter MC, Figuera-Losada M, Rojas C, Slusher BS (2013) Targeting the glutamatergic system for the treatment of HIV-associated neurocognitive disorders. J Neuroimmune Pharmacol 8:594–607

Prasad V, Suomalainen M, Jasiqi Y, Hemmi S, Hearing P, Hosie L, Burgert H-G, Greber UF (2020) The UPR sensor IRE1α and the adenovirus E3-19K glycoprotein sustain persistent and lytic infections. Nat Commun 11:1997

Qie X, Wen D, Guo H, Xu G, Liu S, Shen Q, Liu Y, Zhang W, Cong B, Ma C (2017) Endoplasmic reticulum stress mediates methamphetamine-induced blood-brain barrier damage. Front Pharmacol 8:639

Ranga U, Shankarappa R, Siddappa NB, Ramakrishna L, Nagendran R, Mahalingam M, Mahadevan A, Jayasuryan N, Satishchandra P, Shankar SK, Prasad VR (2004) Tat protein of human immunodeficiency virus type 1 subtype C strains is a defective chemokine. J Virol 78:2586–2590

Rao VR, Neogi U, Talboom JS, Padilla L, Rahman M, Fritz-French C, Gonzalez-Ramirez S, Verma A, Wood C, Ruprecht RM, Ranga U, Azim T, Joska J, Eugenin E, Shet A, Bimonte-Nelson H, Tyor WR, Prasad VR (2013) Clade C HIV-1 isolates circulating in Southern Africa exhibit a greater frequency of dicysteine motif-containing Tat variants than those in Southeast Asia and cause increased neurovirulence. Retrovirology 10:61

Ronsard L, Rai T, Rai D, Ramachandran VG, Banerjea AC (2017) In silico analyses of subtype specific HIV-1 Tat-TAR RNA interaction reveals the structural determinants for viral activity. Front Microbiol 8:1467

Ruiz AP, Ajasin DO, Ramasamy S, DesMarais V, Eugenin EA, Prasad VR (2019) A naturally occurring polymorphism in the HIV-1 tat basic domain inhibits uptake by bystander cells and leads to reduced neuroinflammation. Sci Rep 9:3308

Rumbaugh JA, Tyor W (2015) HIV-associated neurocognitive disorders: five new things. Neurol Clin Pract 5:224–231

Said N, Venketaraman V (2025) Neuroinflammation, blood-brain barrier, and HIV reservoirs in the CNS: an in-depth exploration of latency mechanisms and emerging therapeutic strategies. Viruses 17

Samikkannu T, Atluri VSR, Adriana Y Arias, Kesava VK Rao, Carmen T Mulet, Rahul D Jayant, Madhavan PN Nair (2014) HIV-1 subtypes B and C tat differentially impact synaptic plasticity expression and implicates HIV-associated neurocognitive disorders §. Curr HIV Res 12:397–405

Santerre M, Wang Y, Arjona S, Allen C, Sawaya BE (2019) Differential contribution of HIV-1 subtypes B and C to neurological disorders: mechanisms and possible treatments. AIDS Rev 21:76–83

Schier CJ, Marks WD, Paris JJ, Barbour AJ, McLane VD, Maragos WF, McQuiston AR, Knapp PE, Hauser KF (2017) Selective vulnerability of striatal D2 versus D1 dopamine receptor-expressing medium spiny neurons in HIV-1 Tat transgenic male mice. J Neurosci 37:5758–5769

Sharma AL, Wang H, Zhang Z, Millien G, Tyagi M, Hongpaisan J (2022) HIV promotes neurocognitive impairment by damaging the hippocampal microvessels. Mol Neurobiol 59:4966–4986

Shen J, Prywes R (2004) Dependence of site-2 protease cleavage of ATF6 on prior site-1 protease digestion is determined by the size of the luminal domain of ATF6. J Biol Chem 279:43046–43051

Sil S, Periyasamy P, Thangaraj A, Niu F, Chemparathy DT, Buch S (2021) Advances in the experimental models of HIV-associated neurological disorders. Curr HIV/AIDS Rep 18:459–474

Singh I, Goody R, Dean C et al. (2004) Apoptotic death of striatal neurons induced by HIV-1 Tat and gp120: differential involvement of caspase-3 and endonuclease G. J Neurovirol 10:141–151

Spector C, Mele AR, Wigdahl B, Nonnemacher MR (2019) Genetic variation and function of the HIV-1 Tat protein. Med Microbiol Immunol 208:131–169

Strazza M, Pirrone V, Wigdahl B, Nonnemacher MR (2011) Breaking down the barrier: the effects of HIV-1 on the blood-brain barrier. Brain Res 1399:96–115

Sun Y, Cai M, Liang Y, Zhang Y (2023) Disruption of blood-brain barrier: effects of HIV Tat on brain microvascular endothelial cells and tight junction proteins. J Neurovirol 29:658–668

Sunny A, James RR, Menon SR, Rayaroth S, Daniel A, Thompson NA, Tharakan B (2024) Matrix Metalloproteinase-9 inhibitors as therapeutic drugs for traumatic brain injury. Neurochem Int 172:105642

Takahashi JL, Giuliani F, Power C, Imai Y, Yong VW (2003) Interleukin-1beta promotes oligodendrocyte death through glutamate excitotoxicity. Ann Neurol 53:588–595

Tao W, Chen Q, Wang L, Zhou W, Wang Y, Zhang Z (2015) Brainstem brain-derived neurotrophic factor signaling is required for histone deacetylase inhibitor-induced pain relief. Mol Pharm 87:1035–1041

Thaney VE, Sanchez AB, Fields JA, Minassian A, Young JW, Maung R, Kaul M (2018) Transgenic mice expressing HIV-1 envelope protein gp120 in the brain as an animal model in neuroAIDS research. J Neurovirol 24:156–167

Thangaraj A, Periyasamy P, Liao K, Bendi VS, Callen S, Pendyala G, Buch S (2018) HIV-1 TAT-mediated microglial activation: role of mitochondrial dysfunction and defective mitophagy. Autophagy 14:1596–1619

Toborek M, Lee YW, Pu H, Malecki A, Flora G, Garrido R, Hennig B, Bauer HC, Nath A (2003) HIV-tat protein induces oxidative and inflammatory pathways in brain endothelium. J Neurochem 84:169–179

Tomitaka A, Arami H, Huang Z, Raymond A, Rodriguez E, Cai Y, Febo M, Takemura Y, Nair M (2017) Hybrid magneto-plasmonic liposomes for multimodal image-guided and brain-targeted HIV treatment. Nanoscale 10:184–194

Vigorito M, Connaghan KP, Chang SL (2015) The HIV-1 transgenic rat model of neuroHIV. Brain Behav Immun 48:336–349

Viorel VI, Pastorello Y, Bajwa N, Slevin M (2024) p38-MAPK and CDK5, signaling pathways in neuroinflammation: a potential therapeutic intervention in Alzheimer's disease? Neural Regen Res 19:1649–1650

Wani WY, Gudup S, Sunkaria A, Bal A, Singh PP, Kandimalla RJ, Sharma DR, Gill KD (2011) Protective efficacy of mitochondrial targeted antioxidant MitoQ against dichlorvos induced oxidative stress and cell death in rat brain. Neuropharmacology 61:1193–1201

Williams ME, Cloete R (2022a) Molecular modeling of subtype-specific tat protein signatures to predict tat-TAR interactions that may be involved in HIV-associated neurocognitive disorders. Front Microbiol 13:866611

Williams ME, Ipser JC, Stein DJ, Joska JA, Naudé PJW (2020a) Peripheral immune dysregulation in the ART era of HIV-associated neurocognitive impairments: a systematic review. Psychoneuroendocrinology 118:104689

Williams ME, Stein DJ, Joska JA, Naudé PJW (2021) Cerebrospinal fluid immune markers and HIV-associated neurocognitive impairments: A systematic review. J Neuroimmunol 358:577649

Williams ME, Zulu SS, Stein DJ, Joska JA, Naudé PJW (2020b) Signatures of HIV-1 subtype B and C Tat proteins and their effects in the neuropathogenesis of HIV-associated neurocognitive impairments. Neurobiol Dis 136:104701

Williams ME, Cloete R (2022b) Molecular modeling of subtype-specific Tat protein signatures to predict Tat-TAR interactions that may be involved in HIV-associated neurocognitive disorders. Front Microbiol 13:866611

Williams ME, Naudé PJW (2024) The relationship between HIV-1 neuroinflammation, neurocognitive impairment and encephalitis pathology: A systematic review of studies investigating post-mortem brain tissue. Rev Med Virol 34:e2519

Wu HM, Tzeng NS, Qian L, Wei SJ, Hu X, Chen SH, Rawls SM, Flood P, Hong JS, Lu RB (2009) Novel neuroprotective mechanisms of memantine: increase in neurotrophic factor release from astroglia and anti-inflammation by preventing microglial activation. Neuropsychopharmacology 34:2344–2357

Wu R-F, Ma Z, Liu Z, Terada LS (2010) Nox4-derived H2O2 mediates endoplasmic reticulum signaling through local Ras activation. Mol Cell Biol 30:3553–3568

Yadav-Samudrala BJ, Yadav AP, Patel RP et al. (2025) HIV-1 Tat protein alters medial prefrontal cortex neuronal activity and recognition memory. iScience 28(3)

Yarandi SS, Duggan MR, Sariyer IK (2021) Emerging role of Nef in the development of HIV associated neurological disorders. J Neuroimmune Pharmacol 16:238–250

Ye X, Zhang Y, Xu Q, Zheng H, Wu X, Qiu J, Zhang Z, Wang W, Shao Y, Xing HQ (2017) HIV-1 Tat inhibits EAAT-2 through AEG-1 upregulation in models of HIV-associated neurocognitive disorder. Oncotarget 8:39922–39934

Zaitsev AV, Malkin SL, Postnikova TY, Smolensky IV, Zubareva OE, Romanova IV, Zakharova MV, Karyakin VB, Zavyalov V (2019) Ceftriaxone treatment affects EAAT2 expression and glutamatergic neurotransmission and exerts a weak anticonvulsant effect in young rats. Int J Mol Sci 20

Zou S, Fuss B, Fitting S et al. (2015) Oligodendrocytes are targets of HIV-1 Tat: NMDA- and AMPA receptor-mediated effects on survival and development. J Neurosci 35(32):11384–11398. https://doi.org/10.1523/JNEUROSCI.4740-14.2015

Metabolomics of HIV in the Modern cART Era

8

Levanco. K. Asia, Tshiamo W. Sebigi, Du Toit Loots,
Shayne Mason, Esme Jansen van Vuren,
and Monray. E. Williams

Abstract

HIV remains a global health challenge, affecting millions of individuals worldwide. While combination antiretroviral therapy (cART) significantly reduces viral loads to undetectable levels, it does not eradicate the virus completely, and a compromised host immune system persists. Traditionally, HIV-1 research has largely focused on elucidating the mechanisms of viral replication, characterising host immune responses, specifically inflammation, and developing effective antiretroviral therapies. However, comparatively less emphasis has historically been placed on understanding the metabolic alterations associated with HIV-1 infection. Metabolomics, a powerful tool that provides a functional understanding of the systemic metabolic changes induced by HIV and cART, is increasingly recognised for its potential to reveal the complex interplay between chronic inflammation, metabolic reprogramming, and long-term health outcomes. This chapter explores the critical metabolic pathways disrupted by HIV infection in the modern cART era, focusing on amino acid, glucose, microbiome, and lipid metabolism. Using metabolomic profiling, this chapter highlights persistent immunometabolic dysfunctions in people living with HIV, highlighting pathways of comorbidity risk and offering insights into therapeutic interventions to improve patient outcomes in the modern cART era.

Keywords

HIV · Metabolomics · cART · Metabolome

L. K. Asia · T. W. Sebigi · D. T. Loots · S. Mason ·
M. E. Williams (✉)
Biomedical and Molecular Metabolism Research (BioMMet), North-West University, Potchefstroom, South Africa
e-mail: Monray.Williams@nwu.ac.za

E. Jansen van Vuren
Hypertension in Africa Research Team (HART), North-West University, Potchefstroom, South Africa

South African Medical Research Council Unit for Hypertension and Cardiovascular Disease, North-West University, Potchefstroom, South Africa

8.1 Introduction

Simian immunodeficiency virus is the precursor of human immunodeficiency virus (HIV) as a result of zoonotic activities (hunting, wild markets, etc.) (Hahn et al. 2000). In the 1980s, HIV was identified as the causative agent of acquired immunodeficiency syndrome (AIDS) (Barré-Sinoussi et al. 1983; Gallo et al. 1984). Human immunodeficiency virus (HIV) remains a significant global health concern. In 2023, an estimated

© The Author(s) 2026
A. Shonhai et al. (eds.), *Advances in Biochemistry and Molecular Biology to meet Africa´s Needs*,
Advances in Experimental Medicine and Biology 1507,
https://doi.org/10.1007/978-3-032-24254-9_8

39.9 million [36.1–44.6 million] people were living with HIV (PLWH) worldwide, more than half of whom were women and girls (53%). During the same year, there were approximately 1.3 million [1.0–1.7 million] new HIV infections, reflecting a 40% reduction since 2010. AIDS-related mortality has also declined substantially, with about 630,000 deaths recorded in 2023, representing a 51% decline since 2010 (UNAIDS 2023). HIV exists in two main strains: HIV-1 and HIV-2. While both viruses can cause AIDS, they differ in their pathogenesis and clinical outcomes (Nyamweya et al. 2013). HIV-1 is the more virulent and widespread of the two, accounting for most infections globally (Nyamweya et al. 2013; Marlink et al. 1994; Popper et al. 1999). In contrast, HIV-2 is less transmissible, largely confined to West Africa, and typically associated with slower disease progression (Nyamweya et al. 2013; Campbell-Yesufu and Gandhi 2011; Esbjörnsson et al. 2018). Due to its global prevalence, significant public health impact, and the depth of existing research, this chapter will focus specifically on HIV-1. HIV-1 primarily targets cluster of differentiation (CD)4$^+$ T cells, macrophages, and dendritic cells, leading to progressive immune dysfunction (Weiss 1993; Porichis and Kaufmann 2011). HIV integrates its genome into host deoxyribonucleic acid (DNA), establishing latent reservoirs that persist despite combination antiretroviral therapy (cART) (Weiss 1993; Cohn et al. 2020; Chun et al. 1997). This chronic infection drives sustained immune activation, inflammation, and gradual CD4+ T-cell depletion, disrupting immune homeostasis and contributing significantly to the risk of the host developing other opportunistic infections (Okoye and Picker 2013; Cockerham et al. 2014).

While cART effectively reduces plasma viral loads to undetectable levels, it does not eliminate HIV. Low-level viral persistence and immune activation continue to fuel chronic inflammation, sufficient to disrupt key metabolic pathways, including lipid, glucose, and amino acid metabolism (Massanella et al. 2016; Svensson Akusjärvi et al. 2023; Baer et al. 2021; Wan et al. 2025). Furthermore, cART does not invariably restore immune protection. A significant proportion of PLWH, and especially those who initiate therapy

with low nadir CD4$^+$ counts, experience incomplete CD4+ T-cell count recovery, which correlates with a higher long-term mortality despite virologic control (Martin-Iguacel et al. 2022; Gazzola et al. 2009). In such cases, persistent immune dysfunction and a low CD4/CD8 ratio elevate risks for non-AIDS comorbidities and mortality (Mou et al. 2025). Emerging metabolomic evidence indicates that distinct lipid and bile acid metabolic signatures, including altered primary bile acid pathways and DHA-containing lipids, are closely tied to CD4 recovery trajectories, suggesting metabolic dysregulation may reflect or even contribute to immune reconstitution outcomes (Wan et al. 2025).

These metabolic shifts contribute to the heightened risk of comorbidities such as cardiovascular disease (CVD) (Triant 2013), metabolic syndrome (Nix and Tien 2014), diabetes mellitus (Kalra et al. 2011), cognitive impairment (Williams et al. 2021), and tuberculosis (TB) (Herbert et al. 2023), observed in PLWH. Understanding these metabolic shifts is imperative to identifying non-virological drivers of disease progression and developing new therapeutic strategies. Historically, HIV-1 research has focused on investigating viral dynamics (Zuo et al. 2024; Pybus and Rambaut 2009; Hill et al. 2018), immune cell depletion (Deeks et al. 2015; Février et al. 2011), inflammatory pathways (Deeks et al. 2013) and therapeutic development. In comparison, interest in understanding the metabolome of HIV-1 within clinical settings has been under-prioritised.

In the modern cART era, metabolomics has historically been underutilised as it has been used in isolation from other omics techniques. However, studies are now opting for the integration of metabolomics with other omics, e.g., lipidomics, proteomics, and genomics, to get a better physiological understanding of what is happening beyond just the metabolism (Lee et al. 2025; Mikaeloff et al. 2023; Pei et al. 2021). Integration with other omics approaches provides a more comprehensive understanding of the metabolite levels in conjunction with gene expression, as an example, and a holistic approach to better describing and understanding this disease. As such, metabolomics is increasingly recognised as a crucial tool towards the better understanding of the systemic consequences of HIV infection and cART (Clish 2015; Pei et al. 2021; Mikaeloff et al. 2023). It offers a broader functional insight into how chronic immune activation and long-term therapy reprogram host metabolism, particularly in the modern cART era, where viral suppression is commonly achieved, but comorbidities persist (Akusjärvi et al. 2023; Baer et al. 2021).

This chapter explores the role of metabolomics in elucidating the complex interplay between HIV infection and host metabolic regulation. Here, we focused on understanding metabolic profiles in the modern cART era, defined as the period of HIV treatment characterized by the widespread use of cART that is highly effective, better tolerated, and less toxic than earlier regimens (Sheth et al. 2016). This era generally began in the mid-to-late 2000s and is ongoing. Different cART regimens may have various adverse effects, e.g., nausea, depression, hyperglycaemia, and hyperlipidaemia (Montessori et al. 2004; Ganta and Chaubey 2019). Certain cART classes, such as nucleoside/nucleotide reverse transcriptase inhibitors (NRTIs) and protease inhibitors (PIs), specifically inhibit the HIV replication cycle; they also result in host mitochondrial dysfunction (Brinkman et al. 1998). This mitochondrial dysfunction occurs when these cART classes deplete mitochondrial DNA (mtDNA), inhibit mitochondrial polymerase γ (pol γ), and inhibit mitochondrial peptidase processing (Lewis et al. 2003; Holec et al. 2017; Ganta and Chaubey 2019). Considering that the mitochondria are responsible for the production of the major energy source, Adenosine triphosphate (ATP), disruption of the mitochondria by cART classes such as NRTIs and PIs may influence various metabolic pathways in PLWH. Focusing on the modern cART era, it highlights how metabolomic profiling can uncover persistent immunometabolic dysfunction, identify key metabolic pathways affected by HIV infection and treatment, and provide insights that may inform future therapeutic strategies to improve long-term health outcomes in PLWH.

8.2 Metabolomics: Definitions and Methodologies

Metabolomics is the study of the metabolome, the full complement of small-molecule metabolites in a cell, tissue, or organism, and how these change in response to perturbations (Schrimpe-Rutledge et al. 2016; Han et al. 2023). In particular, metabolomics can detect both endogenous and exogenous metabolites. Endogenous metabolites refer to metabolites that are naturally occurring or biosynthesized within a living organism, whereas exogenous metabolites refer to metabolites that occur from an external source, e.g., food, xenobiotics, and environmental factors (Qiu et al. 2023; Kong and Hernandez-Ferrer 2020). Therefore, cART metabolites, which are exogenous metabolites, may also be detected using metabolomics (Andriguetti et al. 2025; Guan et al. 2025). Subsequently, metabolomics studies allow for a better understanding of how changes to genes/gene expression or the environment influence the physiology of the biological system (Dunn et al. 2011; Beger et al. 2016). Metabolomics is typically classified into two broad approaches: (1) untargeted metabolomics, which provides hypothesis-generating, qualitative analysis and offers a more comprehensive understanding of the whole metabolome in the sample (Schrimpe-Rutledge et al. 2016; Han et al. 2023), and (2) targeted metabolomics, which is hypothesis-driven and allows for the absolute quantification of a pre-defined class/selection of metabolites of interest (Schrimpe-Rutledge et al. 2016). In the modern era of systems and integrative biology, metabolomics is no longer regarded solely as a stand-alone technology. It is increasingly applied in combination with other omics approaches such as genomics, transcriptomics, and proteomics to provide complementary insights. This integrative use of metabolomics helps validate findings from other omics fields and explains how molecular changes translate into the physiological alterations observed within biological systems.

The methodologies employed in metabolomics encompass a range of sophisticated techniques that allow for the identification, quantification, and characterisation of metabolites. Generally, metabolomics encompasses the use of chromatographic separation techniques combined with mass spectrometry (MS)-based identification strategies. The most common chromatographic techniques include gas chromatography (GC) or liquid chromatography (LC) for compound separation (Dunn et al. 2011). The GC platform works by volatilising samples and separating compounds based on their volatility and interaction with the column's stationary phase, making it particularly suitable for analysing thermally stable apolar metabolites (Dunn et al. 2011; Kawamura and Fukusaki 2024). However, using various derivatization steps, GC is now being used to analyse both polar and non-polar molecules, from all compound classes (Sun and Xia 2024). Liquid chromatography (LC), on the other hand, typically separates metabolites in liquid samples, making it well-suited for analysing polar and non-polar metabolites depending on which column is selected (Zhao and Li 2020; Plumb et al. 2023).

Coupling these chromatographic techniques with MS enables a robust identification and quantification of metabolites. MS is particularly valued for its sensitivity and ability to identify metabolites based on their mass-to-charge ratio. MS-based methods may include, but are not limited to, tandem MS (MS/MS), time-of-flight (TOF), Orbitrap, and Fourier-transform ion cyclotron resonance mass spectrometry (FTICR) (Bauermeister et al. 2022; Bierla et al. 2020). Proton nuclear magnetic resonance (^{1}H-NMR) spectroscopy is another utilised analytical platform in metabolomics (Dunn et al. 2011; Emwas et al. 2019).

The ^{1}H-NMR platform uses a strong magnetic field to excite hydrogen atoms in metabolites, allowing their identification based on unique chemical signatures. This method is typically used for detecting more abundant metabolites due to its lower sensitivity, as compared to other MS-based techniques.

Each of these analytical platforms has its own unique strengths and weaknesses. In Table 8.1, some of the major differences between the metabolomic platforms are compared. The

LC-MS and GC-MS has a far higher sensitivity when compared to that of NMR (Bujak et al. 2015), however, ^{1}H-NMR has a far higher comparative reproducibility when compared to the LC-MS and GC-MS, due to minimal changes needed to the sample prior to analysis (Bujak et al. 2015; Emwas et al. 2019), since LC and GC requires additional sample preparation procedures, such as metabolite extraction and derivatisation steps prior to analysis (Nagana Gowda and Raftery 2014).

A major sample preparation step for most metabolomics platforms is the removal of proteins prior to analysis. This can be achieved either chemically (e.g., using acetonitrile or methanol) or physically (e.g., using ultracentrifugal filters) (Vuckovic 2012). Chemical methods are generally more cost-effective, and given that both approaches are reliable, are often preferred in routine metabolomics workflows (Nagana Gowda and Raftery 2014; Emwas 2015). These additional sample preparation steps may affect sample integrity or introduce variability, potentially reducing reproducibility compared to NMR.

In terms of costs, the initial setup for NMR is significantly more expensive than that of GC-MS or LC-MS systems. The use of all these analytical platforms synergistically may provide for a more comprehensive quantification and understanding of the entire metabolome changes indicated by HIV infection and its treatment using cART (Petrella et al. 2021). A comparison of the most common platforms/techniques is presented in Table 8.1.

As highlighted above, GC-MS, LC-MS and NMR are the most commonly used metabolomics analytical platforms (Sun and Xia 2024). Other advancements and complementary techniques to the basic GC-MS, LC-MS, and NMR platforms, include derivatisation strategies to enhance the detection of low-abundance or non-volatile metabolites, stable isotope labelling for more accurate quantification and metabolic flux analysis (Chokkathukalam et al. 2014), and two-dimensional chromatography (2D-GC or 2D-LC), which improves separation efficiency and the resolution of complex mixtures (Singh and Chen 2024). Additionally, high-resolution

Table 8.1 Contrasting the common metabolomics platforms/techniques

	GC-MS	LC-MS	¹H-NMR
Sensitivity	High	High	Lower compared to MS-based techniques
Specificity	Medium	Low	High
Reproducibility	Medium	Medium	High
Cost	Cost-effective	Cost-effective	Initial setup is expensive
Metabolite detection	Volatile non-polar metabolites/volatile polar metabolites after derivatization	Polar and non-polar metabolites	Abundant metabolites
Derivatisation	Required	Sometimes required	Not required

mass spectrometry (HRMS) (Bussey 2024) and tandem mass spectrometry (MS/MS) (Antoniewicz 2013) provide greater sensitivity and structural elucidation capabilities. Integration of ion mobility spectrometry (IMS) with MS (Dodds and Baker 2019) further aids in the separation of isobaric and isomeric compounds. In addition to this, there are metabolomics techniques that are less commonly used in HIV metabolomics research in the modern cART era. These techniques include Fourier transform infrared (FTIR) and Raman spectroscopy (Lilo et al. 2022). Each platform brings distinct strengths, allowing researchers to tackle challenges related to sample complexity, sensitivity, and the diversity of metabolites in each sample.

Several of these tools have been used to characterise the metabolic profiles of HIV, particularly in the modern cART era. Metabolomics has opened new avenues for understanding HIV-1 pathogenesis, offering insights that were not previously possible using other analytical platforms. For example, LC–MS/MS approaches have been particularly effective in identifying lipidomic alterations in cART-treated PLWH, including dyslipidemia and lipid signatures linked to cardiovascular risk (Wan et al. 2024). NMR spectroscopy has been used to capture systemic metabolic shifts, including perturbations in amino acid metabolism and energy-related pathways in HIV infection (Munshi et al. 2013; Thirion et al. 2024b). In addition, GC × GC–TOF MS provides enhanced resolution and sensitivity for detecting complex metabolite mixtures, enabling the discovery of volatile and semi-volatile metabolites as well as oxidative stress–related signatures that may be perturbed in HIV

and cART contexts (Thirion et al. 2024a). These approaches have provided novel insights into key metabolic pathways that may remain persistently dysregulated, even in PLWH who exhibit low and controlled viremia (Wan et al. 2025). This persistent dysregulation, along with the identification of specific metabolic markers, may offer valuable targets for therapeutic intervention, diagnostics, and prognostics. Specifically, these tools allow for biomarker discovery to predict metabolic complications (e.g., CVD, insulin resistance) (Sun et al. 2025; Rodríguez-Gallego et al. 2018), monitoring of ART-associated metabolic side effects, and stratification of PLWH at risk of comorbidities (Cassol et al. 2013).

In the remaining sections of this chapter, we will discuss some of the most affected metabolic pathways influenced by HIV-1 infection and cART, as elucidated using metabolomics.

8.3 Key Metabolic Pathways Affected in cART-Treated PLWH

8.3.1 Amino Acid Metabolism

Amino acid metabolism plays a central role in human immune function, energy production, and cellular signalling (Yang et al. 2023). In the context of HIV-1 infection and treatment with cART, several amino acid pathways are significantly disrupted, reflecting chronic immune activation, chronic low-grade inflammation, and metabolic reprogramming (Ahmed et al. 2018). Notably, changes in tryptophan catabolism, glutamine-glutamate cycling, and branched-chain amino acid (BCAA) levels have been consistently observed in PLWH.

8.3.1.1 Tryptophan-Kynurenine Pathway

Tryptophan is an essential amino acid with multiple metabolic fates, including incorporation into proteins, conversion into serotonin and melatonin, or catabolism through the kynurenine pathway (KP). Under normal physiological conditions, tryptophan is primarily used for protein synthesis, with a smaller fraction entering the serotonin pathway (Fig. 8.1), particularly in neuronal and gastrointestinal cells (Yan et al. 2024; Höglund et al. 2019). However, during chronic immune activation, such as in HIV infection, the KP becomes highly upregulated due to increased activity of various enzymes, including indoleamine 2,3-dioxygenase (IDO) (Boasso and Shearer 2007; Routy et al. 2015). This shift diverts tryptophan away from serotonin production and protein synthesis, leading to tryptophan depletion and the accumulation of immunomodulatory and neuroactive metabolites (Lovelace et al. 2017; Wang et al. 2021).

The Tryptophan–Kynurenine pathway (Trp-Kyn) is the primary route for the catabolism of the essential amino acid tryptophan in mammals, accounting for over 95% of its breakdown (Badawy 2017; Davis and Liu 2015; Fujigaki

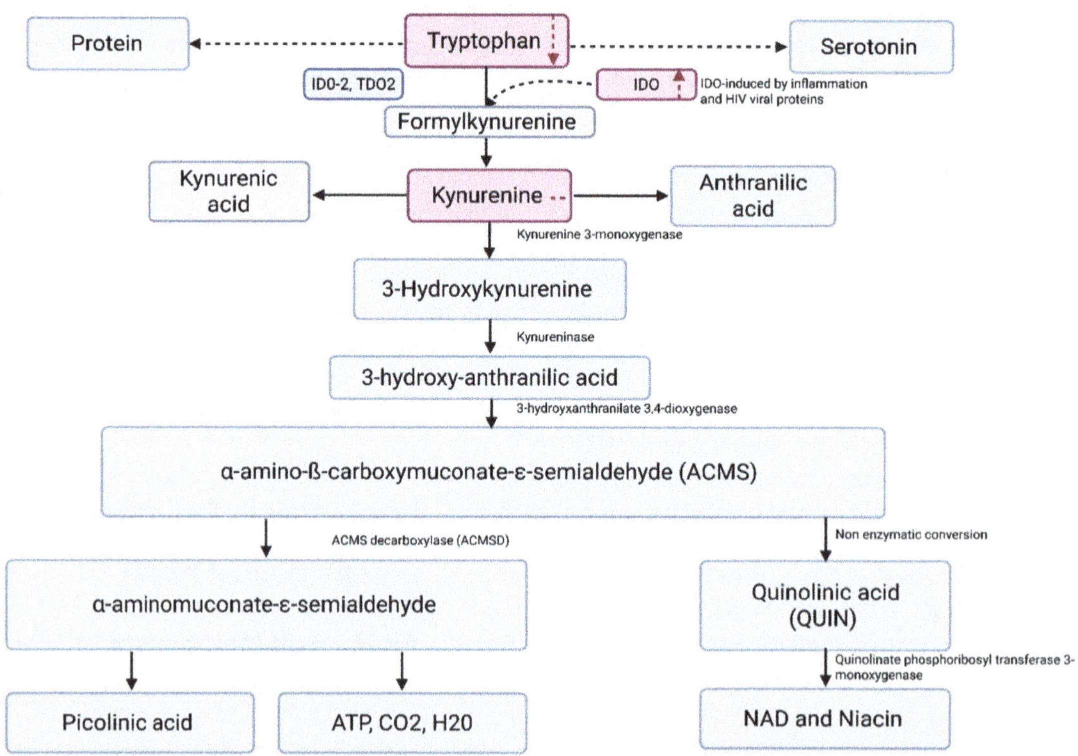

Fig. 8.1 Tryptophan catabolism pathway. Tryptophan is primarily degraded via the kynurenine pathway, initiated by the enzymes IDO or TDO. This pathway produces kynurenine and downstream metabolites such as kynurenic acid, 3-hydroxykynurenine, and quinolinic acid, precursors to NAD+ synthesis. Alternative routes include the serotonin pathway, which leads to neurotransmitter production, and microbial metabolism to indole derivatives. The boxes highlighted in red (Tryptophan, IDO, and kynurenine) and arrows represent the effect that HIV has on the pathway. These pathways are key in immune regulation, neurobiology, and energy metabolism. Abbreviations: IDO Indoleamine 2,3-dioxygenase, TDO Tryptophan 2,3-dioxygenase, ATP Adenosine triphosphate, CO2 Carbon dioxide, H2O Hydrogen dioxide, NAD Nicotinamide adenine dinucleotide. (The figure was adapted from Bipath et al. (2015))

et al. 2017). Tryptophan catabolism is initiated by two rate-limiting enzymes: IDO and tryptophan 2,3-dioxygenase (TDO) (Ball et al. 2014). While both enzymes catalyse the initial conversion of tryptophan to N-formylkynurenine, which is subsequently converted to kynurenine, their main distinction lies in their specific tissue localisation (Ball et al. 2014) (Fig. 8.1). IDO, particularly its isoform IDO-1, is highly inducible by pro-inflammatory cytokines and is expressed in various immune and non-immune cells (Pallotta et al. 2022; Sarkar et al. 2007). Induction of IDO is primarily driven by interferon-gamma (IFN-γ) in the peripheral tissue, and via interleukin-6 (IL-6) within the central nervous system (CNS) (Huang et al. 2020; Sarkar et al. 2007). In contrast, TDO is predominantly expressed in the liver, where it plays a critical role in maintaining systemic tryptophan homeostasis under various physiological conditions (Ball et al. 2014).

After the conversion of tryptophan to N-formylkynurenine, N-formylkynurenine is rapidly converted into kynurenine, the central intermediate of the kynurenine pathway (Bipath et al. 2015). From here, kynurenine can follow several downstream routes: (1) it may be transformed into kynurenic acid via kynurenine aminotransferases (KATs), or into (2) 3-hydroxykynurenine via kynurenine 3-monooxygenase (KMO) (Bipath et al. 2015). The latter is further metabolised to 3-hydroxyanthranilic acid and then converted by 3-hydroxyanthranilate 3,4-dioxygenase to α-amino-β-carboxymuconate-ϵ-semialdehyde (ACMS). ACMS is subsequently converted to α-aminomuconate-ϵ-semialdehyde via the rate-limiting enzyme ACMS decarboxylase (ACMSD). However, when ACMSD activity is low, a portion of ACMS is non-enzymatically converted into quinolinic acid (Deme et al. 2022), a key precursor for NAD^+ and niacin biosynthesis (Bipath et al. 2015). Alternatively, kynurenine can also be directly converted into anthranilic acid via kynureninase, representing another branch of the pathway (Fig. 8.1).

This Trp-Kyn pathway is biologically significant due to its role in immune regulation (Krupa and Kowalska 2021), neuroactive metabolite production (Davis and Liu 2015), and its involvement in various pathological conditions, including but not limited to HIV (Miedema et al. 2013), cancer (Wen et al. 2022), and neurodegenerative disorders (Adamu et al. 2024). The Trp-Kyn pathway has also been reviewed as a key contributor to frailty and inflammaging (Routy et al. 2015; Sultana et al. 2023), immune dysfunction, and the development of neurocognitive disorders in PLWH (Kandanearatchi and Brew 2012). In cART-treated PLWH, the Trp-Kyn pathway is known to be dysregulated (Sebigi et al. 2025). Literature suggests that chronic inflammation in cART-treated PLWH and HIV-derived proteins, Tat and Nef, contribute to the upregulation of IDO to further drive tryptophan degradation and immune modulation during HIV infection despite treatment status (Bipath et al. 2015; Boasso and Shearer 2007; Samikkannu et al. 2009). During HIV infection, IDO-1 activity is strongly upregulated by persistent immune activation and inflammation, e.g., elevated levels of IFN-γ, IL-6, tumour necrosis factor-α (TNF-α), and other pro-inflammatory cytokines (Baer et al. 2021; Favre et al. 2010), ultimately resulting in reduced tryptophan and elevated kynurenine in the cART-treated PLWH (Boasso et al. 2008; Boasso et al. 2009). While several studies have investigated metabolites in the Trp-Kyn, certain metabolites, tryptophan, kynurenine, and IDO, are more frequently examined in the context of cART-treated PLWH compared to HIV-negative controls (Sebigi et al. 2025). While the use of cART is known to reduce the viral load, it influences the tryptophan and kynurenine metabolism and IDO-1 activity, but does not normalize the levels of these metabolites to those seen in healthy controls (Chen et al. 2014; Zangerle et al. 2002). Although cART drugs themselves contain limited direct suppressive activity against IDO-1 activity and the resulting Trp-Kyn metabolites, they do have indirect effects in these by reducing inflammation and viral replication in PLWH (Gelpi et al. 2017; Zangerle et al. 2002). The majority of studies make use of LC/MS to measure this pathway (Sebigi et al. 2025), and studies consistently show that in the blood of cART-treated PLWH, tryptophan levels are significantly lower (Akusjärvi et al. 2023; Babu et al. 2019a; Frias et al. 2024; Fuchs et al. 1991; Sitole et al. 2019; Yang et al. 2023; Yuan et al. 2023) (Fig. 8.1, red box, down arrow), and IDO activity (as measured by the Kyn/Trp ratio) elevated (Fig. 8.1, red box, up arrow) (Babu et al. 2019a; Baer et al. 2021; Chen et al. 2014; Chen et al. 2019; Frias et al. 2024; Jenabian et al. 2013; Somsouk et al. 2015; Yuan et al. 2023). However, kynurenine levels themselves appear to be inconsistent when comparing PLWH to HIV-negative controls (Fig. 8.1, red box, dashed line), with studies reporting higher (Fuchs et al. 1991; Jenabian et al. 2013; Yuan et al. 2023), lower (Akusjärvi et al. 2023) and/or no significant differences (Babu et al. 2019b; Chen et al. 2014; Li et al. 2018; Yang et al. 2023), suggesting these changes in the kynurenine pathways are associated with general infection, and not necessarily specific to HIV per se, however further research is needed.

8.3.1.2 Other Key Amino Acid Metabolic Pathways

The glutamine–glutamate cycling refers to the interplay between glucose and glutamate metabolism (Massucci et al. 2013). Glucose is a primary energy source for cellular functioning (Liu et al. 2025), and glutamate, which can be derived from glucose metabolism, serves as a key metabolic substrate (Zhang et al. 2024). This cycling plays an essential role in cellular metabolism, immune regulation, and neurotransmission (Andersen 2025). Once transported into the mitochondria, glutamine is first deamidated by glutaminase to form glutamate. Glutamate can then undergo transamination or oxidative deamination via glutamate dehydrogenase, producing α-ketoglutarate, a central carbon metabolism intermediate of the tricarboxylic acid (TCA) cycle. In this way, glutamine metabolism directly fuels the TCA cycle, supporting mitochondrial function, ATP generation, and biosynthetic processes (Dohl et al. 2020; Yoo et al. 2020). In the context of neuroscience, this interplay between glucose and glutamate for energy production is especially important because glutamate is not only a metabolic substrate but also the major excitatory neurotransmitter in the CNS (Vázquez-

Santiago et al. 2014). Excess glutamate is then rapidly cleared from the cleft by astrocytes, where it is converted to glutamine via glutamine synthetase. The glutamine is subsequently transported back to neurons, completing the glutamate–glutamine cycle, which ensures both neurotransmitter homeostasis and continuous energy supply. This astrocyte–neuron metabolic coupling is therefore critical for maintaining synaptic transmission, preventing excitotoxicity, and supporting overall brain energy metabolism (Albrecht et al. 2007; McKenna 2013). This pathway is disrupted by HIV-1 infection.

HIV-1 typically infects activated CD4$^+$ T cells, which rely on glutamine for proliferation and execution of their effector functionality (Hegedus et al. 2017). In a clinical study using untargeted ^{1}H-NMR spectroscopy on serum, followed by targeted GC–MS, glutamine levels were found to be significantly lower in PLWH when compared to HIV-negative healthy controls, suggesting an altered amino acid metabolism associated with HIV infection (Sitole et al. 2019). Similarly, another study using GC and LC/MS by Svensson Akusjärvi et al. (2023) reported lower levels of both glutamine and glucose in PLWH compared to HIV-negative healthy controls (Svensson Akusjärvi et al. 2023) (Fig. 8.2a, red box). Notably, glutamine metabolism has been identified as an alternative energy source in HIV-infected macrophages that survive infection, highlighting its importance in cellular adaptation and viral persistence (Castellano et al. 2019). This further suggests that glutamine contributes to energy production required for cellular processes, particularly under HIV-induced metabolic stress.

Glutamate metabolism is also disrupted in the CNS, where HIV-1 and its viral proteins interfere with normal synaptic transmission (Vázquez-Santiago et al. 2014). Thus, the dysregulation of this pathway plays a crucial role in neuronal communication, synaptic plasticity, and metabolic processes (Zhou and Danbolt 2014). Mechanistically, these viral factors, and the accompanying inflammation, impair glutamate release, astrocytic uptake, and glial recycling, causing extracellular glutamate to accumulate

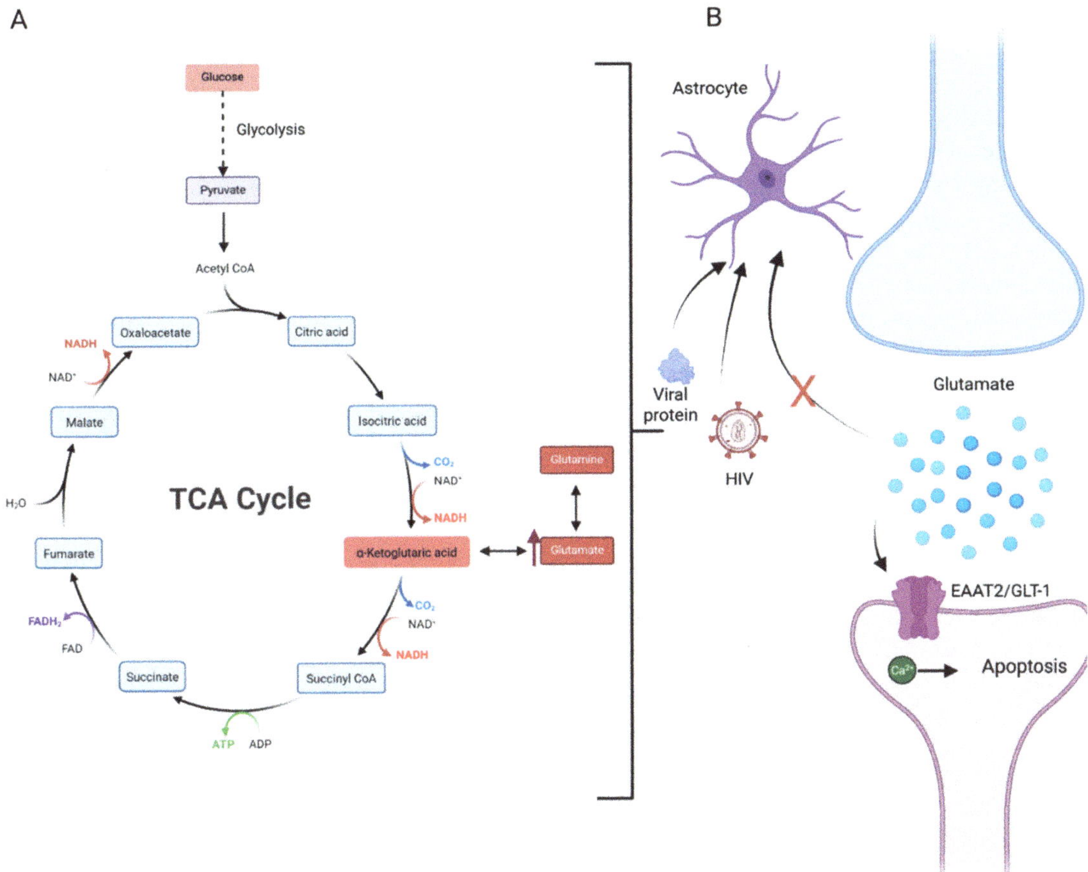

Fig. 8.2 Production of glutamine from glucose. In panel A, glucose undergoes glycolysis and the synthesis of its end product, pyruvate. Pyruvate is typically converted to acetyl-CoA, a substrate for the TCA cycle. The TCA cycle consists of various intermediates, but of particular interest in HIV infection is alpha-ketoglutaric acid. Alpha-ketoglutaric acid is catabolised into glutamate from which glutamine can be synthesized and vice versa. The elevated glutamate and glutamine levels associated with HIV infection may also play a role in the CNS. In panel B, under normal conditions, astrocytes maintain synaptic glutamate homeostasis by clearing excess glutamate from the synaptic cleft via excitatory amino acid transporters EAAT2/GLT-1 receptors. However, HIV-1 viral proteins, including Tat and gp120, disrupt this process by reducing EAAT2/GLT-1 expression or impairing their activity. This inhibition leads to the accumulation of extracellular glutamate, resulting in prolonged activation of glutamate receptors, particularly N-methyl-D-aspartate receptors, on the adjacent neurons. Chronic overstimulation of these receptors causes excessive calcium influx, triggering a cascade of events leading to neuronal apoptosis. Abbreviations: ATP Adenosine triphosphate, ADP Adenosine diphosphate, NAD Nicotinamide adenine dinucleotide, FAD Flavin adenine dinucleotide, CO_2 Carbon dioxide. (The figure was created in BioRender. Mason, S. (2026). https://BioRender.com/6nh31qv)

(Fig. 8.2b) (Melendez et al. 2016; Ru and Tang 2017). Excess glutamate drives excitotoxicity by overactivation of glutamate receptors, often potentiated by viral proteins, which leads to neuronal injury and cell death (Fig. 8.2b) (Dong et al. 2009). Because this pathway underpins neuronal communication, synaptic plasticity, and brain energy metabolism, its dysregulation is closely linked to neurocognitive impairment, including in treatment-experienced PLWH (Cassol et al. 2013).

HIV-induced neuroinflammation further amplifies glutamatergic dysregulation, disrupts the blood-brain barrier, and perpetuates CNS injury. Importantly, disturbed glutamate homeostasis skews the balance between excitatory (glutamate) and inhibitory (GABA) neurotransmission, increasing neuronal excitability and neurotoxicity (Sears and Hewett 2021). Consequently, alterations in the glutamine-glutamate axis not only impair cellular energy homeostasis in PLWH but also interfere with immune signalling and neuronal communication. While cART has no direct modulatory effect on this pathway, some regimens have been linked to mitochondrial dysfunction and oxidative stress, which may indirectly exacerbate glutamatergic imbalance (Sitole et al. 2022; Cassol et al. 2014).

In addition to glutamine-glutamate, Svensson Akusjärvi and colleagues identified reduced arginine levels in PLWH when compared to healthy controls (Svensson Akusjärvi et al. 2023). Arginine is metabolised into nitric oxide and citrulline and can also be hydrolysed into ornithine and urea via arginase enzymes (Fig. 8.3).

Conversely, using LC–MS/MS/HPLC-based quantification, Dirajlal-Fargo et al. reported higher arginine levels in PLWH than in people without HIV (Dirajlal-Fargo et al. 2017). In PLWH, several metabolites showed significant associations with various immune markers: lysine was inversely associated with TNFRI, D-dimer, and sCD163; arginine was inversely associated with hsCRP, TNFRI, and D-dimer; citrulline was inversely associated with D-dimer; sDMA was positively associated with IL-6, TNFRI, TNFRII, D-dimer, and VCAM; ADMA was positively associated with TNFRII and VCAM; and the global arginine bioavailability ratio (GABR) was inversely associated with TNFRI, TNFRII, D-dimer, and VCAM. Ornithine showed no significant association with VCAM (Dirajlal-Fargo et al. 2017) (Fig. 8.3). These findings suggest a potential immunometabolic role of

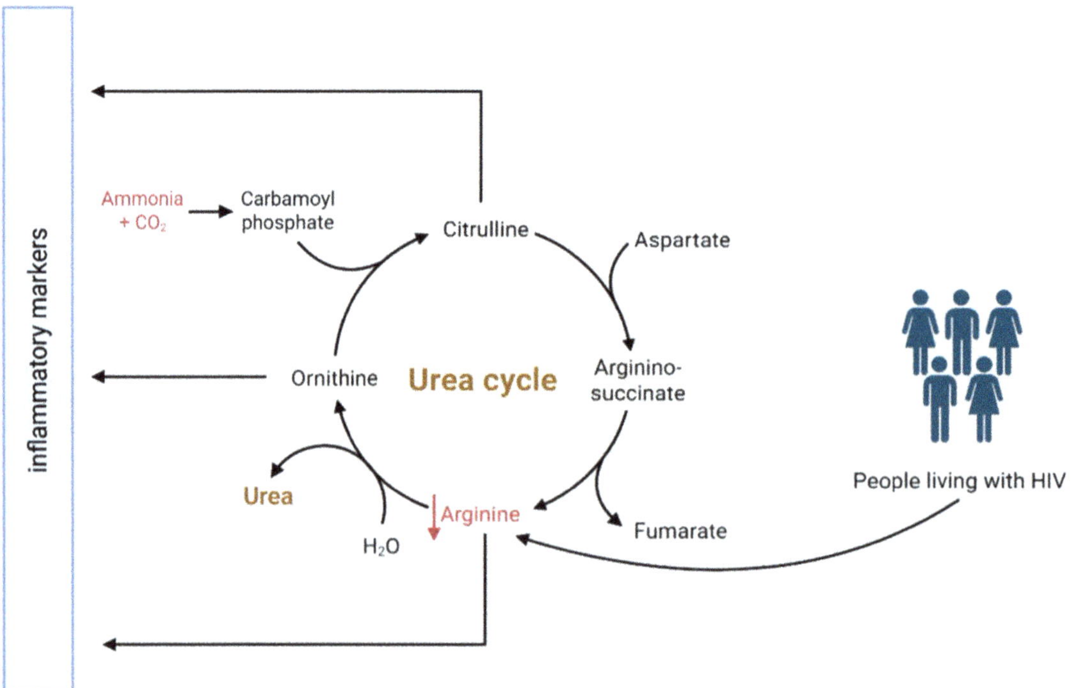

Fig. 8.3 The urea cycle in PLWH. The urea cycle's function is to remove ammonia, a natural end-product of amino acid catabolism, which is considered to be neurotoxic. During HIV infection, some studies suggest lowered levels of arginine hinder the ability of the urea cycle to remove ammonia, resulting in subsequent liver dysfunction and/or hepatic encephalopathy. Furthermore, studies also suggest associations between metabolites of the urea cycle with immune-related markers in PLWH. (The figure was created in BioRender. Mason, S. (2026). https://BioRender.com/9dgzt2q)

arginine metabolism dysregulation, which could contribute to the inflammatory and immune alterations observed in HIV infection.

BCAAs, including leucine, isoleucine, and valine, are essential amino acids that play a vital role in muscle metabolism, immune function, and overall energy balance (Nie et al. 2018). In PLWH, BCAA metabolism is notably altered due to both the viral infection itself and the chronic inflammation it induces. HIV-associated immune activation may, in turn, lead to increased protein catabolism, thereby depleting BCAA levels (Ziegler et al. 2017). Moreover, ART further changes amino acid metabolism by altering liver function and mitochondrial metabolism. A study using ion-exchange chromatography with post-column detection to measure plasma amino acid concentrations in youth living with HIV indicated that, regardless of their CD4+ T cell count, PLWH had significantly reduced levels of total, essential, branched-chain, and sulphur-containing amino acids when compared to healthy controls (Ziegler et al. 2017). These changes were not associated with changes to their inflammatory markers, suggesting that the altered amino acid profiles are a consistent feature of HIV infection, rather than a reflection of general immune activation (Ziegler et al. 2017). Another study using metabolomics in PLWH on long-term cART revealed significant sex-based differences. Women with HIV had lower plasma levels of isoleucine, leucine, and valine when compared to men, alongside greater insulin sensitivity and a more favourable metabolic profile, even after adjusting for body fat and cART duration. These findings suggest that HIV and cART regimens may alter BCAA metabolism in a sex-specific manner, contributing to differences in metabolic disease risk (Koethe et al. 2016).

8.3.2 Glucose Metabolism

Glucose metabolism is essential for providing energy and biosynthetic intermediates required for cellular functions (Liu et al. 2025). In normal cells, glucose is primarily metabolised through glycolysis to produce pyruvate, which enters the TCA cycle and the Nicotinamide adenine dinucleotide (NADH) and Flavin adenine dinucleotide($FADH_2$) that are produced are then utilised in the mitochondria for oxidative phosphorylation to generate ATP in the presence of oxygen (Liu et al. 2025; MacDonald 1993). However, in rapidly proliferating cells, including cancerous and immune cells, glucose metabolism is often reprogrammed to favour glycolysis alone, even in the presence of oxygen, rather than the more efficient oxidative phosphorylation pathway used by normal cells. This leads to high glucose uptake and lactate production, a phenomenon known as the Warburg effect (Vaupel et al. 2019; Warburg 1956; Liberti and Locasale 2016). This shift allows for faster production of ATP and generates key intermediates for biosynthesis, supporting cell growth and immune activation. In PLWH, this metabolic shift is observed in activated immune cells, where the increased demand for energy and biosynthetic precursors supports both viral replication and immune response (Kang and Tang 2020; Hegedus et al. 2017; Ayrga and Koorsen 2025). Consequently, glucose metabolism becomes crucial for maintaining immune function and viral persistence, while also contributing to metabolic disturbances associated with HIV infection.

To replicate, the virus must first attach to the host's CD4 immune cells and subsequently depend on the host metabolites to support its replication. Glucose can enter CD4 immune cells via glucose (GLUT1) transporters, where it is metabolised for energy to fuel viral replication (Fig. 8.4) (Palmer et al. 2016; Macintyre et al. 2014). Once HIV establishes itself in the host's immune cells, it consequently disrupts normal host cell glucose metabolism.

In cART-experienced PLWH, few studies have identified or targeted dysregulated glucose metabolism. A study investigating the metabolic profiles of PLWH found a reduction in glucose and an increase in beta-glucose (a glucose isomer) (Svensson Akusjärvi et al. 2023). Although this study involved a cART-treated cohort, the observed changes in glucose and its isomer indicate additional mechanisms at play contributing to glucose dysregulation during HIV infection.

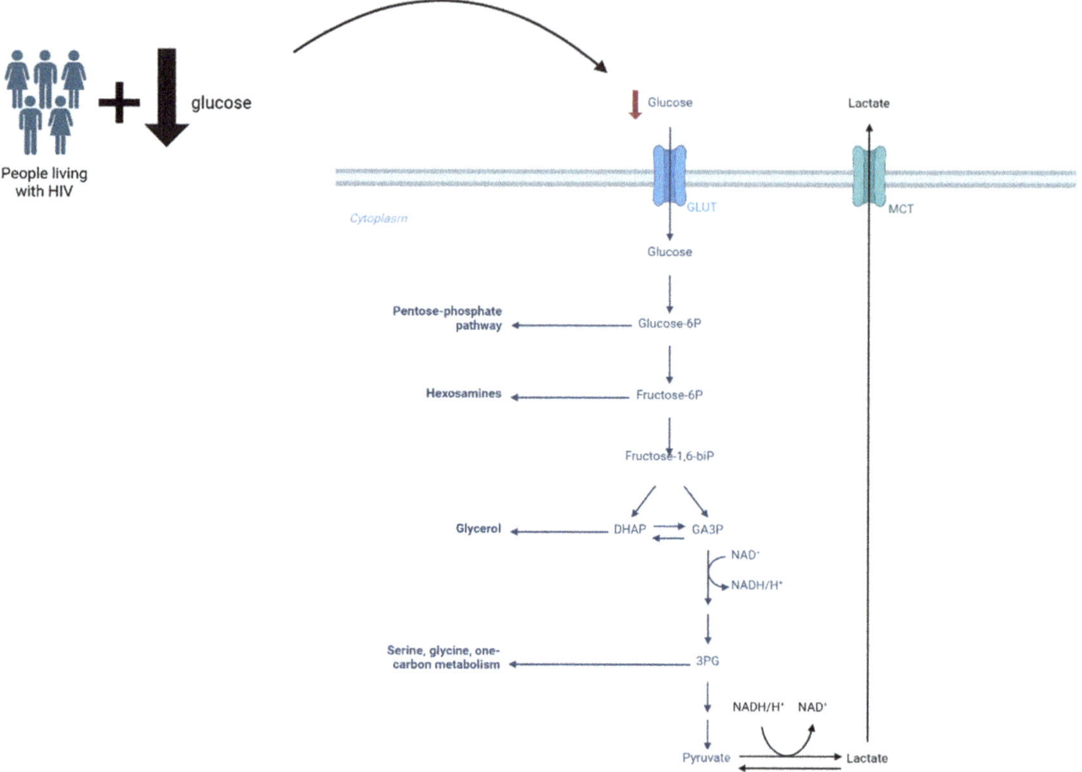

Fig. 8.4 Warburg effect in CD4+ T cells of PLWH on cART. Clinical studies have reported reduced circulating glucose levels in PLWH who are receiving cART. In these individuals, glucose enters CD4+ T cells via the glucose transporter GLUT1; however, despite the presence of oxygen, glucose is preferentially metabolised through glycolysis alone, and the intermediates are not channelled to oxidative phosphorylation (the Warburg effect), resulting in the production of pyruvate and, subsequently, lactate. This metabolic shift supports increased ATP production, which may fuel ongoing HIV replication within the CD4+ T cells. However, this enhanced metabolic demand can deplete systemic glucose levels and ultimately contribute to immune cell exhaustion and death, impairing host immune function. Abbreviations: DHAP Dihydroxyacetone phosphate, GA3P Glyceraldehyde 3-phosphate, NAD Nicotinamide adenine dinucleotide, 3PG 3-Phosphoglycerate, GLUT Glucose transporter, MCT Monocarboxylate transporter. (The figure was adapted from Sun et al. (2023) and created with BioRender. Mason, S. (2026). https://BioRender.com/ tn2q9pa)

Additionally, studies support the notion that the levels of glucose are dysregulated in cART-treated PLWH compared to HIV uninfected participants (Gabazana and Sitole 2021; Roux et al. 2024). Furthermore, the study by Svensson Akusjärvi et al. (2023) revealed an increased number of GLUT1 transporters on the classical monocytes in peripheral blood mononuclear cells (PBMCs) of PLWH when compared to healthy controls. This suggests that while glucose influx into immune cells is elevated, HIV may manipulate and disrupt glucose metabolism, even in PLWH who have been on cART for an average of 7 years, indicating persistent immune dysregulation, despite cART. Additionally, IL-7 upregulates GLUT1 expression during HIV infection (Wofford et al. 2008; Loisel-Meyer et al. 2012). Chronic inflammation often presents in PLWH despite cART and suppressed viral loads (Kuller et al. 2008; Hunt et al. 2014), further contributing to this dysregulation. Therefore, the relationship between immunometabolism and HIV infection plays a significant role in glucose dysregulation, even in individuals on cART treatment.

8.3.3 Microbiome-Derived Metabolites

Microbiome-derived metabolites are small molecules produced by the gut microbiota (or other host-associated microbiomes) through the digestion of dietary components, host-derived metabolites, or microbial interactions (Zhang et al. 2023; Nicholson et al. 2012). These metabolites can have significant systemic effects on host physiology, immunity, and disease progression (Zhang et al. 2023; Nicholson et al. 2012). In the context of HIV infection, the gut microbiome undergoes significant alterations (dysbiosis), leading to changes in the production of microbial metabolites (Brenchley et al. 2006). These shifts contribute to systemic inflammation, immune activation, and HIV-associated comorbidities (Gáspár et al. 2024; Brenchley et al. 2006).

Vujkovic-Cvijin and colleagues observed that in progressive HIV infection, the microbiome of PLWH is enriched with mucosal-adherent bacterial communities capable of catabolising tryptophan, notably through the production of the rate-limiting enzyme IDO-1, a key driver of kynurenine pathway activation and immune dysregulation as previously described (Vujkovic-Cvijin et al. 2013). The study used high-resolution bacterial community profiling to investigate the role of gut microbiota in HIV-associated immune dysfunction, and analysed mucosal-adherent colonic bacteria in PLWH, including those on cART, and found a dysbiotic community enriched in Proteobacteria and depleted in Bacteroidia. This dysbiosis was associated with markers of mucosal immune disruption, systemic inflammation (e.g., IL-6), and the aforementioned Trp-Kyn pathway (Vujkovic-Cvijin et al. 2013). They also identified specific gut bacteria capable of producing kynurenine, suggesting a microbial contribution to HIV immunopathogenesis despite viral suppression. These findings were further reinforced by an additional metabolomic analysis, which detected the KP byproduct 3-hydroxyanthranilate in gut bacteria (Serrano-Villar et al. 2016). The researchers analysed gut microbial metabolism of PLWH and compared it to participants with systemic lupus erythematosus, Clostridium difficile-associated diarrhoea, and healthy controls (Serrano-Villar et al. 2016). Using metabolomic profiling, they additionally found that HIV uniquely alters gut microbial activity and the metabolism of four specific amino acids, indicating that changes in microbial metabolism are disease-specific. These findings suggest potential targets for nutritional interventions in PLWH (Serrano-Villar et al. 2016).

Short-chain fatty acids (SCFAs) have also previously been implicated in HIV pathogenesis (Dillon et al. 2017). SCFAs are the main fermentation products of gut microbiota from dietary fibre (Akhtar et al. 2022; Morrison and Preston 2016). Acetate (two carbons), propionate (three carbons), and butyrate (four carbons) are three of the most abundantly produced SCFAs (Akhtar et al. 2022; Koh et al. 2016). Interesting links have been found between gut microbiota, SCFAs and the host's physiology. Previous studies have shown that in treated PLWH, several key butyrate-producing bacterial genera, such as *Roseburia, Coprococcus, Faecalibacterium,* and *Eubacterium*, are significantly reduced when compared to those of HIV-uninfected controls (Dillon et al. 2017). This reduction is associated with altered SCFA profiles, particularly lower butyrate levels (Mutlu et al. 2014; Vázquez-Castellanos et al. 2015; Sun et al. 2016).

Trimethylamine-N-oxide (TMAO), a gut microbiota-derived metabolite of choline (Sun et al. 2016; Wang et al. 2011), has been studied in the context of HIV infection, though findings remain inconsistent. In one study, plasma TMAO levels were measured in 50 untreated and 50 cART-treated PLWH (Haissman et al. 2017). TMAO was not associated with platelet aggregation or overall platelet reactivity in either group. However, in untreated HIV infection, TMAO levels were reported to correlate with sCD14, a marker of microbial translocation and monocyte activation (Haissman et al. 2017). When compared to untreated individuals, those on cART had lower plasma carnitine and betaine but similar TMAO levels. Notably, the cART-treated group showed higher TMAO-to-carnitine and TMAO-to-betaine ratios, suggesting a possible cART-related influence on TMAO metabolism

(Haissman et al. 2017). However, several other studies found no significant differences in TMAO levels between PLWH and uninfected individuals, nor did they observe notable associations between TMAO levels and HIV-related factors such as cART use or CD4 counts (Serrano-Villar et al. 2016; Miller et al. 2016; Srinivasa et al. 2015).

8.3.4 Lipid Metabolism

8.3.4.1 Lipidomics

Lipidomics refers to the isolation, separation and identification of lipids in biological samples. Lipids are categorised into fatty acyls, glycerolipids, glycerophospholipids, sphingolipids, sterol lipids, prenol lipids, saccharolipids, and polyketides, based on their chemically functional backbones, with fatty acids serving as key building blocks for many of these lipid classes (Fahy et al. 2005). Citrate from the TCA cycle may form these fatty acid synthesis intermediates (Sáez-Cirión and Sereti 2021). This metabolic process is crucial for HIV replication. During HIV replication, fatty acids create an ideal environment at the membrane of CD4+ T cells to assist with the entry and budding of the virus (Sáez-Cirión and Sereti 2021). This ensures that the virus can successfully bud out of the CD4[+] T cell and go on to infect other healthy immune cells. Considering that these fatty acids serve as the backbone for lipids, several studies have reported lipid dysregulation in PLWH. Apart from the aforementioned changes to various SCFA associated with an altered microbiome in PLWH, a study by Sperk and colleagues used untargeted plasma metabolomics by LC–MS/MS and identified significant differences in lipid metabolism between the following groups: PLWH who exhibit slow disease progression despite having high levels of viral load (i.e., viraemic progressors) vs. PLWH who is able to maintain undetectable viral loads for at least 12 months despite not having started cART (i.e., elite controllers), viraemic progressors vs. healthy controls, and elite controllers vs. healthy controls (Sperk et al. 2021). This study was conducted in a cART-naïve group, which indicates that HIV dysregulates lipid metabolism, independ-

dent of cART. A subsequent study by Peltenburg and colleagues used targeted LC/MS and observed a trend of having increased lipid changes, compared to baseline samples from cART-naïve individuals, particularly for cholesterol esters, lysophosphatidylcholines, phosphatidylcholines (PCs), plasmalogen PCs, phosphatidylethanolamines (PE), plasmalogen PEs, and sphingomyelins (Peltenburg et al. 2018), indicating that cART also influences the lipid profiles in PLWH. In another study using LC–MS/MS, a large set of lipid species that differed between cART-treated PLWH and people without HIV was reported (Wan et al. 2025). All PLWH were virally suppressed and stratified by CD4 count (<350 vs ≥350 cells/mm^3) and both strata still showed numerous lipid differences when compared to people without HIV (Wan et al. 2025). A further study using LC/MS, in which a proportion of participants reported stimulant use, differences were observed in specific lipids, particularly dihexosylceramide (d18:1/24:1), triacylglyceride (20:3_34:0), and lysophosphatidylcholines (Cherenack et al. 2025). However, another study, although in a HIV paediatric cohort using UHPLC–qTOF, found no difference in lipidomics analysis when comparing groups that initiated cART early (0–12 weeks) and late (12–50 weeks) (Tarancón-Diez et al. 2021). Both groups were under virological control for at least 65 weeks.

8.4 Metabolic Complications

It is evident that metabolic dysregulation can persist in cART-treated PLWH and subsequently contribute to associated diseases and comorbidities. As a result of these dysregulated metabolic pathways, cART-treated PLWH remain at risk for metabolic complications. These include CVD, with PLWH showing a higher prevalence of non-calcified coronary plaques and coronary artery calcium (Shah et al. 2018; Soares et al. 2021). Inflammation, which is linked to metabolic dysregulation (Baer et al. 2021; Svensson Akusjärvi et al. 2023; Wan et al. 2025), is also associated with CVD (Castley et al. 2016; Dolan et al. 2005; Gager et al. 2020). Moreover, studies report associations between metabolites and inflammatory markers (Baer et al. 2021; Svensson Akusjärvi et al. 2023; Wan et al. 2025), suggesting an indirect mechanism contributing to CVD risk.

Dyslipidemia further contributes to CVD (Grand et al. 2020). It is well established that dyslipidaemia occurs during HIV infection, and in the modern cART era, treatment itself may increase total cholesterol and triglyceride levels (Cummins 2022). Certain cART regimens can also interact with statins used to manage hypercholesterolemia, necessitating regimen-appropriate choices (Cummins 2022). Protease inhibitors have been shown to impair glucose transporters, thereby increasing the risk of insulin resistance (Cummins 2022; Noor et al. 2006). Weight gain and disrupted glucose homeostasis have also been observed with some contemporary regimens, warranting monitoring for diabetes and metabolic syndrome (Eckard and McComsey 2020).

Metabolic dysfunction–associated steatotic liver disease and fibrosis risk may be heightened through the combined effects of lipid and amino acid pathway disruption alongside chronic inflammation (Fittipaldi et al. 2025). Chronic inflammation and exposure to certain antiretrovirals can also affect renal and bone health, contributing to chronic kidney disease and reduced bone mineral density (Debeb et al. 2021). In addition, adipose tissue dysfunction, including residual lipodystrophy and ectopic fat deposition, reflects underlying mitochondrial and lipid pathway alterations in some individuals. Collectively, these complications underscore the need for integrated risk stratification and management, incorporating lifestyle modification, thoughtful regimen selection, and appropriate lipid- and glucose-lowering therapies (Martens et al. 2025).

8.5 Viral and Lifestyle Factors Influencing Metabolic Health in PLWH

Beyond cART, both virologic and lifestyle factors substantially shape the metabolic landscape of PLWH. Metabolomic studies demonstrate that viral load stratification is closely tied to metabolic phenotypes in PLWH. For example, chronic progressors with high viremia show altered glucose, lipid, and amino acid metabolites compared to low-viraemic controllers (Gómez-Archila et al. 2023). Similarly, elite controllers with undetectable viral loads display distinct lipid and antioxidant signatures compared to viraemic progressors (Sperk et al. 2021). In addition, CSF and plasma metabolomics reveal unique metabolic alterations in untreated high viral load individuals relative to ART-suppressed PLWH (Cassol et al. 2014). Together, these findings highlight that higher viral burden is consistently associated with broader metabolic dysregulation.

Likewise, low nadir or persistently reduced CD4 counts are linked to incomplete immune recovery and distinct metabolic alterations. For example, Wan et al. reported that metabolites such as aspartyl-threonine, ribothymidine, and thymine correlated positively with CD4+ T-cell counts, while 3-hydroxyoctanoic acid showed a negative correlation, even with successful ART use (Wan et al. 2025). Similarly, Feng et al. observed lower dimethylglycine levels in women with higher CD4 counts, highlighting how immune reconstitution status is mirrored in metabolic signatures (Feng et al. 2025).

The duration of untreated HIV infection also leaves a metabolic 'imprint' in untreated PLWH. Dysregulated lipid and biogenic amine metabolism persists even after 12 months of successful cART (Peltenburg et al. 2018), suggesting heightened long-term risks of CVD, neurocognitive decline, and metabolic syndrome.

Superimposed on these viral influences are lifestyle factors that compound metabolic risk. Diet significantly influences metabolic outcomes in PLWH. Greater saturated- and total-fat intake has been associated with higher triglycerides and cholesterol levels among PLWH with metabolic abnormalities (Joy et al. 2007). Concurrently, micronutrient deficiencies, including zinc, B-vitamins, and others, are prevalent in PLWH and are known to impair glucose and lipid metabolism by disrupting key enzymatic and energy pathways (Osuna-Padilla et al. 2024). Physical inactivity accelerates visceral adiposity, insulin resistance, and loss of muscle mass, whereas regular exercise can partially restore mitochondrial efficiency and lipid metabolism. Additionally, lifestyle factors may directly influence metabolism. Lifestyle factors such as alcohol use, smoking, substance use, etc., may influence metabolic levels. Substance use, including alcohol, tobacco, and stimulants such as cocaine and methamphetamines, exacerbates oxidative stress, liver injury, and cardiovascular risk (Ande et al. 2015; Amegashie et al. 2025). PLWH often exhibit premature immunosenescence, maintaining signs of immune aging, such as chronic inflammation and T-cell exhaustion, even under successful ART-driven viral control (Chauvin and Sauce 2022). Concurrently, both HIV infection and cART contribute to mitochondrial dysfunction, undermining oxidative metabolism and accelerating age-related metabolic disruptions (Schank et al. 2021). Indeed, older PLWH show impaired mitochondrial fatty acid β-oxidation, reflecting a persistent metabolic 'scar' that overlaps with natural aging processes (Akusjärvi and Neogi 2023). Socioeconomic stressors including food insecurity, have been associated with heightened immune activation (e.g., elevated sCD14, sCD27, sCD163) in PLWH, a pathway that overlaps with metabolic dysregulation and CVD risk (Tamargo et al. 2021). Concurrently, chronic psychosocial stress elevates cortisol levels via HPA-axis activation, promoting gluconeogenesis, insulin resistance, and lipolysis/glycogenesis imbalances, thereby directly impairing glucose and fat metabolism (Levy et al. 2021). Taken together, viral and lifestyle factors interact to amplify the metabolic burden in HIV, underscoring the need for holistic approaches that integrate virologic control with behavioural and lifestyle interventions to optimize long-term health outcomes.

8.6 Integration with Other 'Omics'

Metabolomics provides a dynamic snapshot of cellular function, but its maximum potential is realised when integrated with other omics layers such as transcriptomics, proteomics, genomics and microbiome profiling (O'Donnell et al. 2019; Jendoubi 2021). This multi-omics approach enables a systems-level understanding of HIV pathogenesis by capturing the complex interplay among gene expression, protein activity, microbial composition, and metabolic outcomes. For example, combining transcriptomics with metabolomics allows researchers to link gene-level changes to measurable biochemical alterations. This has been shown in previous studies related to HIV (Yuan et al. 2025). The study integrated transcriptomic and metabolomic analyses to examine changes in HIV-1-infected macrophages and subsequently identified differentially expressed genes and associated changes to amino acid metabolism, glycerophospholipids, and linoleic acid metabolism. Their findings subsequently provided insights into the metabolic disruptions caused by HIV-1 infection and highlighted its potential for identifying therapeutic targets (Yuan et al. 2025). Another study has also integrated metabolomics and transcriptomics to investigate HIV-related comorbidities, including cardiometabolic disease and CVD (Sun et al. 2025). These multi-omics approaches have identified key metabolic and transcriptomic alterations linked to inflammation, endothelial dysfunction, and other pathological processes in PLWH (Sun et al. 2025).

Proteomics elucidates changes in enzyme abundance and activity, which can explain or predict shifts in metabolite levels (Chandramouli and Qian 2009), and can be used as a complement to metabolomics. In PLWH, proteomic studies have shown alterations in mitochondrial proteins, redox regulators, and immune-related molecules (Grabowska et al. 2021). These changes correspond with metabolomic findings such as disruptions in energy metabolism, increased oxidative stress, and depletion of key antioxidants like glutathione (Grabowska et al. 2021). The proteomics and metabolomics of HIV-associated neurocognitive disorders have also been widely reviewed (Williams et al. 2021; Pendyala and Fox 2010). Together, in combination, proteomic and metabolomic data provides deeper insight into the biochemical consequences of chronic viral infection and ART. The integration of microbiome data with metabolomics has also revealed important insights into HIV-associated immune dysfunction, namely that HIV infection is often accompanied by gut microbial dysbiosis, characterised by a loss of beneficial bacterial species and an overrepresentation of pro-inflammatory microbes (Wang and Qi 2019).

As seen above, by combining these various omics platforms, researchers can achieve a more comprehensive understanding of host-virus interactions. Multi-omics integration supports the discovery of robust biomarkers, the identification of therapeutic targets, and the stratification of individuals for personalised interventions with a far higher level of certainty. This approach not only strengthens mechanistic insight into HIV pathogenesis but also offers new avenues for clinical management by revealing molecular signatures that may guide more effective, individualised treatments.

8.7 Chemometrics for Metabolomics

8.7.1 The Role of Chemometrics in HIV Metabolomics

Metabolomics, or the study of metabolites, reveals the end products of metabolic processes, which provide a direct readout of an individual's physiological state. Chemometrics helps make sense of the overwhelming amount of data collected from analytical instruments such as NMR spectroscopy and MS.

8.7.2 Key Chemometric Methods

- Principal component analysis (PCA): An unsupervised technique used to reduce data dimensionality. It is often employed in early-stage exploratory analysis to identify patterns and similarities among samples, such as

grouping healthy individuals versus PLWH (Jolliffe and Cadima 2016).

- Partial least squares discriminant analysis (PLS-DA): A supervised method for classification that uses a regression approach to find a relationship between the metabolic profile and a variable of interest, such as disease status (Gromski et al. 2015). Studies have used PLS-DA and similar methods like Orthogonal Projections to Latent Structures-Discriminant Analysis (OPLS-DA) to successfully distinguish PLWH from healthy controls (Sitole et al. 2014).

- Linear discriminant analysis (LDA): Another supervised classification technique that has been used to effectively sort samples. In one study, LDA applied to NMR spectra achieved 100% accuracy in classifying healthy individuals, PLWH not on therapy, and PLWH on ART (McKnight et al. 2014).

- Regression analysis: Techniques such as Partial Least Squares Regression (PLSR) can be used to correlate metabolic data with continuous clinical parameters, such as CD4 cell count and viral load, to identify specific metabolites associated with disease progression (Tinarwo et al. 2019; Dickens et al. 2015).

- Network-based systems biology approaches: Advanced methods can be used to identify system-level metabolic alterations by constructing networks of metabolite interactions (Yan et al. 2018). This provides deeper insight into the affected biological pathways.

8.8 Challenges and Future Directions

8.8.1 Need for Standardization Across Studies

Metabolomics can be done in either a targeted or an untargeted approach. However, for each of these approaches, there is not one standard method that is utilised. The chemical nature of the metabolites may determine the type of metabolomics analytical platform to use (Schrimpe-Rutledge et al. 2016). Additionally, there is a lack of a standard reference material for many metabolites, which may create variability in comparing studies investigating the same metabolites using different standards and/or internal standards (Schrimpe-Rutledge et al. 2016). The use of different reference standards may affect batch corrections, calibrations, and ultimately quantification, influencing the interpretation of the results when comparing these metabolomics studies. However, there is a growing initiative of using a reference standardization for accurate quantification in metabolomics (Liu et al. 2020). This may be especially beneficial for large-scale metabolomics. The chromatographic column used during metabolomics may also influence the quantification of metabolites (Hemmer et al. 2024). The use of different columns may affect the separation of metabolites in the sample type and the overall quantification of the metabolites if they are not separated well. However, columns are made to cater to specific metabolomics approaches, providing specialised metabolite separation based on the nature of the metabolites investigated (Hemmer et al. 2024). This aids with the standardisation of particular approaches for metabolomics. The column choice may also determine the mobile phases needed to efficiently separate the metabolite for quantification. Also, the choice of protein removal methods is not routine. However, most analysts make use of either a chemical solvent such as acetonitrile and/or methanol or ultracentrifugal filters for physical protein removal (Nagana Gowda and Raftery 2014). There is currently no gold standard for metabolomics analysis; however, increasing efforts are being made to introduce standardisation into the field.

Beyond technical aspects, HIV metabolomics studies are also constrained by study design and clinical heterogeneity. Many existing investigations are limited by small sample sizes, which reduces statistical power and reproducibility across cohorts (Sebigi et al. 2025). The absence of standardized protocols for sample collection, preparation, and data analysis further complicates comparisons between studies. Additionally, variability in cART regimens introduces metabolic heterogeneity, since different drug classes (e.g., protease inhibitors, integrase inhibitors, NNRTIs) exert distinct influences on lipid and glucose metabolism (Lagathu et al. 2019). This makes it challenging to disentangle whether observed metabolic alterations are attributable to HIV infection itself or are secondary to ART exposure. Most current datasets are cross-sectional rather than longitudinal, which further limits causal inference.

8.8.2 Development of HIV-Specific Metabolite Reference Databases

Currently, most metabolite databases are generalized and lack HIV-specific contextual information. Developing dedicated HIV-focused metabolite reference libraries would enable more accurate identification and annotation of metabolites altered during infection and treatment. Such databases could capture unique metabolic fingerprints of viral proteins, ART regimens, and HIV-associated comorbidities, thereby enhancing reproducibility and comparability across studies. The lack of HIV-specific metabolite reference datasets hampers benchmarking of findings, as current databases such as Human Metabolome Database (HMDB) do not capture HIV- or ART-specific ranges. Addressing these gaps will require multicentre studies with harmonized protocols and the development of HIV-focused metabolomics resources to enable more robust biomarker discovery and clinical translation.

8.8.3 Integration of Metabolomics with Immunological and Clinical Profiling

Metabolomics alone provides a biochemical snapshot, but its impact is maximized when integrated with immunological markers (e.g., cytokines, T-cell subsets) and clinical measures (e.g., viral load, CD4 counts, neurocognitive assessments). This systems-level approach can uncover direct links between metabolic alterations, immune dysregulation, and patient outcomes (Fu et al. 2023). Ultimately, such integration would facilitate precision medicine strategies by identifying patient subgroups at higher risk of HIV-related complications.

8.8.4 Longitudinal Study Designs to Monitor Treatment Effects and Disease Progression

Cross-sectional metabolomics studies provide limited insights into temporal changes. Longitudinal cohort designs are essential to track how metabolic pathways evolve over the course of infection, treatment initiation, ART adherence, and aging. This approach would allow researchers to disentangle short-term drug effects from long-term disease progression and to identify early metabolic indicators of comorbidities such as CVD, neurocognitive decline, and metabolic syndrome in PLWH (Hudson et al. 2025).

8.8.5 Combining Untargeted and Targeted Metabolomics to Both Discover and Validate Biomarkers

Untargeted metabolomics is well-suited for hypothesis generation, enabling the discovery of novel metabolic changes in HIV infection. However, findings from untargeted approaches must be validated using targeted assays to ensure accuracy and clinical applicability. Combining both strategies allows researchers to first cast a wide net to identify potential biomarkers, and then confirm their relevance, robustness, and sensitivity for clinical translation, particularly in the context of HIV-related comorbidities (Di Minno et al. 2021).

8.8.6 Underrepresentation of Low-Income Settings in Metabolomics Research

Metabolomics remains largely a preserve of high-income societies (González-Alcaide et al. 2020). In relation to HIV research, this calls for more metabolomics research to be done in low-income settings such as sub-Saharan Africa, which is a highly endemic region for HIV. The lack of metabolomics research in these low-income settings shows a bias in our understanding of HIV metabolomics towards high-income settings. There are different types of HIV subtypes around the world that are attributed to amino acid sequence variations, which may contribute to the differences seen in clinical outcomes in PLWH (Hemelaar et al. 2019; Asia et al. 2023). Introducing HIV metabolomics into more regions of the globe, for investigating metabolic profiles of PLWH, may provide more representative data and give better insights into the disease mechanisms and response to treatment, especially considering that factors such as climate, environment, genetics, etc., most likely have an influence on these and other factors. In sub-Saharan Africa, South Africa is leading metabolomics research, and most likely in conjunction with international collaborations with other high-income regions (González-Alcaide et al. 2020). Thus, there is also a need for more collaboration between South Africa and other international bodies with other regions of sub-Saharan African countries.

8.8.7 Precision Medicine Approaches and How Metabolomics Can Inform Therapeutic Decisions

Precision medicine may refer to disease prevention and treatment catered to a specific group. Here we refer to precision medicine in the context of HIV. Metabolomics can be used to identify metabolic indicators of disease (e.g., diabetes, cancer) (Mayers et al. 2014; Rhee et al. 2011). Considering the literature of HIV metabolomics in the modern cART era, there is a need to better understand the dysregulation of metabolic pathways and their association with clinical parameters. These may then potentially serve as a metabolic indicator for HIV. For example a study done by Li et al. (2018) indicated that butyrylcarnitine, in combination with myristic acid, from plasma in treatment-naïve patients could predict dyslipidaemia caused by cART therapies later on (Li et al. 2018). Hence, certain metabolic states in cART-treated PLWH may serve as potential biomarkers for HIV and related diseases based on their association with clinical parameters of HIV infection, e.g., CD4 counts and viral load.

8.9 Conclusions

This chapter underscores the importance of metabolomics in advancing our understanding of the systemic effects of HIV infection and cART. While significant progress has been made in controlling the viral load with cART, the metabolic disruptions caused by chronic immune activation and low-level viral persistence remain to be fully understood. Key metabolic pathways, including those involved in amino acid metabolism, glucose regulation, lipid metabolism, and microbiome-derived metabolites, are typically disrupted in HIV, leading to a higher risk of comorbidities such as CVD, metabolic dysfunction, diabetes mellitus, and neurocognitive disorders. Metabolomics provides crucial insights into these ongoing metabolic disturbances, offering a comprehensive view of how HIV and cART shape host metabolism beyond the virological effects. By identifying specific biomarkers of metabolic dysregulation, this chapter also highlights the potential for precision medicine in HIV care, guiding therapeutic strategies that can better address the long-term health of PLWH. The integration of metabolomics with other 'omics' technologies promises to unlock new avenues for understanding HIV pathogenesis and improving clinical outcomes, emphasising the need for stan-

dardised research practices and continued investigation into the non-virological drivers of disease progression in PLWH.

Conflict of Interest Author LKA declares that he has no conflict of interest. Author TS declares that he has no conflict of interest. Author DTL declares that he has no conflict of interest. Author SM declares that he has no conflict of interest. Author EJvV declares that she has no conflict of interest. Author MEW declares that he has no conflict of interest.

Ethical Approval This chapter does not contain any studies with human participants performed by any of the authors.

References

Ahmed D, Roy D, Cassol E (2018) Examining relationships between metabolism and persistent inflammation in HIV patients on antiretroviral therapy. Mediat Inflamm 2018:6238978. https://doi.org/10.1155/2018/6238978

Adamu A, Li S, Gao F et al (2024) Neuroinflammation in neurodegenerative diseases: therapeutic targets. Front Aging Neurosci 16:1347987

Akhtar M, Chen Y, Ma Z et al (2022) Gut microbiota-derived short chain fatty acids are potential mediators in gut inflammation. Anim Nutr 8:350–360. https://doi.org/10.1016/j.aninu.2021.11.005

Akusjärvi SS, Krishnan S, Ambikan AT et al (2023) Role of myeloid cells in system-level immunometabolic dysregulation during prolonged successful HIV-1 treatment. AIDS 37(7):1023–1033. https://doi.org/10.1097/qad.0000000000003512

Akusjärvi SS, Neogi U (2023) Biological aging in people living with HIV on successful antiretroviral therapy: do they age faster? Curr HIV/AIDS Rep 20(2):42–50. https://doi.org/10.1007/s11904-023-00646-0

Albrecht J, Sonnewald U, Waagepetersen HS et al (2007) Glutamine in the central nervous system: function and dysfunction. Front Biosci 12:332–343. https://doi.org/10.2741/2067

Amegashie EA, Sikeola RO, Tagoe EA et al (2025) Oxidative stress in people living with HIV: are diverse supplement sources the solution? Health Sci Rep 8(5):e70824. https://doi.org/10.1002/hsr2.70824

Ande A, Sinha N, Rao PS et al (2015) Enhanced oxidative stress by alcohol use in HIV+ patients: possible involvement of cytochrome P450 2E1 and antioxidant enzymes. AIDS Res Ther 12:29. https://doi.org/10.1186/s12981-015-0071-x

Andersen JV (2025) The glutamate/GABA–glutamine cycle: insights and advances. J Neurochem 169(3):e70029

Andriguetti NB, Barratt DT, Tucci J et al (2025) Lamivudine and tenofovir pharmacokinetic variability in people with HIV in Papua New Guinea. Br J Clin Pharmacol. https://doi.org/10.1002/bcp.70188

Antoniewicz MR (2013) Tandem mass spectrometry for measuring stable-isotope labeling. Curr Opin Biotechnol 24(1):48–53. https://doi.org/10.1016/j.copbio.2012.10.011

Asia LK, Jansen Van Vuren E, Williams ME (2023) The influence of viral protein R amino acid substitutions on clinical outcomes in people living with HIV: a systematic review. Eur J Clin Investig 53(5):e13943. https://doi.org/10.1111/eci.13943

Ayrga S, Koorsen G (2025) Metabolic reprogramming in HIV+ CD4(+) T-cells: implications for immune dysfunction and therapeutic targets in M. Tuberculosis co-infection. Meta 15(5). https://doi.org/10.3390/metabo15050285

Babu H, Sperk M, Ambikan AT et al (2019a) Plasma metabolic signature and abnormalities in HIV-infected individuals on long-term successful antiretroviral therapy. Meta 9(10):210

Babu H, Sperk M, Ambikan AT et al (2019b) Plasma metabolic signature and abnormalities in HIV-infected individuals on long-term successful antiretroviral therapy. Meta 9(10). https://doi.org/10.3390/metabo9100210

Badawy AA (2017) Kynurenine pathway of tryptophan metabolism: regulatory and functional aspects. Int J Tryptophan Res 10:1178646917691938. https://doi.org/10.1177/1178646917691938

Baer SL, Colombo RE, Johnson MH et al (2021) Indoleamine 2,3 dioxygenase, age, and immune activation in people living with HIV. J Investig Med 69(6):1238–1244. https://doi.org/10.1136/jim-2021-001794

Ball HJ, Jusof FF, Bakmiwewa SM et al (2014) Tryptophan-catabolizing enzymes–party of three. Front Immunol 5:485

Barré-Sinoussi F, Chermann JC, Rey F et al (1983) Isolation of a T-lymphotropic retrovirus from a patient at risk for acquired immune deficiency syndrome (AIDS). Science 220(4599):868–871. https://doi.org/10.1126/science.6189183

Bauermeister A, Mannochio-Russo H, Costa-Lotufo LV et al (2022) Mass spectrometry-based metabolomics in microbiome investigations. Nat Rev Microbiol 20(3):143–160. https://doi.org/10.1038/s41579-021-00621-9

Beger RD, Dunn W, Schmidt MA et al (2016) Metabolomics enables precision medicine: "A white paper, community perspective". Metabolomics 12(9):149. https://doi.org/10.1007/s11306-016-1094-6

Bierla K, Chiappetta G, Vinh J et al (2020) Potential of fourier transform mass spectrometry (orbitrap and ion cyclotron resonance) for speciation of the selenium metabolome in selenium-rich yeast. Front Chem 8:612387. https://doi.org/10.3389/fchem.2020.612387

Bipath P, Levay PF, Viljoen M (2015) The kynurenine pathway activities in a sub-Saharan HIV/AIDS population. BMC Infect Dis 15(1):346. https://doi.org/10.1186/s12879-015-1087-5

Boasso A, Hardy AW, Anderson SA et al (2008) HIV-induced type I interferon and tryptophan catabolism drive T cell dysfunction despite phenotypic activation. PLoS One 3(8):e2961

Boasso A, Shearer G, Chougnet C (2009) Immune dysregulation in human immunodeficiency virus infection: know it, fix it, prevent it? J Intern Med 265(1):78–96

Boasso A, Shearer GM (2007) How does indoleamine 2, 3-dioxygenase contribute to HIV-mediated immune dysregulation. Curr Drug Metab 8(3):217–223

Brenchley JM, Price DA, Schacker TW et al (2006) Microbial translocation is a cause of systemic immune activation in chronic HIV infection. Nat Med 12(12):1365–1371. https://doi.org/10.1038/nm1511

Brinkman K, ter Hofstede HJ, Burger DM et al (1998) Adverse effects of reverse transcriptase inhibitors: mitochondrial toxicity as common pathway. AIDS 12(14):1735–1744. https://doi.org/10.1097/00002030-199814000-00004

Bujak R, Struck-Lewicka W, Markuszewski MJ et al (2015) Metabolomics for laboratory diagnostics. J Pharm Biomed Anal 113:108–120. https://doi.org/10.1016/j.jpba.2014.12.017

Bussey IRO (2024) Applications of ultra high-performance liquid chromatography coupled with high-resolution mass spectrometry. In: Núñez O (ed) Relevant applications of high-performance liquid chromatography in food, environmental, clinical and biological fields. IntechOpen, Rijeka. https://doi.org/10.5772/intechopen.1006638

Campbell-Yesufu OT, Gandhi RT (2011) Update on human immunodeficiency virus (HIV)-2 infection. Clin Infect Dis 52(6):780–787. https://doi.org/10.1093/cid/ciq248

Cassol E, Misra V, Dutta A et al (2014) Cerebrospinal fluid metabolomics reveals altered waste clearance and accelerated aging in HIV patients with neurocognitive impairment. AIDS 28(11):1579–1591. https://doi.org/10.1097/qad.0000000000000303

Cassol E, Misra V, Holman A et al (2013) Plasma metabolomics identifies lipid abnormalities linked to markers of inflammation, microbial translocation, and hepatic function in HIV patients receiving protease inhibitors. BMC Infect Dis 13(1):203. https://doi.org/10.1186/1471-2334-13-203

Castley A, Williams L, James I et al (2016) Plasma CXCL10, sCD163 and sCD14 levels and cardiovascular risk in HIV. PLoS One 11(6):e0158169

Castellano P, Prevedel L, Valdebenito S et al (2019) HIV infection and latency induce a unique metabolic signature in human macrophages. Sci Rep 9(1):3941. https://doi.org/10.1038/s41598-019-39898-5

Chandramouli K, Qian PY (2009) Proteomics: challenges, techniques and possibilities to overcome biological sample complexity. Hum Genom Proteom 2009. https://doi.org/10.4061/2009/239204

Chauvin M, Sauce D (2022) Mechanisms of immune aging in HIV. Clin Sci (Lond) 136(1):61–80. https://doi.org/10.1042/cs20210344

Chen J, Shao J, Cai R et al (2014) Anti-retroviral therapy decreases but does not normalize indoleamine 2,

3-dioxygenase activity in HIV-infected patients. PLoS One 9(7):e100446

Chen J, Xun J, Yang J et al (2019) Plasma indoleamine 2, 3-dioxygenase activity is associated with the size of the human immunodeficiency virus reservoir in patients receiving antiretroviral therapy. Clin Infect Dis 68(8):1274–1281

Cherenack EM, Larson ME, Murray K et al (2025) Stimulant use, HIV, and plasma metabolites among men. J Neuroimmune Pharmacol 20(1):68. https://doi.org/10.1007/s11481-025-10223-4

Chokkathukalam A, Kim DH, Barrett MP et al (2014) Stable isotope-labeling studies in metabolomics: new insights into structure and dynamics of metabolic networks. Bioanalysis 6(4):511–524. https://doi.org/10.4155/bio.13.348

Chun TW, Stuyver L, Mizell SB et al (1997) Presence of an inducible HIV-1 latent reservoir during highly active antiretroviral therapy. Proc Natl Acad Sci USA 94(24):13193–13197. https://doi.org/10.1073/pnas.94.24.13193

Clish CB (2015) Metabolomics: an emerging but powerful tool for precision medicine. Cold Spring Harb Mol Case Stud 1(1):a000588. https://doi.org/10.1101/mcs.a000588

Cockerham LR, Siliciano JD, Sinclair E et al (2014) CD4+ and CD8+ T cell activation are associated with HIV DNA in resting CD4+ T cells. PLoS One 9(10):e110731. https://doi.org/10.1371/journal.pone.0110731

Cohn LB, Chomont N, Deeks SG (2020) The biology of the HIV-1 latent reservoir and implications for cure strategies. Cell Host Microbe 27(4):519–530. https://doi.org/10.1016/j.chom.2020.03.014

Cummins NW (2022) Metabolic complications of chronic HIV infection: a narrative review. Pathogens 11(2)

Davis I, Liu A (2015) What is the tryptophan kynurenine pathway and why is it important to neurotherapeutics? Expert Rev Neurother 15(7):719–721. https://doi.org/10.1586/14737175.2015.1049999

Debeb SG, Muche AA, Kifle ZD et al (2021) Tenofovir-associated renal dysfunction in people living with HIV. HIV AIDS (Auckl) 13:491–503

Deeks SG, Overbaugh J, Phillips A et al (2015) HIV infection. Nat Rev Dis Primers 1:15035. https://doi.org/10.1038/nrdp.2015.35

Deeks SG, Tracy R, Douek DC (2013) Systemic effects of inflammation on health during chronic HIV infection. Immunity 39(4):633–645. https://doi.org/10.1016/j.immuni.2013.10.001

Deme P, Rubin LH, Yu D et al (2022) Immunometabolic reprogramming in response to HIV infection is not fully normalized by suppressive antiretroviral therapy. Viruses 14(6):1313

Di Minno A, Gelzo M, Stornaiuolo M et al (2021) The evolving landscape of untargeted metabolomics. Nutr Metab Cardiovasc Dis 31(6):1645–1652. https://doi.org/10.1016/j.numecd.2021.01.008

Dickens AM, Anthony DC, Deutsch R et al (2015) Cerebrospinal fluid metabolomics implicate bioen-ergetic adaptation as a neural mechanism regulating shifts in cognitive states of HIV-infected patients. AIDS 29(5):559–569. https://doi.org/10.1097/qad.0000000000000580

Dillon SM, Kibbie J, Lee EJ et al (2017) Low abundance of colonic butyrate-producing bacteria in HIV infection is associated with microbial translocation and immune activation. AIDS 31(4):511–521. https://doi.org/10.1097/qad.0000000000001366

Dirajlal-Fargo S, Alam K, Sattar A et al (2017) Comprehensive assessment of the arginine pathway and its relationship to inflammation in HIV. AIDS 31(4):533–537. https://doi.org/10.1097/qad.0000000000001363

Dodds JN, Baker ES (2019) Ion mobility spectrometry: fundamental concepts, instrumentation, applications, and the road ahead. J Am Soc Mass Spectrom 30(11):2185–2195. https://doi.org/10.1007/s13361-019-02288-2

Dohl J, Passos MEP, Földi J et al (2020) Glutamine depletion disrupts mitochondrial integrity and impairs myoblast proliferation and differentiation. Nutr Res 84:42–52

Dong XX, Wang Y, Qin ZH (2009) Molecular mechanisms of excitotoxicity and their relevance to pathogenesis of neurodegenerative diseases. Acta Pharmacol Sin 30(4):379–387. https://doi.org/10.1038/aps.2009.24

Dolan SE, Hadigan C, Killilea KM et al (2005) Increased cardiovascular disease risk indices in HIV-infected women. J Acquir Immune Defic Syndr 39(1):44–54

Dunn WB, Broadhurst DI, Atherton HJ et al (2011) Systems level studies of mammalian metabolomes: the roles of mass spectrometry and nuclear magnetic resonance spectroscopy. Chem Soc Rev 40(1):387–426

Eckard AR, McComsey GA (2020) Weight gain and integrase inhibitors. Curr Opin Infect Dis 33(1):10–19

Emwas AH (2015) The strengths and weaknesses of NMR spectroscopy and mass spectrometry with particular focus on metabolomics research. Methods Mol Biol 1277:161–193. https://doi.org/10.1007/978-1-4939-2377-9_13

Emwas AH, Roy R, McKay RT et al (2019) NMR spectroscopy for metabolomics research. Meta 9(7). https://doi.org/10.3390/metabo9070123

Esbjörnsson J, Månsson F, Kvist A et al (2018) Long-term follow-up of HIV-2-related AIDS and mortality in Guinea-Bissau: a prospective open cohort study. Lancet HIV. https://doi.org/10.1016/s2352-3018(18)30254-6

Fahy E, Subramaniam S, Brown HA et al (2005) A comprehensive classification system for lipids. J Lipid Res 46(5):839–861. https://doi.org/10.1194/jlr.E400004-JLR200

Favre D, Mold J, Hunt PW et al (2010) Tryptophan catabolism by indoleamine 2,3-dioxygenase 1 alters the balance of TH17 to regulatory T cells in HIV disease. Sci Transl Med 2(32):32ra36

Feng A, Zhao H, Qiu C et al (2025) Gut microbiota metabolites impact immunologic responses to antiretroviral therapy in HIV-infected men who have sex with men.

Infect Dis Poverty 14(1):21. https://doi.org/10.1186/s40249-025-01291-y

Février M, Dorgham K, Rebollo A (2011) CD4+ T cell depletion in human immunodeficiency virus (HIV) infection: role of apoptosis. Viruses 3(5):586–612. https://doi.org/10.3390/v3050586

Fittipaldi J, Cardoso SW, Nunes EP et al (2025) Metabolic dysfunction-associated liver disease and fibrosis in people with HIV. AIDS 39(12)

Frias MA, Pagano S, Bararpour N et al (2024) People living with HIV display increased anti-apolipoprotein A1 auto-antibodies, inflammation, and kynurenine metabolites: a case–control study. Front Cardiovasc Med 11:1343361

Fu J, Zhu F, Xu CJ et al (2023) Metabolomics meets systems immunology. EMBO Rep 24(4):e55747. https://doi.org/10.15252/embr.202255747

Fuchs D, Möller AA, Reibnegger G et al (1991) Increased endogenous interferon-gamma and neopterin correlate with increased degradation of tryptophan in human immunodeficiency virus type 1 infection. Immunol Lett 28(3):207–211

Fujigaki H, Yamamoto Y, Saito K (2017) L-tryptophan-kynurenine pathway enzymes are therapeutic target for neuropsychiatric diseases: focus on cell type differences. Neuropharmacol 112(Pt B):264–274. https://doi.org/10.1016/j.neuropharm.2016.01.011

Gabazana Z, Sitole L (2021) Raman-based metabonomics unravels metabolic changes related to a first-line tenofovir-based treatment in a small cohort of South African HIV-infected patients. Spectrochim Acta A Mol Biomol Spectrosc 248:119256. https://doi.org/10.1016/j.saa.2020.119256

Gager GM, Biesinger B, Hofer F et al (2020) Interleukin-6 level is a predictor of long-term cardiovascular mortality in acute coronary syndrome. Vascul Pharmacol 135:106806

Gallo RC, Salahuddin SZ, Popovic M et al (1984) Frequent detection and isolation of cytopathic retroviruses (HTLV-III) from patients with AIDS and at risk for AIDS. Science 224(4648):500–503. https://doi.org/10.1126/science.6200936

Ganta KK, Chaubey B (2019) Mitochondrial dysfunctions in HIV infection and antiviral drug treatment. Expert Opin Drug Metab Toxicol 15(12):1043–1052. https://doi.org/10.1080/17425255.2019.1692814

Gáspár Z, Nagavci B, Szabó BG et al (2024) Gut microbiome alteration in HIV/AIDS and the role of antiretroviral therapy-a scoping review. Microorganisms 12(11). https://doi.org/10.3390/microorganisms12112221

Gazzola L, Tincati C, Bellistré GM et al (2009) The absence of CD4+ T cell count recovery despite receipt of virologically suppressive highly active antiretroviral therapy: clinical risk, immunological gaps, and therapeutic options. Clin Infect Dis 48(3):328–337. https://doi.org/10.1086/695852

Gelpi M, Hartling HJ, Ueland PM et al (2017) Tryptophan catabolism and immune activation in HIV infection. BMC Infect Dis 17:349

Gómez-Archila LG, Palomino-Schätzlein M, Zapata-Builes W et al (2023) Plasma metabolomics by nuclear magnetic resonance reveals biomarkers and metabolic pathways associated with the control of HIV-1 infection/progression. Front Mol Biosci 10:1204273. https://doi.org/10.3389/fmolb.2023.1204273

González-Alcaide G, Menchi-Elanzi M, Nacarapa E et al (2020) HIV/AIDS research in Africa and the Middle East: participation and equity in North-South collaborations and relationships. Glob Health 16(1):83. https://doi.org/10.1186/s12992-020-00609-9

Grabowska K, Harwood E, Ciborowski P (2021) HIV and proteomics: what we have learned from high throughput studies. Proteomics Clin Appl 15(1):e2000040. https://doi.org/10.1002/prca.202000040

Grand M, Bia D, Diaz A (2020) Cardiovascular risk assessment in people living with HIV: a systematic review. Curr HIV Res 18(1):5–18

Gromski PS, Muhamadali H, Ellis DI et al (2015) A tutorial review: metabolomics and partial least squares-discriminant analysis--a marriage of convenience or a shotgun wedding. Anal Chim Acta 879:10–23. https://doi.org/10.1016/j.aca.2015.02.012

Guan Y, Wei L, Tao Y et al (2025) Compare antiretroviral drug concentrations in hair and plasma across EFV-based regimens in China. Biosci Trends. https://doi.org/10.5582/bst.2025.01157

Hahn BH, Shaw GM, De Cock KM et al (2000) AIDS as a zoonosis: scientific and public health implications. Science 287(5453):607–614. https://doi.org/10.1126/science.287.5453.607

Haissman JM, Haugaard AK, Ostrowski SR, Berge RK, Hov JR, Trøseid M, Nielsen SD (2017) Microbiota-dependent metabolite and cardiovascular disease marker trimethylamine-N-oxide (TMAO) is associated with monocyte activation but not platelet function in untreated HIV infection. BMC Infect Dis 17(1):445. https://doi.org/10.1186/s12879-017-2547-x

Han W, Ward JL, Kong Y et al (2023) Editorial: targeted and untargeted metabolomics for the evaluation of plant metabolites in response to the environment. Front Plant Sci 14:1167513. https://doi.org/10.3389/fpls.2023.1167513

Hegedus A, Kavanagh Williamson M, Khan MB et al (2017) Evidence for altered glutamine metabolism in human immunodeficiency virus type 1 infected primary human CD4(+) T cells. AIDS Res Hum Retrovir 33(12):1236–1247. https://doi.org/10.1089/aid.2017.0165

Hemelaar J, Elangovan R, Yun J et al (2019) Global and regional molecular epidemiology of HIV-1, 1990-2015: a systematic review, global survey, and trend analysis. Lancet Infect Dis 19(2):143–155. https://doi.org/10.1016/s1473-3099(18)30647-9

Hemmer S, Manier SK, Wagmann L et al (2024) Comparison of reversed-phase, hydrophilic interaction, and porous graphitic carbon chromatography columns for an untargeted toxicometabolomics study in pooled human liver microsomes, rat urine, and rat plasma. Metabolomics 20(3):49. https://doi.org/10.1007/s11306-024-02115-0

Herbert C, Luies L, Loots DT et al (2023) The metabolic consequences of HIV/TB co-infection. BMC Infect Dis 23(1):536. https://doi.org/10.1186/s12879-023-08505-4

Hill AL, Rosenbloom DIS, Nowak MA et al (2018) Insight into treatment of HIV infection from viral dynamics models. Immunol Rev 285(1):9–25. https://doi.org/10.1111/imr.12698

Höglund E, Øverli Ø, Winberg S (2019) Tryptophan metabolic pathways and brain serotonergic activity: a comparative review. Front Endocrinol (Lausanne) 10:158. https://doi.org/10.3389/fendo.2019.00158

Holec AD, Mandal S, Prathipati PK et al (2017) Nucleotide reverse transcriptase inhibitors: a thorough review, present status and future perspective as HIV therapeutics. Curr HIV Res 15(6):411–421. https://doi.org/10.2174/1570162x15666171120110145

Huang YS, Ogbechi J, Clanchy FI et al (2020) IDO and kynurenine metabolites in peripheral and CNS disorders. Front Immunol 11:388. https://doi.org/10.3389/fimmu.2020.00388

Hudson A, Wu P, Kroll KW et al (2025) Longitudinal analysis of rhesus macaque metabolome during acute SIV infection reveals disruption in broad metabolite classes. J Virol 99(3):e0163424. https://doi.org/10.1128/jvi.01634-24

Hunt PW, Sinclair E, Rodriguez B et al (2014) Gut epithelial barrier dysfunction and innate immune activation predict mortality in treated HIV infection. J Infect Dis 210(8):1228–1238. https://doi.org/10.1093/infdis/jiu238

Jenabian M-A, Patel M, Kema I et al (2013) Distinct tryptophan catabolism and Th17/Treg balance in HIV progressors and elite controllers. PLoS One 8(10):e78146

Jendoubi T (2021) Approaches to integrating metabolomics and multi-omics data: a primer. Meta 11(3). https://doi.org/10.3390/metabo11030184

Jolliffe IT, Cadima J (2016) Principal component analysis: a review and recent developments. Philos Trans A Math Phys Eng Sci 374(2065):20150202. https://doi.org/10.1098/rsta.2015.0202

Joy T, Keogh HM, Hadigan C et al (2007) Dietary fat intake and relationship to serum lipid levels in HIV-infected patients with metabolic abnormalities in the HAART era. AIDS 21(12):1591–1600. https://doi.org/10.1097/QAD.0b013e32823644ff

Kalra S, Kalra B, Agrawal N et al (2011) Understanding diabetes in patients with HIV/AIDS. Diabetol Metab Syndr 3(1):2. https://doi.org/10.1186/1758-5996-3-2

Kandanearatchi A, Brew BJ (2012) The kynurenine pathway in HIV-associated neurocognitive disorders. FEBS J 279(8):1366–1374

Kang S, Tang H (2020) HIV-1 infection and glucose metabolism reprogramming of T cells: another approach toward functional cure and reservoir eradication. Front Immunol 11:572677. https://doi.org/10.3389/fimmu.2020.572677

Kawamura K, Fukusaki E (2024) Novel sampling and gas-phase derivatization strategy: proof-of-concept by profiling ionic polar metabolites using gas chromatography-mass spectrometry. J Biosci Bioeng 138(5):462–468. https://doi.org/10.1016/j.jbiosc.2024.07.019

Koethe JR, Jenkins CA, Petucci C et al (2016) Superior glucose tolerance and metabolomic profiles, independent of adiposity, in HIV-infected women compared with men on antiretroviral therapy. Medicine (Baltimore) 95(19):e3634. https://doi.org/10.1097/md.0000000000003634

Koh A, De Vadder F, Kovatcheva-Datchary P et al (2016) From dietary fiber to host physiology: short-chain fatty acids as key bacterial metabolites. Cell 165(6):1332–1345. https://doi.org/10.1016/j.cell.2016.05.041

Kong SW, Hernandez-Ferrer C (2020) Assessment of coverage for endogenous metabolites and exogenous chemical compounds using an untargeted metabolomics platform. Pac Symp Biocomput 25:587–598

Krupa A, Kowalska I (2021) The kynurenine pathway—new linkage between innate and adaptive immunity in autoimmune endocrinopathies. Int J Mol Sci 22(18). https://doi.org/10.3390/ijms22189879

Kuller LH, Tracy R, Belloso W et al (2008) Inflammatory and coagulation biomarkers and mortality in patients with HIV infection. PLoS Med 5(10):e203. https://doi.org/10.1371/journal.pmed.0050203

Lagathu C, Béréziat V, Gorwood J et al (2019) Metabolic complications affecting adipose tissue, lipid and glucose metabolism associated with HIV antiretroviral treatment. Expert Opin Drug Saf 18(9):829–840. https://doi.org/10.1080/14740338.2019.1644317

Lee KS, Su X, Huan T (2025) Metabolites are not genes - avoiding the misuse of pathway analysis in metabolomics. Nat Metab 7(5):858–861. https://doi.org/10.1038/s42255-025-01283-0

Levy ME, Waters A, Sen S et al (2021) Psychosocial stress and neuroendocrine biomarker concentrations among women living with or without HIV. PLoS One 16(12):e0261746. https://doi.org/10.1371/journal.pone.0261746

Lewis W, Day BJ, Copeland WC (2003) Mitochondrial toxicity of nrti antiviral drugs: an integrated cellular perspective. Nat Rev Drug Discov 2(10):812–822. https://doi.org/10.1038/nrd1201

Li X, Wu T, Jiang Y et al (2018) Plasma metabolic changes in Chinese HIV-infected patients receiving lopinavir/ritonavir based treatment: implications for HIV precision therapy. Cytokine 110:204–212. https://doi.org/10.1016/j.cyto.2018.05.001

Liberti MV, Locasale JW (2016) The Warburg effect: how does it benefit cancer cells? Trends Biochem Sci 41(3):211–218. https://doi.org/10.1016/j.tibs.2015.12.001

Lilo T, Morais CLM, Shenton C et al (2022) Revising Fourier-transform infrared (FT-IR) and Raman spectroscopy towards brain cancer detection. Photodiagn

Photodyn Ther 38:102785. https://doi.org/10.1016/j.pdpdt.2022.102785

Liu H, Wang S, Wang J et al (2025) Energy metabolism in health and diseases. Signal Transduct Target Ther 10(1):69

Liu KH, Nellis M, Uppal K et al (2020) Reference standardization for quantification and harmonization of large-scale metabolomics. Anal Chem 92(13):8836–8844. https://doi.org/10.1021/acs.analchem.0c00338

Loisel-Meyer S, Swainson L, Craveiro M et al (2012) Glut1-mediated glucose transport regulates HIV infection. Proc Natl Acad Sci USA 109(7):2549–2554. https://doi.org/10.1073/pnas.1121427109

Lovelace MD, Varney B, Sundaram G et al (2017) Recent evidence for an expanded role of the kynurenine pathway of tryptophan metabolism in neurological diseases. Neuropharmacol 112(Pt B):373–388. https://doi.org/10.1016/j.neuropharm.2016.03.024

MacDonald MJ (1993) Glucose enters mitochondrial metabolism via both carboxylation and decarboxylation of pyruvate in pancreatic islets. Metabolism 42(10):1229–1231. https://doi.org/10.1016/0026-0495(93)90118-8

Macintyre AN, Gerriets VA, Nichols AG et al (2014) The glucose transporter Glut1 is selectively essential for CD4 T cell activation and effector function. Cell Metab 20(1):61–72. https://doi.org/10.1016/j.cmet.2014.05.004

Marlink R, Kanki P, Thior I et al (1994) Reduced rate of disease development after HIV-2 infection as compared to HIV-1. Science 265(5178):1587–1590. https://doi.org/10.1126/science.7915856

Martens F, Visseren FLJ, Westerink J et al (2025) ESC guidelines for cardiovascular disease management in diabetes. Neth Heart J 33(7–8):216–225

Martin-Iguacel R, Reyes-Urueña J, Bruguera A et al (2022) Determinants of long-term survival in late HIV presenters: the prospective PISCIS cohort study. E Clin Med 52:101600. https://doi.org/10.1016/j.eclinm.2022.101600

Massanella M, Fromentin R, Chomont N (2016) Residual inflammation and viral reservoirs: alliance against an HIV cure. Curr Opin HIV AIDS 11(2):234–241. https://doi.org/10.1097/coh.0000000000000230

Massucci FA, DiNuzzo M, Giove F et al (2013) Energy metabolism and the glutamate–glutamine cycle in the brain: a stoichiometric modeling perspective. BMC Syst Biol 7:103

Mayers JR, Wu C, Clish CB et al (2014) Elevation of circulating branched-chain amino acids is an early event in human pancreatic adenocarcinoma development. Nat Med 20(10):1193–1198. https://doi.org/10.1038/nm.3686

McKenna MC (2013) Glutamate pays its own way in astrocytes. Front Endocrinol (Lausanne) 4:191. https://doi.org/10.3389/fendo.2013.00191

McKnight TR, Yoshihara HA, Sitole LJ et al (2014) A combined chemometric and quantitative NMR analysis of HIV/AIDS serum discloses metabolic alterations associated with disease status. Mol BioSyst 10(11):2889–2897. https://doi.org/10.1039/c4mb00347k

Melendez RI, Roman C, Capo-Velez CM et al (2016) Decreased glial and synaptic glutamate uptake in the striatum of HIV-1 gp120 transgenic mice. J Neurovirol 22:358–365

Miedema F, Hazenberg MD, Tesselaar K et al (2013) Immune activation and collateral damage in AIDS pathogenesis. Front Immunol 4:298

Mikaeloff F, Gelpi M, Benfeitas R et al (2023) Network-based multi-omics integration reveals metabolic at-risk profile within treated HIV-infection. elife 12. https://doi.org/10.7554/eLife.82785

Miller PE, Haberlen SA, Brown TT et al (2016) Brief report: intestinal microbiota-produced trimethylamine-: N:-oxide and its association with coronary stenosis and HIV serostatus. JAIDS 72(1):114–118

Montessori V, Press N, Harris M et al (2004) Adverse effects of antiretroviral therapy for HIV infection. CMAJ 170(2):229–238

Morrison DJ, Preston T (2016) Formation of short chain fatty acids by the gut microbiota and their impact on human metabolism. Gut Microbes 7(3):189–200. https://doi.org/10.1080/19490976.2015.1134082

Mou T, Gao KC, Chen X et al (2025) Clinical events associated with poor CD4(+) T-cell recovery in people living with HIV following ART: a systematic review and meta-analysis. J Infect 90(2):106414. https://doi.org/10.1016/j.jinf.2025.106414

Munshi SU, Rewari BB, Bhavesh NS et al (2013) Nuclear magnetic resonance based profiling of biofluids reveals metabolic dysregulation in HIV-infected persons and those on anti-retroviral therapy. PLoS One 8(5):e64298. https://doi.org/10.1371/journal.pone.0064298

Mutlu EA, Keshavarzian A, Losurdo J et al (2014) A compositional look at the human gastrointestinal microbiome and immune activation parameters in HIV infected subjects. PLoS Pathog 10(2):e1003829. https://doi.org/10.1371/journal.ppat.1003829

Nagana Gowda GA, Raftery D (2014) Quantitating metabolites in protein precipitated serum using NMR spectroscopy. Anal Chem 86(11):5433–5440. https://doi.org/10.1021/ac5005103

Nicholson JK, Holmes E, Kinross J et al (2012) Host-gut microbiota metabolic interactions. Science 336(6086):1262–1267. https://doi.org/10.1126/science.1223813

Nie C, He T, Zhang W et al (2018) Branched chain amino acids: beyond nutrition metabolism. Int J Mol Sci 19(4). https://doi.org/10.3390/ijms19040954

Nix LM, Tien PC (2014) Metabolic syndrome, diabetes, and cardiovascular risk in HIV. Curr HIV/AIDS Rep 11(3):271–278. https://doi.org/10.1007/s11904-014-0219-7

Noor MA, Flint OP, Maa JF et al (2006) Effects of protease inhibitors on glucose uptake and insulin sensitivity. AIDS 20(14):1813–1821

Nyamweya S, Hegedus A, Jaye A et al (2013) Comparing HIV-1 and HIV-2 infection: lessons for viral immunopathogenesis. Rev Med Virol 23(4):221–240. https://doi.org/10.1002/rmv.1739

O'Donnell ST, Ross RP, Stanton C (2019) The progress of multi-omics technologies: determining function in lactic acid bacteria using a systems level approach. Front Microbiol 10:3084. https://doi.org/10.3389/fmicb.2019.03084

Okoye AA, Picker LJ (2013) CD4(+) T-cell depletion in HIV infection: mechanisms of immunological failure. Immunol Rev 254(1):54–64. https://doi.org/10.1111/imr.12066

Osuna-Padilla IA, Rodríguez-Moguel NC, Aguilar-Vargas A et al (2024) Zinc and selenium supplementation on treated HIV-infected individuals induces changes in body composition and on the expression of genes responsible of naïve CD8+ T cells function. Front Nutr 11:1417975. https://doi.org/10.3389/fnut.2024.1417975

Pallotta MT, Rossini S, Suvieri C et al (2022) Indoleamine 2,3-dioxygenase 1 (IDO1): an up-to-date overview of an eclectic immunoregulatory enzyme. FEBS J 289(20):6099–6118. https://doi.org/10.1111/febs.16086

Palmer CS, Hussain T, Duette G et al (2016) Regulators of glucose metabolism in CD4(+) and CD8(+) T cells. Int Rev Immunol 35(6):477–488. https://doi.org/10.3109/08830185.2015.1082178

Pei L, Fukutani KF, Tibúrcio R et al (2021) Plasma metabolomics reveals dysregulated metabolic signatures in HIV-associated immune reconstitution inflammatory syndrome. Front Immunol 12:693074. https://doi.org/10.3389/fimmu.2021.693074

Peltenburg NC, Schoeman JC, Hou J et al (2018) Persistent metabolic changes in HIV-infected patients during the first year of combination antiretroviral therapy. Sci Rep 8(1):16947. https://doi.org/10.1038/s41598-018-35271-0

Pendyala G, Fox HS (2010) Proteomic and metabolomic strategies to investigate HIV-associated neurocognitive disorders. Genome Med 2(3):22. https://doi.org/10.1186/gm143

Petrella G, Montesano C, Lentini S et al (2021) Personalized metabolic profile by synergic use of NMR and HRMS. Molecules 26(14). https://doi.org/10.3390/molecules26144167

Plumb RS, Gethings LA, Rainville PD et al (2023) Advances in high throughput LC/MS based metabolomics: a review. TrAC 160:116954. https://doi.org/10.1016/j.trac.2023.116954

Popper SJ, Sarr AD, Travers KU et al (1999) Lower human immunodeficiency virus (HIV) type 2 viral load reflects the difference in pathogenicity of HIV-1 and HIV-2. J Infect Dis 180(4):1116–1121. https://doi.org/10.1086/315010

Porichis F, Kaufmann DE (2011) HIV-specific CD4 T cells and immune control of viral replication. Curr Opin HIV AIDS 6(3):174–180. https://doi.org/10.1097/COH.0b013e3283454058

Pybus OG, Rambaut A (2009) Evolutionary analysis of the dynamics of viral infectious disease. Nat Rev Genet 10(8):540–550. https://doi.org/10.1038/nrg2583

Qiu S, Cai Y, Yao H et al (2023) Small molecule metabolites: discovery of biomarkers and therapeutic targets. Signal Transduct Target Ther 8(1):132. https://doi.org/10.1038/s41392-023-01399-3

Rhee EP, Cheng S, Larson MG et al (2011) Lipid profiling identifies a triacylglycerol signature of insulin resistance and improves diabetes prediction in humans. J Clin Invest 121(4):1402–1411. https://doi.org/10.1172/jci44442

Rodríguez-Gallego E, Gómez J, Domingo P et al (2018) Circulating metabolomic profile can predict dyslipidemia in HIV patients undergoing antiretroviral therapy. Atherosclerosis 273:28–36. https://doi.org/10.1016/j.atherosclerosis.2018.04.008

Routy JP, Mehraj V, Vyboh K et al (2015) Clinical relevance of kynurenine pathway in HIV/AIDS: an immune checkpoint at the crossroads of metabolism and inflammation. AIDS Rev 17(2):96–106

Roux CG, Mason S, du Toit LDV et al (2024) Comparative effects of Efavirenz and Dolutegravir on metabolomic and inflammatory profiles, and platelet activation of people living with HIV: a pilot study. Viruses 16(9). https://doi.org/10.3390/v16091462

Ru W, Tang SJ (2017) HIV-associated synaptic degeneration. Mol Brain 10(1):40. https://doi.org/10.1186/s13041-017-0321-z

Sáez-Cirión A, Sereti I (2021) Immunometabolism and HIV-1 pathogenesis: food for thought. Nat Rev Immunol 21(1):5–19. https://doi.org/10.1038/s41577-020-0381-7

Samikkannu T, Saiyed ZM, Rao K et al (2009) Differential regulation of indoleamine 2,3-dioxygenase by HIV-1 Tat. AIDS Res Hum Retroviruses 25(3):329–335

Sarkar SA, Wong R, Hackl SI et al (2007) Induction of indoleamine 2,3-dioxygenase by interferon-gamma in human islets. Diabetes 56(1):72–79. https://doi.org/10.2337/db06-0617

Schank M, Zhao J, Moorman JP et al (2021) The impact of HIV- and ART-induced mitochondrial dysfunction in cellular senescence and aging. Cells 10(1). https://doi.org/10.3390/cells10010174

Schrimpe-Rutledge AC, Codreanu SG, Sherrod SD et al (2016) Untargeted metabolomics strategies-challenges and emerging directions. J Am Soc Mass Spectrom 27(12):1897–1905. https://doi.org/10.1007/s13361-016-1469-y

Sears SM, Hewett SJ (2021) Influence of glutamate and GABA transport on brain excitatory/inhibitory balance. Exp Biol Med (Maywood) 246(9):1069–1083. https://doi.org/10.1177/1535370221989263

Sebigi TW, Asia LK, January GG et al (2025) The tryptophan-kynurenine pathway in people living with HIV: a systematic review. Infection. https://doi.org/10.1007/s15010-025-02557-1

Serrano-Villar S, Rojo D, Martínez-Martínez M et al (2016) HIV infection results in metabolic alterations in the gut microbiota different from those induced by other diseases. Sci Rep 6(1):26192. https://doi.org/10.1038/srep26192

Shah ASV, Stelzle D, Lee KK et al. (2018) Global burden of atherosclerotic cardiovascular disease in people living with HIV: a systematic review and meta-analysis. Circulation 138(11):1100–1112

Sheth AN, Ofotokun I, Buchacz K et al (2016) Antiretroviral regimen durability and success in treatment-naive and treatment-experienced patients by year of treatment initiation, United States, 1996-2011. J Acquir Immune Defic Syndr 71(1):47–56. https://doi.org/10.1097/qai.0000000000000813

Singh Y, Chen S (2024) Two-dimensional liquid chromatography advancing metabolomics research. In: Núñez O (ed) Relevant applications of high-performance liquid chromatography in food, environmental, clinical and biological fields. IntechOpen, Rijeka. https://doi.org/10.5772/intechopen.1006558

Sitole L, Fortuin R, Tugizimana F (2022) Metabolic profiling of HIV infected individuals on an AZT-based antiretroviral treatment regimen reveals persistent oxidative stress. J Pharm Biomed Anal 220:114986

Sitole L, Steffens F, Krüger TP et al (2014) Mid-ATR-FTIR spectroscopic profiling of HIV/AIDS sera for novel systems diagnostics in global health. OMICS 18(8):513–523. https://doi.org/10.1089/omi.2013.0157

Sitole LJ, Tugizimana F, Meyer D (2019) Multi-platform metabonomics unravel amino acids as markers of HIV/combination antiretroviral therapy-induced oxidative stress. J Pharm Biomed Anal 176:112796. https://doi.org/10.1016/j.jpba.2019.112796

Somsouk M, Estes JD, Deleage C et al (2015) Gut epithelial barrier and systemic inflammation during chronic HIV infection. AIDS 29(1):43–51

Soares C, Samara A, Yuyun MF et al (2021) Coronary artery calcification in people living with HIV: a systematic review. J Am Heart Assoc 10(19):e019291

Sperk M, Mikaeloff F, Svensson-Akusjärvi S et al (2021) Distinct lipid profile, low-level inflammation, and increased antioxidant defense signature in HIV-1 elite control status. iScience 24(2). https://doi.org/10.1016/j.isci.2021.102111

Srinivasa S, Fitch KV, Lo J et al (2015) Plaque burden in HIV-infected patients is associated with serum intestinal microbiota-generated trimethylamine. AIDS 29(4):443–452

Sultana S, Elengickal A, Bensreti H et al (2023) The kynurenine pathway in HIV, frailty and inflammaging. Front Immunol 14:1244622. https://doi.org/10.3389/fimmu.2023.1244622

Sun Y, Cai M, Liang Y, Zhang Y (2023) Disruption of blood-brain barrier: effects of HIV tat on brain microvascular endothelial cells and tight junction proteins. J Neurovirol 29:658–668

Sun J, Xia Y (2024) Pretreating and normalizing metabolomics data for statistical analysis. Genes Dis 11(3):100979. https://doi.org/10.1016/j.gendis.2023.04.018

Sun L, Shi C, Li X et al (2025) Metabolomic and transcriptomic features of the cardiometabolic disease spectrum in people living with HIV. J Infect Dis Immun. https://doi.org/10.1097/id9.0000000000000164

Sun Y, Ma Y, Lin P et al (2016) Fecal bacterial microbiome diversity in chronic HIV-infected patients in China. Emerg Microbes Infect 5(1):1–7

Tamargo JA, Hernandez-Boyer J, Teeman C et al (2021) Immune activation: a link between food insecurity and chronic disease in people living with human immunodeficiency virus. J Infect Dis 224(12):2043–2052. https://doi.org/10.1093/infdis/jiab257

Tarancón-Diez L, Rull A, Herrero P et al (2021) Early antiretroviral therapy initiation effect on metabolic profile in vertically HIV-1-infected children. J Antimicrob Chemother 76(11):2993–3001. https://doi.org/10.1093/jac/dkab277

Thirion A, Loots DT, Williams ME et al (2024a) An exploratory investigation of the CSF metabolic profile of HIV in a South African paediatric cohort using GCxGC-TOF/MS. Metabolomics 20(2):33

Thirion A, Loots DT, Williams ME et al (2024b) 1H-NMR metabolomics investigation of CSF from children with HIV reveals altered neuroenergetics due to persistent immune activation. FFront Neurosci 18:1270041

Tinarwo P, Zewotir T, Yende-Zuma N et al (2019) An evaluation to determine the strongest CD4 count covariates during HIV disease progression in women in South Africa. Infect Dis Ther 8(2):269–284. https://doi.org/10.1007/s40121-019-0235-4

Triant VA (2013) Cardiovascular disease and HIV infection. Curr HIV/AIDS Rep 10(3):199–206. https://doi.org/10.1007/s11904-013-0168-6

UNAIDS GJUNPoHA (2023) The path that ends AIDS:UNAIDS Global AIDS Update 2023

Vaupel P, Schmidberger H, Mayer A (2019) The Warburg effect: essential part of metabolic reprogramming and central contributor to cancer progression. Int J Radiat Biol 95(7):912–919. https://doi.org/10.1080/09553000.2019.1589653

Vázquez-Castellanos J, Serrano-Villar S, Latorre A et al (2015) Altered metabolism of gut microbiota contributes to chronic immune activation in HIV-infected individuals. Mucosal Immunol 8(4):760–772

Vázquez-Santiago FJ, Noel RJ Jr, Porter JT et al (2014) Glutamate metabolism and HIV-associated neurocognitive disorders. J Neurovirol 20(4):315–331. https://doi.org/10.1007/s13365-014-0258-2

Vuckovic D (2012) Current trends and challenges in sample preparation for global metabolomics using liquid chromatography–mass spectrometry. Anal Bioanal Chem 403(6):1523–1548

Vujkovic-Cvijin I, Dunham RM, Iwai S et al (2013) Dysbiosis of the gut microbiota is associated with HIV disease progression and tryptophan catabolism. Sci Transl Med 5(193):193ra191. https://doi.org/10.1126/scitranslmed.3006438

Wan LY, Lam SM, Huang HH et al (2025) Multi-omics dissection of metabolic dysregulation associated with immune recovery in people living with HIV-1. J Transl Med 23(1):143. https://doi.org/10.1186/s12967-025-06168-0

Wan Z, Su J, Zhu X et al (2024) Distinct Lipidomic profiles between people living with HIV treated with E/C/F/TAF or B/F/TAF: an open-label prospective cohort study. Infect Dis Ther 13(4):727–744. https://doi.org/10.1007/s40121-024-00943-0

Wang X, Mehra S, Kaushal D et al (2021) Abnormal tryptophan metabolism in HIV and mycobacterium tuberculosis infection. Front Microbiol 12:666227. https://doi.org/10.3389/fmicb.2021.666227

Wang Z, Klipfell E, Bennett BJ et al (2011) Gut flora metabolism of phosphatidylcholine promotes cardiovascular disease. Nature 472(7341):57–63. https://doi.org/10.1038/nature09922

Wang Z, Qi Q (2019) Gut microbial metabolites associated with HIV infection. Future Virol 14(5):335–347. https://doi.org/10.2217/fvl-2019-0002

Warburg O (1956) On the origin of cancer cells. Science 123(3191):309–314. https://doi.org/10.1126/science.123.3191.309

Weiss RA (1993) How does HIV cause AIDS? Science 260(5112):1273–1279. https://doi.org/10.1126/science.8493571

Wen Y, Zhu Y, Zhang C et al (2022) Chronic inflammation, cancer development and immunotherapy. Front Pharmacol 13:1040163

Williams ME, Naudé PJW, van der Westhuizen FH (2021) Proteomics and metabolomics of HIV-associated neurocognitive disorders: a systematic review. J Neurochem 157(3):429–449. https://doi.org/10.1111/jnc.15295

Wofford JA, Wieman HL, Jacobs SR et al (2008) IL-7 promotes Glut1 trafficking and glucose uptake via STAT5-mediated activation of Akt to support T-cell survival. Blood 111(4):2101–2111. https://doi.org/10.1182/blood-2007-06-096297

Yan J, Chen D, Ye Z et al (2024) Molecular mechanisms and therapeutic significance of Tryptophan metabolism and signaling in cancer. Mol Cancer 23(1):241. https://doi.org/10.1186/s12943-024-02164-y

Yan J, Risacher SL, Shen L et al (2018) Network approaches to systems biology analysis of complex disease: integrative methods for multi-omics data. Brief Bioinform 19(6):1370–1381. https://doi.org/10.1093/bib/bbx066

Yang L, Chu Z, Liu M et al (2023) Amino acid metabolism in immune cells: essential regulators of the effector functions, and promising opportunities to enhance cancer immunotherapy. J Hematol Oncol 16(1):59. https://doi.org/10.1186/s13045-023-01453-1

Yoo HC, Yu YC, Sung Y et al (2020) Glutamine reliance in cell metabolism. Exp Mol Med 52(9):1496–1516

Yuan Z, Mo C, Kang Y et al (2025) Integration of transcriptomics and metabolomics reveals metabolism dysregulation in HIV-1-infected macrophages. Curr Microbiol 82(5):232. https://doi.org/10.1007/s00284-025-04204-2

Yuan ZW, Gan HL, Jin HL et al (2023) Evaluation of characteristic metabolites of aromatic amino acids in patients with HIV infection at different stages of disease. J Clin Lab Anal 37(1):e24795

Zangerle R, Widner B, Quirchmair G et al (2002) Antiretroviral therapy reduces tryptophan degradation in HIV-1 infection. Clin Immunol 104(3):242–2476

Zhang Y, Chen R, Zhang D et al (2023) Metabolite interactions between host and microbiota during health and disease: which feeds the other? Biomed Pharmacother 160:114295. https://doi.org/10.1016/j.biopha.2023.114295

Zhang D, Hua Z, Li Z (2024) Glutamate and glutamine metabolism in nerve cells. CNS Neurosci Ther30(2):e14617

Zhao S, Li L (2020) Chemical derivatization in LC-MS-based metabolomics study. TrAC 131:115988. https://doi.org/10.1016/j.trac.2020.115988

Zhou Y, Danbolt NC (2014) Glutamate as a neurotransmitter in the healthy brain. J Neural Transm (Vienna) 121(8):799–817. https://doi.org/10.1007/s00702-014-1180-8

Ziegler TR, Judd SE, Ruff JH et al (2017) Amino acid concentrations in HIV-infected youth compared to healthy controls and associations with CD4 counts and inflammation. AIDS Res Hum Retrovir 33(7):681–689. https://doi.org/10.1089/aid.2015.0369

Zuo K, Gao W, Wu Z et al (2024) Evolution of virology: science history through milestones and technological advancements. Viruses 16(3). https://doi.org/10.3390/v16030374

The Role of Lactic Acid Bacteria in Colon Cancer Mitigation: Insights into Mechanisms of Action

Mduduzi P. Mokoena, Paul K. Chelule, Ivy Rukasha, Xolani H. Makhoba, Zukile Mbita, and Ademola O. Olaniran

Abstract

Lactic acid bacteria (LAB) have garnered significant attention for their potential role in mitigating cancer through various mechanisms of action, such as immunoregulation. The LAB, primarily from the genera *Lactobacillus* and *Bifidobacterium*, contribute to gut health by maintaining microbial balance, enhancing gut barrier function, and producing short-chain fatty acids (SCFAs). Colon cancer and colorectal cancer are some of the leading causes of cancer-related morbidity and mortality worldwide, with their incidence rising in many parts of the world. This review explores the ecology and the contributions of LAB species in maintaining the health of the human gut. We further discuss the specific mechanisms through which gut LAB may reduce the risk of colon cancer and/or colorectal cancer, including the modulation of inflammatory pathways, induction of apoptosis in cancer cells, regulation of gut microbiota composition, metabolic activity, and immunomodulation. By understanding these mechanisms, we could better appreciate the potential of LAB as a preventive strategy against cancer and inform future drug strategies and dietary recommendations.

M. P. Mokoena (✉) · I. Rukasha
Department of Pathology, University of Limpopo, Sovenga, South Africa
e-mail: mduduzi.mokoena@ul.ac.za

P. K. Chelule
Department of Public Health, Sefako Makgatho Health Sciences University, Pretoria, South Africa

X. H. Makhoba
Department of Life and Consumer Sciences, University of South Africa (UNISA), Roodepoort, South Africa

Z. Mbita
Department of Biochemistry Microbiology and Biotechnology, University of Limpopo, Sovenga, South Africa

A. O. Olaniran
Discipline of Microbiology, University of KwaZulu-Natal (Westville Campus), Durban, South Africa

Keywords

Lactic acid bacteria · Cancer · Gut microbiota · Immune system

9.1 Introduction

Colon cancer, a growth of cells that begins in a part of the large intestine called the colon, is one of the leading causes of cancer-related morbidity and mortality worldwide, with its incidence rising in many parts of the world (Arnold et al. 2017; Awedew et al. 2022). Like all cancers, early detection of colon and colorectal cancer improves patient survival rates, whereas for late presentation, the prognosis is dismal, with very

© The Author(s) 2026
A. Shonhai et al. (eds.), *Advances in Biochemistry and Molecular Biology to meet Africa's Needs*, Advances in Experimental Medicine and Biology 1507,
https://doi.org/10.1007/978-3-032-24254-9_9

high medical costs. Thus, prevention measures, including diet containing lactic acid bacteria, are of great necessity. The development of cancer is a complex process influenced by a myriad of factors, including genetic predisposition, lifestyle choices, dietary habits, and the composition and health of the gut microbiota (Appunni et al. 2021; Tamayo et al. 2024). The gut microbiota, a diverse community of microorganisms residing in the gastrointestinal tract, plays a crucial role in maintaining host health by aiding digestion, synthesizing vitamins, and regulating immune responses. Disruptions in this microbial balance, known as dysbiosis, are linked to a variety of gastrointestinal disorders, including inflammatory bowel disease and colorectal cancer, a cancer that develops in the colon and/or the rectum (Fan et al. 2021; Santana et al. 2022; Walker and Lawley 2013).

Recent studies have begun to elucidate the significant role that specific gut bacteria can play in cancer prevention and progression (Dougherty and Jobin 2023; El Tekle and Garrett 2023; Sánchez-Alcoholado et al. 2020). Among these, lactic acid bacteria (LAB), primarily belonging to the genera *Lactobacillus* and *Bifidobacterium*, have garnered considerable attention for their potential health benefits. LAB are well-known probiotics that produce lactic acid as a primary fermentation product, contributing to the maintenance of a healthy gut environment (Ağagündüz et al. 2021; Jurášková et al. 2022). LAB produce some antimicrobials (such as bacteriocins), modulate the immune system, and enhance the gut barrier function, thereby positioning themselves as suppressors of infections and reducers of cancer risks (Zhao et al. 2023a, b).

The manifold protective effects of LAB against cancer involve various biological mechanisms (Rytsyk et al. 2020). These include the modulation of inflammatory pathways, induction of apoptosis in cancer cells, regulation of gut microbiota composition, production of short-chain fatty acids (SCFAs), and immunomodulatory effects (Yao et al. 2022). Understanding how these key events regulate gut health is critical for developing effective dietary strategies and therapeutic interventions aimed at reducing cancer

risk. This review aims to provide a comprehensive overview of the mechanisms through which gut lactic acid bacteria may mitigate cancer, emphasizing their potential as a preventive measure in public health.

9.2 Sources and Value of Lactic Acid Bacteria in Human Health

Lactic acid bacteria (LAB) are microorganisms that produce lactic acid as their main fermentation product (Masood et al. 2011). They constitute a large group including *Lactobacillus, Streptococcus, Enterococcus, Lactococcus, Bifidobacterium,* and *Leuconostoc*. Despite their presence in many habitats, they have limited biosynthetic abilities and need a continuous supply of purines, pyrimidines, vitamins, and amino acids from their growth environment (Miranda et al. 2021). LAB are facultative anaerobic, non-spore forming, non-motile organisms and obtain energy through fermenting sugar supplied from the diet, and are widely distributed as normal flora in the intestinal tracts of various animals (Dicks 2023).

The largest genus of LAB is *Lactobacillus*, containing almost 80 species, which are employed in producing different food products such as pickle, sauerkraut, beer, wine, juices, cheese, yogurt, and sausage (Ruiz-Rodríguez et al. 2017). Due to its highly diverse nature, this genus is now re-classified into *Lactobacillus, Paralactobacillus* and 23 novel genera. The twenty three (23) novel genera include: *Amylolactobacillus, Acetilactobacillus, Agrilactobacillus, Apilactobacillus, Bombilactobacillus, Companilactobacillus, Dellaglioa, Fructilactobacillus, Furfurilactobacillus, Holzapfelia, Lacticaseibacillus, Lactiplantibacillus, Lapidilactobacillus, Latilactobacillus, Lentilactobacillus, Levilactobacillus, Ligilactobacillus, Limosilactobacillus, Liquorilactobacillus, Loigolactobacilus, Paucilactobacillus, Schleiferilactobacillus, and Secundilactobacillus* (Chen et al. 2022).

Lactic acid bacteria are ubiquitous in nature, occupying niches in fermented dairy, meat, and vegetable products, the gastrointestinal and urogenital tracts of humans and animals, and soil and water (Li et al. 2024). Ecologically, LAB have transitioned over time from their soil and plant habitats to colonize and adapt in the gut of mammals (George et al. 2018). Consequently, the mammalian intestine is a source of approximately 100 trillion microorganisms that are regarded as part of this organ's microbiota. The microorganisms that colonize the gastrointestinal tract mediate gut health by enhancing metabolism, facilitating digestion, boosting the immune system, and rendering probiotic effects (Koboziev et al. 2014; Kim et al. 2024). The adaptation of microbiota to the mammalian gut is based primarily on three factors, which include adhesion to intestinal cells, ability to bypass host barriers, and substrate fermentation in the gut (Lin et al. 2024).

9.3 Dominant Lactic Acid Bacteria in the Human Intestinal Microbiota

Of the approximately 100 trillion microbiota, about ten billion bacteria belong to 500 species that coexist in the human gastrointestinal tract, and arise from 20 dominant genera, including lactic acid bacteria (LAB) (Masood et al. 2011). Some of these genera are *Bacteroids, Lactobacillus, Clostridium, Fusobacterium, Bifidobacterium, Eubacterium, Peptococcus, Peptostreptococcus, Escherichia*, and *Veillonella*. Microbial balance is critical for maintaining intestinal homeostasis. Live LAB intake through dairy products has several beneficial effects on the human gastrointestinal tract. These range from correction of lactose malabsorption, alleviation of viral and drug-induced diarrhoea, post-operative pouchitis, irritable bowel syndrome, inflammatory bowel syndrome, anti-neoplastic effects on human cell lines, maintenance of normal insulin levels in blood and enhanced absorption of fatty acids through the intestinal wall (Masood et al. 2011). LAB produce these beneficial effects by restoration of normal intestinal flora, elimination of intestinal pathogens, reinforcement of intestinal barrier capacity to foreign antigens, stimulation of nonspecific immunity such as phagocytosis, stimulation of humoral immunity and production of anti-inflammatory products (Hove et al. 1999; Harish and Varghese 2006). Table 9.1 summarizes the genera of lactic acid bacteria dominant in the human and mouse intestine.

The *Lactobacillus/Enterococcus* group represents 1% of the total population of faecal bacteria. *Lactobacillus gasseri* and *Lactobacillus reuteri* are considered true GI commensals while other species, such as *Lactobacillus plantarum, Lactobacillus rhamnosus*, and *Lactobacillus paracasei*, appear to be transient species. Metabolism and growth of these bacteria are accompanied by high carbon fluxes, rapid acidification, and low growth yield. However, species such as *Lactobacillus delbrueckii* and *Streptococcus thermophilus* are highly adapted to the fermentation of dairy substrates and demonstrate advanced specialization towards the utilization of lactose and dairy proteins as sources of amino acids (Derrien and van Hylckama 2015).

A study by Gryaznova et al. (2022) demonstrated that when mice received probiotic supplements, this significantly altered the composition

Table 9.1 Summary of dominant lactic acid bacteria in the intestine and their roles

Sample	Genera	Role	References
Human intestine	*Bacteroids, Lactobacillus, Clostridium, Fusobacterium, Bifidobacterium, Eubacterium, Peptococcus, Peptostreptococcus, Escherichia, Veillonella*	Restoration of normal intestinal flora, elimination of intestinal pathogens, reinforcement of intestinal barrier capacity to foreign antigens, stimulation of nonspecific immunity	Masood et al. (2011)), Harish and Varghese (2006)
Mouse Intestine	*Actinobacteriota, Bacteroidota, Verrucomicrobia, Proteobacteria*	Positive effects on health	Gryaznova et al. (2022)

of the microbiome (the microorganisms in a particular environment, including the body or a part of the body) at the phylum level. Notable changes included an increase in the levels of Actinobacteriota, Bacteroidota, Verrucomicrobia and Proteobacteria, all of which have potentially positive effects on gut health. A decrease in the abundance of bacteria involved in inflammatory processes such as *Nocardioides*, *Helicobacter* and *Mucispirillum* genus was observed in mice fed with lactic acid bacteria (LAB), most likely because of their susceptibility to the products of LAB, such as bacteriocins and lactic acid. For the group of mice that was fed with bifidobacteria, a decrease was seen in the number of members of the *Tyzzerella* and *Akkermansia* genus. Probiotics that are based on members of the Lactobacillaceae family have a more positive effect on the gut microbiome than probiotics that are based on Bifidobacteria.

Unlike in mice, the estimated proportion of LAB in the human proximal small intestine is always subdominant (6%), while their relative abundance in the colon is mostly below 0.5% or non-detectable (George et al. 2018). LAB have also been detected in low quantities in the stomach of humans although the actual numbers are subject to huge variations depending on individuals and their health status, the methodology used, and the type of samples (George et al. 2018). Importantly, LAB that frequently inhabit niches rich in plant-derived carbohydrates produce enzymes that hydrolyze complex plant polymers such as hemicellulose and xylan. These enzyme systems may also contribute to adaptation to the gut environment where plant polymers serve as sole fermentation substrates, explaining the frequent detection of these species in human faecal samples (Derrien and van Hylckama 2015).

Lactobacilli and enterococci are first colonizers and dominant bacteria at infancy; they then become subdominant at toddler and adult ages, mainly due to changes in diet, including the removal of fermented foods. According to Rossi et al. (2016), a metagenomic study identified nearly 60 distinct species of lactobacilli in human faecal samples, not exceeding 0.04% of all bacteria present, corresponding to nearly 8.5 log cells per g from an average bacterial load of 12 log. Within this diversity, one or two major species (*L. rhamnosus* and *L. acidophilus*) were estimated at 8 log. Of the six most represented species, those estimated from 7.5 log were identified as *L. rhamnosus, L. ruminis, L. acidophilus, L. delbrueckii, L. casei, and L. plantarum*, while others were detected below 5 log cells per gram, close to the detection threshold (George et al. 2018).

The abundance of enterococci in the human gut, notably *E. faecium* and *E. faecalis*, is estimated between 4 and 6 log bacteria per gram wet weight (Layton et al. 2010). This accounts for 1% of the relative abundance, which is up to 100 times higher than that of lactobacilli. Although commensal LAB are subdominant, their contribution to gut physiology is significant as they maintain the essential gut microbiome through the fermentation of non-digestible carbohydrates, nutrient absorption, and biosynthesis of some vitamins, and modulation of host immunity (Wong and Yu 2023). However, the interactions between LAB and non-LAB within the gut still remain to be clearly elucidated, with some authors suggesting that LAB fermentations can yield substrates that are converted by non-LAB into butyrate (George et al. 2018).

Burakova et al. (2022) evaluated the bacterial composition of the gut of obese patients before and after 2 weeks of lactic acid bacteria (*Lactobacillus acidophilus, Lactiplantibacillus plantarum, Limosilactobacillus fermentum*, and *Lactobacillus delbrueckii*) intake. An increase in the population of members of the phylum Actinobacteriota was noted in the group taking nutritional supplements, while the number of phylum Bacteroidota decreased in comparison with the control group. There was an increase in potentially beneficial groups of *Bifidobacterium, Blautia, Eubacterium, Anaerostipes, Lactococcus, Lachnospiraceae* ND3007, *Streptococcus, Escherichia-Shigella, and Lachnoclostridium*. A decrease was noted in the genera of *Faecalibacterium, Pseudobutyrivibrio, Subdoligranulum, Faecalibacterium, Clostridium* sensu stricto 1 and 2, *Catenibacterium, Megasphaera, Phascolarctobacterium*, and the

Oscillospiraceae NK4A214 group, which contribute to the development of various metabolic disorders. Thus, modulation of the gut microbiota (the unique combination of microorganisms that exist in a specific environment) by lactic acid bacteria may be one of the ways to treat obesity.

9.4 Probiotic Gut Microbiome and Its Impact on Colon Cancer Mitigation

The onset of cancer in patients is preceded by inflammatory and carcinogenic stimuli, which can influence changes in the composition of gut microbiota (Drago 2019). Hence, the establishment of the tumorigenic state depends on the overabundance of cancerogenic bacteria, which is likely to be the case under dysbiosis conditions. While it is widely demonstrated that epigenetic changes and gene regulations can also occur during the development of colon cancer (CC), it is not clear how some members of the gut microbiome expedite CC pathogenesis in the host. That said, there is a growing consensus that factors such as diet, lifestyle, genetics, oncogenic infection, specific microorganisms or the variability of the microbiome are associated with tumour formation (Drago 2019).

Some of the known cancerogenic bacteria include *Fusobacterium nucleatum* and *Porphyromonas gingivalis*. Their sequences are overabundant in tumours versus matched normal control tissue, and they are positively associated with lymph node metastasis, suggesting that these microorganisms can represent a risk factor for disease progression from adenoma to cancer, possibly affecting patient survival outcomes. On the other hand, some gut microbiota have probiotic effects such as colon cancer prevention, through various ways to maintain gut health and attenuation of dysbiosis, reduction in carcinogenic compound-producing bacteria (e.g., *Escherichia coli* and *Clostridium perfringens*) and carcinogen degradation, the enhanced production of some beneficial molecules (e.g., short-chain fatty acids), and immune modulation (Binmama et al. 2022). Bacterial metabolites of the probiotic gut microbiome are considered bioprotective. Butyrate in particular, produced when specific members of the *Firmicutes* phylum ferment dietary fibre and resistant starches, can modulate inflammation, epithelial proliferation, and apoptosis. Butyrate was recognized by the host colonic receptors GPR109 and GPR43, and mice that were deficient in GPR109 exhibited increased tumour susceptibility and colon tumorigenesis in two different colon cancer mouse models (Ganapathy et al. 2013; Drago 2019).

9.5 Mechanisms of Action of Lactic Acid Bacteria in Colon Cancer Mitigation

Lactic acid bacteria are known to impact the microbial composition of the gut and mitigate colon cancer using several strategies, as depicted in Fig. 9.1. These are described in detail in the sub-sections that follow hereafter.

9.5.1 Modulation of Inflammatory Pathways by Lactic Acid Bacteria

The gut microbiota broadly supports immune system development and function, but studies differ in their emphasis on specific mechanisms. Ferreira et al. (2014) highlight microbiota-driven stimulation of mucin production and lymphoid tissue formation as foundational for gut immune defence, whereas Yoo et al. (2020) and Chu et al. (2023) expand this by illustrating microbiota interactions with enteroendocrine cells (EECs) and toll-like receptors (TLRs) to induce cytokine responses. The latter studies emphasize a signalling cascade involving pro-inflammatory cytokine peptides, which contrast with Bhatia et al. (2022), where LAB reduce inflammation by downregulating pro-inflammatory cytokines (IL-6, TNF-α) and upregulating anti-inflammatory cytokines. This suggests a dual role of gut microbes: modulating baseline immune activation via TLRs and restraining excessive inflammation through cytokine balance.

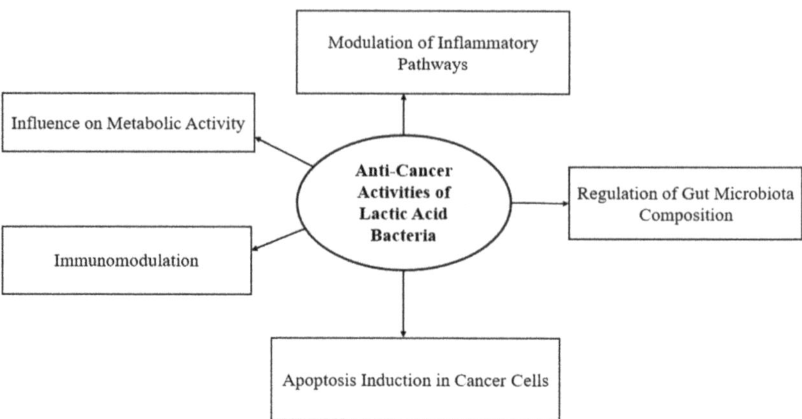

Fig. 9.1 A summary of potential mechanisms utilized by lactic acid bacteria in cancer mitigation

Moreover, the immunomodulatory effects vary depending on bacterial species. Lactobacillus promotes T regulatory cells (Tregs), *Clostridium perfringens* enhances both Tregs and Th 17 cells (increasing IL-17), and Bifidobacterium stimulates B cell sIgA production (Kaur and Ali 2022). This contrasts with LAB strains used in metabolic disease models (Fabersani et al. 2017), where their influence on adipokine secretion, specifically leptin, and macrophage-adipocyte crosstalk reveals immune modulation beyond the gut, indicating species- and strain-specific effects that extend to systemic immune-metabolic regulation.

Menard et al. (2004) used conditioned media from *Bifidobacterium breve* and *Streptococcus thermophiles* in vitro to demonstrate inhibition of TNF-α secretion and NFκB activation in immune cells, with metabolites resistant to digestive enzymes and capable of crossing intestinal barriers. This in vitro epithelial and immune cell co-culture model contrasts with studies employing porcine intestinal epithelial (PIE) cells (Laino et al. 2016; Bhandary et al. 2023), which elucidate the role of LAB exopolysaccharides (EPS) in modulating TLR signalling via RP105/MD1 complex, highlighting host-receptor specificity.

Meanwhile, Fabersani et al. (2017) utilized macrophage-adipocyte co-cultures to capture the complex interplay between immune and metabolic cells, revealing that co-culture systems better mimic in vivo interactions than single cultures.

This diversity in models, from epithelial cell lines to co-cultures and in vivo animal models, demonstrates that while LAB effects are consistently anti-inflammatory, the mechanisms are context-dependent and may vary with cell types and system complexity.

A clear contrast exists between studies focusing on small-molecule metabolites and those emphasizing macromolecular effectors. Menard et al. (2004) identified low-molecular-weight (<300 da) metabolites capable of crossing the epithelial layer to inhibit inflammation. In contrast, Saadat et al. (2019) underscore the role of EPS interacting with LTRs to downregulate inflammatory signalling, with EPS-mediated immunoregulation involving distinct complexes like RP105/MD1.

Further expanding metabolite diversity, Chen et al. (2023) highlight cyclic peptides from LAB, which stand out due to their stability under extreme pH and temperature conditions. These peptides inhibit LPS-induced cytokine production in macrophages and also promote wound healing, indicating a multifunctional role that may surpass traditional small metabolites or EPS in therapeutic potential.

LAB's role varies across disease models. In colon cancer, chronic inflammation is a key driver, and LAB-mediated reduction of pro-inflammatory cytokines may help prevent tumorigenesis (Bhatia et al. 2022). Meanwhile, LAB show promise in inflammatory bowel disease

(IBD), with studies (Caioni et al. 2021; Li et al. 2024) demonstrating enhanced intestinal barrier function and microbiota modulation. However, the differential responses of ulcerative colitis (UC) and Crohn's disease (CD) to LAB treatment suggest that disease-specific microbiota and immune milieu must be considered for effective therapy.

LeBlanc et al. (2020) and Chen et al. (2023) present a therapeutic angle where engineered LAB strains produce antioxidant enzymes and anti-inflammatory cytokines, adding another layer of complexity to their use in IBD, potentially offering personalized therapy or probiotic formulations. Conversely, Fabersani et al. (2017) demonstrate LAB's immunomodulatory potential extends to metabolic diseases through the effects on leptin and macrophage-adipocyte interactions, and mechanisms distinct from gut-centred inflammatory modulation but relevant to systemic immune regulation.

LAB's interaction with traditional herbal medicines represents an emerging avenue for immune modulation. Tiwari et al. (2018) review how LAB fermentation modifies herbal compounds, generating new immunomodulatory metabolites that activate both innate and adaptive immunity with low toxicity. This combinatorial approach contrasts with probiotic monotherapy, suggesting synergy in targeting multiple immune pathways.

In summary, across these studies, LAB and gut microbiota display multifaceted immuno-modulatory effects that depend on bacterial species, metabolites produced, host cell targets, and disease contexts. Experimental models range from simple in vitro systems to complex co-cultures and in vivo models, each revealing unique aspects of LAB-host interactions. These contrasts emphasize the importance of tailored approaches in developing LAB-based therapies for inflammation-related diseases, including colon cancer, IBD, and metabolic syndrome, and highlight promising directions such as engineered probiotics and combined herbal-probiotic strategies.

9.5.2 Pro-apoptotic Activities of Lactic Acid Bacteria

Recent studies converge on the anticancer potential of lactic acid bacteria (LAB), but diverge in targeted cancer types, mechanisms of action, and bioactive delivery formats. Regarding target specificity, Adiyoga et al. (2022) focused on colon cancer (WiDr cell line), using Indonesian meat-based LAB strains (*L. plantarun* IIA5 and *L. acodophilus* IIA-2B4). These strains displayed dose-dependent inhibition of cell viability, with extracts exhibiting stronger cytotoxicity than intracellular ones. Similarly, Salek et al. (2023) observed that cell-free supernatant (CFS) from *Enterococcus faecalis* inhibited both HT-29 (colon) and AGS (gastric) cancer cell lines in a proteinaceous, dose-dependent manner, suggesting peptide/protein components as key effectors.

In contrast, Budu et al. (2024) broadened the scope to melanoma (A375) and breast cancer (MCF-7), showing that *L. plantarum* triggers apoptosis via modulation of intrinsic pathways, upregulating Bcl-2. These findings indicate potential systemic anticancer effects of LAB, not limited to the gut. Concerning mechanistic diversity, a mechanistic divergence is evident when comparing whole-cell components with purified molecular structures. Wang et al. (2023) demonstrated that whole peptidoglycan (WPG) from *L. paracasei* induced apoptosis in HT-29 colon cancer cells by disrupting mitochondrial membrane potential, triggering cytochrome c release, and activating caspase-3 via ROS-mediated mitochondrial pathways. This highlights a contrast with Adiyoga et al. (2022), where the cytotoxicity was not linked to apoptosis pathways but was thought to involve proteinaceous metabolites.

When it comes to the form and function of the bioactives, while Adiyoga et al. and Salek et al. focused on extracellular extracts or CFS (secreted metabolites), Wong et al. emphasized structural cell wall components (WPG), suggesting distinct delivery forms of LAB bioactives: secreted, structural, or both. Niamah et al. (2024) emphasize drug development potential but caution that stability, dosage, and clinical applicability remain challenges.

9.5.3 Regulation of Gut Microbiota Composition by Lactic Acid Bacteria

Beyond anticancer effects, LAB are also central to regulating gut microbiota composition, with implications for both intestinal and systemic health. LAB in fermented foods modulate the gut microbiome and enhance host immunity, as shown in studies like Lin et al. (2023), which used LAB-fermented formula milk (LFM) to enhance gut development and immune function in weaned piglets. LFM altered gut microbiota by promoting short-chain fatty acid (SCFA)-producing genera and regulated immune genes through TLR and IL-17 pathways. These findings are echoed by Musungwe and Dong (2023), who also observed microbiota-driven modulation of innate and adaptive immune signalling, including NF-κB and miRNA pathways in weaned piglets. The animal models used in both studies provide translational insights, though differences in species, diet, and developmental stage must be considered when extending results to humans.

Shen et al. (2022) extended the scope of LAB's gut microbiota to obesity management, noting that LAB-produced metabolites, such as SCFAs, linoleic acid, and GABA, regulate lipid metabolism through AMPK and PPAR signalling. This presents a sharp contrast to the cancer studies, where immune modulation or apoptosis was central. Instead, the mechanism here hinges on metabolic regulation via microbial metabolite-host interactions.

Finally, Zhan et al. (2024) synthesized these insights, proposing that LAB could indirectly lower cancer risk by reshaping gut microbiota composition by suppressing carcinogenic bacteria like some *E.coli* strains. This echoes earlier cancer studies (e.g., Adiyoga et al. 2022), but introduces a microbiota-mediated preventive approach rather than direct tumour suppression.

While all these studies demonstrate beneficial effects of LAB, they vary significantly in models used, bioactive forms, mechanisms of action, and target outcomes. Importantly, the cellular specificity of LAB effects remains a recurring strength: across cancer studies (Salek et al. 2023; Wang et al. 2022), LAB selectively inhibit tumour cells while sparing normal cells and have an advantage over conventional chemotherapy. However, as noted by Niamah et al. (2024), clinical translation remains limited, with key gaps in understanding dosage, stability, and delivery.

The collective body of research underscores LAB's multi-modal potential in disease prevention and treatment, ranging from cancer suppression via apoptosis and bacteriocins to gut microbiota regulation for metabolic and immune health. However, distinctions in experimental approaches and bioactive forms lead to divergent findings, underlining the need for standardized comparative studies and clinical trials to validate therapeutic potential. The integration of LAB into functional foods, adjuvant therapies, or even standalone interventions holds promise, but must overcome formulation and delivery challenges from laboratory to clinic.

9.5.4 Influence of Lactic Acid Bacteria on Metabolic Activity

Phenolic compounds in plant-based foods are well-documented for their contribution to sensory characteristics such as flavour, astringency, and colour, while also offering antioxidant-mediated health benefits (Wang et al. 2022; Rodríguez et al. 2009). These properties make them important not only for food quality but also for their potential protective role against carcinogens and mutagenesis. However, the interaction between phenolic compounds and lactic acid bacteria (LAB) remains underexplored, particularly regarding the impact on LAB growth, viability, and metabolic pathways involved in phenolic degradation. While Rodríguez et al. (2009) suggest that *Lactobacillus plantarum* can degrade phenolic compounds to enhance antioxidant capacity and aroma, the mechanistic pathways remain largely undefined.

In contrast, Tang et al. (2023) broaden the focus from phenolics to LAB-derived metabolites such as organic acids, SCFAs, bacteriocins, and vitamins, underlining their multifaceted roles in gut health, immune modulation, and host metabolism. This study emphasizes LAB's systemic health contributions beyond their interaction with dietary polyphenols. Importantly, these beneficial roles contrast sharply with recent findings by Colbert et al. (2023), which reveal that certain LAB-like species, particularly *Lactobacillus inners*, may have detrimental roles in cancer therapy resistance. Specifically, *L. inners* was implicated in metabolic rewiring within the tumour microenvironment, decreasing survival rates and increasing resistance to chemo- and radiotherapy in cervical and other cancers.

This juxtaposition reveals a significant context-dependent duality in LAB function. While some strains like *L. plantarum* are linked to health promotion, others like *L. iners* may foster tumour persistence and therapy resistance, suggesting that not all LAB confer uniformly beneficial effects. The potential for LAB to produce lactate, which can modulate immune responses and metabolic pathways in colorectal cancer (CRC), adds another layer of complexity (Sun et al. 2023). These findings challenge the conventional view of LAB as universally beneficial and underscore the need to distinguish between strains and their ecological contexts.

Further expanding on this complexity, postbiotics, which are non-viable bacterial products or metabolic byproducts, have emerged as promising adjuncts to cancer therapies. Rafique et al. (2023) and Wasiak et al. (2024) demonstrated that postbiotics from *L. plantarum* and *L. rhamnosus* significantly enhanced the cytotoxic effects of tamoxifen and experimental drugs on MCF-7 breast cancer cells. Unlike the ambiguous effects of live LAB stains like *L. iners*, these postbiotics increased apoptosis without affecting normal cells, indicating selective bioactivity with reduced risk of promoting resistance or adverse tumour interactions.

The preventive role of LAB in colorectal cancer has also drawn both support and skepticism. Rafter (2003) and Zhong et al. (2014) argue that while direct clinical evidence is lacking, indirect mechanistic data suggest LAB may reduce CRC risk via SCFA production, modulation of histone deacetylase activity, and elicit favourable changes in gene expression linked to cell differentiation. These proposed benefits echo findings from Tang et al. (2023), yet stand in partial contradiction to Colbert et al.'s (2023) report, which implicates lactate, also a product of LAB metabolism, in tumour progression and therapeutic resistance. Therefore, contextual factors, including bacterial strain, site of colonization, and disease stage, appear to critically influence LAB effects. The foregoing contrasts are summarized in Table 9.2. By shifting the focus to comparative mechanisms and outcomes, this analysis underscores the strain-specific, disease-specific, and context-dependent nature of LAB and their derivatives. Future research must refine our understanding of these complexities to harness the therapeutic potential of LAB while mitigating unintended consequences in disease contexts.

Table 9.2 Summary of the metabolic roles of LAB in colorectal cancer

Aspect	Positive/beneficial findings	Contrasting or adverse findings
Phenolic interaction	*L. plantarum* enhances aroma and antioxidant activity (Rodríguez et al. 2009)	Mechanistic pathways unclear Generalizability limited
Health effects of LAB	LAB supports immune function, metabolic health (Tang et al. 2023)	*L. iners* is linked to therapy resistance and poor prognosis (Colbert et al. 2023)
Lactate role	SCFA production linked to anticancer gene regulation (Rafter 2003)	Lactate fuels cancer metabolism and immune evasion (Sun et al. 2023)
Postbiotics	Induce apoptosis, enhance drug efficacy, safe for normal cells (Wasiak et al. 2024)	No adverse effects reported; promising adjunct therapy
Cancer prevention evidence	Indirect support for CRC risk reduction (Zhong et al. 2014)	No direct human evidence Potential strain-specific risks

9.5.5 Immunomodulation by Lactic Acid Bacteria

Despite advances in treatment, cancer remains a global health challenge, with current therapies often limited by toxicity, resistance, and economic inaccessibility (Bedada et al. 2020; Chin et al. 2024). These limitations have driven the search for safer and more sustainable alternatives. Among emerging candidates, probiotics and their derivatives, including heat-killed bacteria as well as postbiotics such as SCFAs, proteins, and ferrichrome, are gaining attention for their anticancer potential (Bedada et al. 2020; Nair et al. 2014; Bahuguna and Dubey 2023). However, the efficacy and safety of such interventions in human clinical settings remain underexplored, with a clear need for rigorous clinical validation.

The value of lactic acid bacteria (LAB) lies in their multifaceted mechanisms of action and generally safety, which contrasts starkly with the cytotoxic profile of traditional chemotherapeutics. LAB exhibit strain-specific abilities to modulate the gut microbiota, influence immune responses, and inhibit tumour proliferation without damaging normal tissues (Delesa 2017; Zhao et al. 2023a, b). This immunomodulatory potential is particularly compelling when viewed against the backdrop of conventional treatments, which often compromise immune function and tissue integrity.

Importantly, strain-specificity is critical when evaluating LAB for cancer therapy. Studies show that not all LAB strains exert beneficial effects, and the context of host-microbe interactions is essential for determining outcomes. For instance, Herich and Levkut (2002) highlighted the role of LAB in mucosal immunity, emphasizing their ability to support gut barrier integrity, stimulate immune responses, and regulate systemic immune function. However, these benefits are strain-dependent, and universal claims about LAB effectiveness are premature without mechanistic specificity.

A more targeted exploration of immunological pathways is offered by Saito et al. (2024), who demonstrated that heat-killed *Lactococcus* *lactis* subsp. *cremoris* C60 (HK-C60) could enhance antigen presentation and CD8+ T cell responses via upregulation of the 20S immunoproteasome in dendritic cells. This mechanistic clarity, particularly in enhancing MHC class I-restricted antigen presentation, offers a significant advancement over general probiotic studies, which often rely on observational or correlational data. HK-C60 was shown to attenuate tumour growth in the melanoma mouse model, suggesting that non-viable LAB can exert immunotherapeutic effects comparable to, or even exceeding, those of some live strains.

These findings are complemented by the research of Wang et al. (2020) and Khaledi et al. (2023), who demonstrated that LAB can stimulate immunoglobulin production, activate dendritic cells, and inhibit cancer metastasis and proliferation. However, unlike Saito et al.'s work, these studies are less specific in their molecular mechanisms, pointing to the need for standardized, mechanistically driven research models. Table 9.3 presents a summary of findings of the studies exploring the immunomodulation by LAB.

9.6 Conclusion

Lactic acid bacteria (LAB) are a diverse group of microorganisms that primarily ferment sugars into lactic acid. They are ubiquitous in fermented foods and known for their health benefits, particularly in promoting gut health. LAB have emerged as multifunctional agents in immune modulation, disease prevention, and therapeutic intervention. While a general consensus supports their anti-inflammatory and immunostimulatory roles, recent research highlights substantial variability depending on specific bacterial strain, disease context, and experimental conditions. For instance, *Lactobacillus plantarum* has consistently been associated with beneficial outcomes such as enhanced antioxidant activity, apoptosis induction in cancer cells, and improved gut barrier function. In contrast *Lactobacillus iners* has been linked to adverse outcomes, including metabolic reprogramming of tumours and resistance to

Table 9.3 Consistencies and divergences of findings regarding LAB immunomodulation in cancer

Aspect	Supportive evidence	Limitations or contrasts
Therapeutic potential	LAB and postbiotics reduce tumour growth (Bedada et al. 2020; Saito et al. 2024)	Lack or paucity of clinical trials in humans
Immune modulation	LAB enhance dendritic cell activation and CD8+ T-cell response (Saito et al. 2024)	Many studies lack mechanistic specificity (Khaledi et al. 2023)
Toxicity and safety	LAB exhibit low systemic toxicity and do not harm normal cells (Delesa 2017)	Strain-specific effects not fully characterized
Mode of delivery	Both live and heat-killed LAB effective (Nair et al. 2014; Saito et al. 2024)	Few comparative studies on live vs. heat-killed LAB efficacy
Scope of action	Target both innate and adaptive immunity, as well as microbiota balance (Wang et al. 2020)	Variable host response depending on gut ecology and strain

cancer therapies. This divergence challenges the notion of LAB as uniformly beneficial, emphasizing the importance of precise strain selection and mechanistic understanding when designing probiotic-based therapeutic interventions.

Mechanistically, LAB operate through both low- and high-molecular-weight bioactives, such as short-chain fatty acids (SCFAs), cyclic peptides, and exopolysaccharides, which influence toll-like receptor signalling, cytokine profiles, and mitochondrial apoptotic pathways. Their ability to both stimulate the immune responses and suppress chronic inflammation positions them as potential therapeutic tools in conditions like inflammatory bowel disease, metabolic syndrome, and colorectal cancer. However, findings vary across model systems, from in vitro cell cultures to complex in vivo and co-culture models, underscoring the need for translational research that bridges laboratory findings with clinical relevance.

That said, to fully harness LAB's therapeutic potential, future work must focus on standardized methodologies, disease-specific targeting, and safe, effective delivery formats, whether as live cultures, heat-killed, or postbiotic compounds. Continued collaboration among researchers, clinicians, and public health advocates will be critical in harnessing the full potential of lactic acid bacteria in the fight against colon cancer.

Conflict of Interest Author MPM declares that he has no conflict of interest. Author PKC declares that he has no conflict of interest. Author IR declares that she has no conflict of interest. Author XHM declares that he has no conflict of interest. Author ZM declares that he has no conflict of interest. Author AOO declares that he has no conflict of interest.

Ethical Approval This chapter does not contain any studies with human participants performed by any of the authors.

Funding This study was funded by the University of Limpopo, Sefako Makgatho Health Sciences University, the University of South Africa, and the University of KwaZulu-Natal.

References

Adiyoga R, Arief II, Budiman C, Abidin Z (2022) In vitro anticancer potentials of *Lactobacillus plantarum* IIA-1A5 and *Lactobacillus acidophilus* IIA-2B4 extracts against WiDr human colon cancer cell line. Food Sci Technol 42:e87221

Ağagündüz D, Yılmaz B, Şahin TÖ, Güneşlio BE, Ayte Ş, Russo P, Özogul F (2021) Dairy lactic acid bacteria and their potential function in dietetics: the food–gut–health axis. Foods 10(12):3099

Amer MN, Elmaghraby MM, Abdellatif AA, Salem SR, Salama MM, Maksoud OAA, Nasr RM, Amer MN, Marghany MM, Awad HM (2024) Probiotics as promoters of human health. Nov Res Microbiol J 8(5):2580–2603

Appunni S, Rubens M, Ramamoorthy V, Tonse R, Saxena A, McGranaghan P, Kotecha R (2021) Emerging evidence on the effects of dietary factors on the gut microbiome in colorectal cancer. Front Nutr 8:718389

Arnold M, Sierra MS, Laversanne M, Soerjomataram I, Jemal A, Bray F (2017) Global patterns and trends in colorectal cancer incidence and mortality. Gut 66(4):683–691

Awedew AF, Asefa Z, Belay WB (2022) Burden and trend of colorectal cancer in 54 countries of Africa 2010–2019: a systematic examination for Global Burden of Disease. BMC Gastroenterol 22(1):204

Azad MAK, Sarker M, Li T, Yin J (2018) Probiotic species in the modulation of gut microbiota: an overview. Biomed Res Int 2018(1):9478630

Bahuguna A, Dubey SK (2023) Overview of the mechanistic potential of probiotics and prebiotics in cancer chemoprevention. Mol Nutr Food Res 67(19):2300221

Barbara G, Barbaro MR, Fuschi D, Palombo M, Falangone F, Cremon C et al (2021) Inflammatory and microbiota-related regulation of the intestinal epithelial barrier. Front Nutr 8:718356

Bedada TL, Feto TK, Awoke KS, Garedew AD, Yifat FT, Birri DJ (2020) Probiotics for cancer alternative prevention and treatment. Biomed Pharmacother 129:110409

Bhandary T, Kurian C, Muthu M, Anand A, Anand T, Paari KA (2023) Exopolysaccharides derived from probiotic bacteria and their health benefits. J Pure Appl Microbiol 17(1):35

Bhatia R, Sharma S, Bhadada SK, Bishnoi M, Kondepudi KK (2022) Lactic acid bacterial supplementation ameliorated the lipopolysaccharide-induced gut inflammation and dysbiosis in mice. Front Microbiol 13:930928

Binmama S, Dang CP, Visitchanakun P, Hiengrach P, Somboonna N, Cheibchalard T, Leelahavanichkul A (2022) Beta-glucan from *S. cerevisiae* protected AOM-induced colon cancer in cGAS-deficient mice partly through dectin-1-manipulated macrophage cell energy. Int J Mol Sci 23(18):10951

Budu O, Mioc A, Soica C, Caruntu F, Milan A, Oprean C, Banciu C (2024) *Lactiplantibacillus plantarum* induces apoptosis in melanoma and breast cancer cells. Microorganisms 12(1):182

Burakova I, Smirnova Y, Gryaznova M, Syromyatnikov M, Chizhkov P, Popov E, Popov V (2022) The effect of short-term consumption of lactic acid bacteria on the gut microbiota in obese people. Nutrients 14(16):3384

Caioni G, Viscido A, d'Angelo M, Panella G, Castelli V, Merola C, Benedetti E (2021) Inflammatory bowel disease: new insights into the interplay between environmental factors and PPARγ. Int J Mol Sci 22(3):985

Caldarelli M, Rio P, Marrone A, Giambra V, Gasbarrini A, Gambassi G, Cianci R (2024) Inflammaging: the next challenge—exploring the role of gut microbiota environmental factors and sex differences. Biomedicine 12(8):1716

Chen B, Loo BZL, Cheng YY, Song P, Fan H, Latypov O, Kittelmann S (2022) Genome-wide high-throughput signal peptide screening via plasmid pUC256E improves protease secretion in *Lactiplantibacillus plantarum* and *Pediococcus acidilactici*. BMC Genomics 23(1):48

Chen Y, Gao H, Zhao J, Ross RP, Stanton C, Zhang H, Yang B (2023) Exploiting lactic acid bacteria for inflammatory bowel disease: a recent update. Trends Food Sci Technol 138:126–140

Chin KW, Khoo SC, Paul RPM, Luang-In V, Lam SD, Ma NL (2024) Potential of synbiotics and probiotics as chemopreventive agent. In: Probiotics antimicrob proteins, pp 1–17

Chu J, Feng S, Guo C, Xue B, He K, Li L (2023) Immunological mechanisms of inflammatory diseases caused by gut microbiota dysbiosis: A review. Biomed Pharmacother 164:114985

Colbert LE, El Alam MB, Wang R, Karpinets T, Lo D, Lynn EJ, Klopp AH (2023) Tumor-resident *Lactobacillus iners* confer chemoradiation resistance through lactate-induced metabolic rewiring. Cancer Cell 41(11):1945–1962

De Filippis F, Pasolli E, Ercolini D (2020) The food-gut axis: lactic acid bacteria and their link to food the gut microbiome and human health. FEMS Microbiol Rev 44(4):454–489

Delesa DA (2017) Bacteriocin as an advanced technology in food industry. Int J Adv Res Biol Sci 4(12):178–190

Derrien M, van Hylckama VJE (2015) Fate activity and impact of ingested bacteria within the human gut microbiota. Trends Microbiol 23(6):354–366

Dicks LMT (2023) Interactions between gut microbiota and the central nervous system with emphasis on quorum sensing between commensal lactic acid bacteria and human cells. Stellenbosch University

Dougherty MW, Jobin C (2023) Intestinal bacteria and colorectal cancer: etiology and treatment. Gut Microbes 15(1):2185028

Drago L (2019) Probiotics and colon cancer. Microorganisms 7(3):66

El Tekle G, Garrett WS (2023) Bacteria in cancer initiation promotion and progression. Nat Rev Cancer 23(9):600–618

Fabersani E, Abeijon-Mukdsi MC, Ross R, Medina R, González S, Gauffin-Cano P (2017) Specific strains of lactic acid bacteria differentially modulate the profile of adipokines in vitro. Front Immunol 8:266

Fan X, Jin Y, Chen G, Ma X, Zhang L (2021) Gut microbiota dysbiosis drives the development of colorectal cancer. Digestion 102(4):508–515

Ferreira CM, Vieira AT, Vinolo MAR, Oliveira FA, Curi R, Martins FDS (2014) The central role of the gut microbiota in chronic inflammatory diseases. J Immunol Res 1:689492

Ganapathy V, Thangaraju M, Prasad PD, Martin PM, Singh N (2013) Transporters and receptors for short-chain fatty acids as the molecular link between colonic bacteria and the host. Curr Opin Pharmacol 13(6):869–874

George F, Daniel C, Thomas M, Singer E, Guilbaud A, Tessier FJ, Foligné B (2018) Occurrence and dynamism of lactic acid bacteria in distinct ecological niches: a multifaceted functional health perspective. Front Microbiol 9:2899

Gryaznova M, Dvoretskaya Y, Burakova I, Syromyatnikov M, Popov E, Kokina A, Popov V (2022) Dynamics of changes in the gut microbiota of healthy mice fed with lactic acid bacteria and bifidobacteria. Microorganisms 10(5):1020

Harish K, Varghese T (2006) Probiotics in humans–evidence based review. Calicut Med J 4(4):e3

Herich R, Levkut M (2002) Lactic acid bacteria probiotics and immune system. Vet Med 47(6):169–180

Hove H, Nørgaard H, Brøbech Mortensen P (1999) Lactic acid bacteria and the human gastrointestinal tract. Eur J Clin Nutr 53(5):339–350

Islam MR, Akash S, Rahman MM, Nowrin FT, Akter T, Shohag S, Simal-Gandara J (2022) Colon cancer and colorectal cancer: prevention and treatment by potential natural products. Chem Biol Interact 368:110170

Jurášková D, Ribeiro SC, Silva CC (2022) Exopolysaccharides produced by lactic acid bacteria: from biosynthesis to health-promoting properties. Foods 11(2):156

Kaur H, Ali SA (2022) Probiotics and gut microbiota: mechanistic insights into gut immune homeostasis through TLR pathway regulation. Food Funct 13(14):7423–7447

Khaledi K, Hoseini R, Gharzi A (2023) Effects of aerobic training and vitamin D supplementation on glycemic indices and adipose tissue gene expression in type 2 diabetic rats. Sci Rep 13(1):10218

Kim J, Lee HK (2022) Potential role of the gut microbiome in colorectal cancer progression. Front Immunol 12:807648

Kim DY, Lee SY, Lee JY, Whon TW, Lee JY, Jeon CO, Bae JW (2024) Gut microbiome therapy: fecal microbiota transplantation vs live biotherapeutic products. Gut Microbes 16(1):2412376

Koboziev I, Webb CR, Furr KL, Grisham MB (2014) Role of the enteric microbiota in intestinal homeostasis and inflammation. Free Radic Biol Med 68:122–133

Kustrimovic N, Balkhi S, Bilato G, Mortara L (2024) Gut microbiota and immune system dynamics in Parkinson's and Alzheimer's diseases. Int J Mol Sci 25(22):12164

Layton BA, Walters SP, Lam LH, Boehm AB (2010) Enterococcus species distribution among human and animal hosts using multiplex PCR. J Appl Microbiol 109(2):539–547

LeBlanc JG, Levit R, Savoy de Giori G, de Moreno de LeBlanc A (2020) Application of vitamin-producing lactic acid bacteria to treat intestinal inflammatory diseases. Appl Microbiol Biotechnol 104:3331–3337

Li W, Wu Q, Kwok LY, Zhang H, Gan R, Sun Z (2024) Population and functional genomics of lactic acid bacteria an important group of food microorganism: current knowledge challenges and perspectives. Food Front 5(1):3–23

Lin A, Yan X, Xu R, Wang H, Su Y, Zhu W (2023) Effects of lactic acid bacteria-fermented formula milk supplementation on colonic microbiota and mucosal transcriptome profile of weaned piglets. Animal 17(9):100959

Lin Q, Lin S, Fan Z, Liu J, Ye D, Guo P (2024) A review of the mechanisms of bacterial colonization of the mammal. Gut Microorganisms 12(5):1026

Masood MI, Qadir MI, Shirazi JH, Khan IU (2011) Beneficial effects of lactic acid bacteria on human beings. Crit Rev Microbiol 37(1):91–98

Menard S, Candalh C, Bambou JC, Terpend K, Cerf-Bensussan N, Heyman M (2004) Lactic acid bacteria secrete metabolites retaining anti-inflammatory properties after intestinal transport. Gut 53(6):821–828

Miranda C, Contente D, Igrejas G, Câmara SP, Dapkevicius MDLE, Poeta P (2021) Role of exposure to lactic acid bacteria from foods of animal origin in human health. Foods 10(9):2092

Musungwe L, Dong L (2023) Nutritional regulation of gut health through the gastrointestinal tract microbiota in weaning piglets. S Afr J Anim Sci 53(2):139–147

Nair N, Kasai T, Seno M (2014) Bacteria: prospective savior in battle against cancer. Anticancer Res 34(11):6289–6296

Niamah AK, Al-Sahlany STG, Verma DK, Shukla RM, Patel AR, Tripathy S, Aguilar CN (2024) Emerging lactic acid bacteria bacteriocins as anti-cancer and anti-tumor agents for human health. Heliyon 10(17):e37054

Pessione E (2012) Lactic acid bacteria contribution to gut microbiota complexity: lights and shadows. Front Cell Infect Microbiol 2:86

Rafique N, Jan SY, Dar AH, Dash KK, Sarkar A, Shams R, Hussain SZ (2023) Promising bioactivities of postbiotics: a comprehensive review. J Agric Food Res 14:100708

Rafter J (2003) Probiotics and colon cancer. Best Pract Res Clin Gastroenterol 17(5):849–859

Rodríguez H, Curiel JA, Landete JM, de las Rivas B, de Felipe FL, Gómez-Cordovés C, Muñoz R (2009) Food phenolics and lactic acid bacteria. Int J Food Microbiol 132(2–3):79–90

Rossi M, Martínez-Martínez D, Amaretti A, Ulrici A, Raimondi S, Moya A (2016) Mining metagenomic whole genome sequences revealed subdominant but constant Lactobacillus population in the human gut microbiota. Environ Microbiol Rep 8(3):399–406

Rossino G, Marchese E, Galli G, Verde F, Finizio M, Serra M, Collina S (2023) Peptides as therapeutic agents: challenges and opportunities in the green transition era. Molecules 28(20):7165

Ruiz-Rodríguez L, Bleckwedel J, Eugenia Ortiz M, Pescuma M, Mozzi F (2017) Lactic acid bacteria. In: Industrial biotechnology: microorganisms, vol 1. Wiley-VCH Verlag GmbH & Co. KGaA, pp 395–451

Rytsyk O, Soroka Y, Shepet I, Vivchar Z, Andriichuk I, Lykhatskyi P, Lisnychuk N (2020) Experimental evaluation of the effectiveness of resveratrol as an antioxidant in colon cancer prevention. Nat Prod Commun 15(6):1934578X20932742

Saadat YR, Khosroushahi AY, Gargari BP (2019) A comprehensive review of anticancer immunomodulatory and health beneficial effects of the lactic acid bacteria exopolysaccharides. Carbohydr Polym 217:79–89

Saito S, Cao DY, Maekawa T, Tsuji NM, Okuno A (2024) Lactococcus lactis subsp cremoris C60 upregulates macrophage function by modifying metabolic preference in enhanced anti-tumor immunity. Cancer 16(10):1928

Salek F, Mirzaei H, Khandaghi J, Javadi A, Nami Y (2023) Apoptosis induction in cancer cell lines and anti-inflammatory and anti-pathogenic properties of proteinaceous metabolites secreted from potential probiotic *Enterococcus faecalis* KUMS-T48. Sci Rep 13(1):7813

Sánchez-Alcoholado L, Ramos-Molina B, Otero A, Laborda-Illanes A, Ordóñez R, Medina JA, Queipo-Ortuño MI (2020) The role of the gut microbiome in colorectal cancer development and therapy response. Cancer 12(6):1406

Santana PT, Rosas SLB, Ribeiro BE, Marinho Y, de Souza HS (2022) Dysbiosis in inflammatory bowel disease: pathogenic role and potential therapeutic targets. Int J Mol Sci 23(7):3464

Sharma H, Ozogul F, Bartkiene E, Rocha JM (2023) Impact of lactic acid bacteria and their metabolites on the techno-functional properties and health benefits of fermented dairy products. Crit Rev Food Sci Nutr 63(21):4819–4841

Shen H, Zhao Z, Zhao Z, Chen Y, Zhang L (2022) Native and engineered probiotics: promising agents against related systemic and intestinal diseases. Int J Mol Sci 23(2):594

Sun Q, Wu J, Zhu G, Li T, Zhu X, Ni B, Li J (2023) Lactate-related metabolic reprogramming and immune regulation in colorectal cancer. Front Endocrinol 13:1089918

Tamayo M, Olivares M, Ruas-Madiedo P, Margolles A, Espín JC, Medina I, Sanz Y (2024) How diet and lifestyle can fine-tune gut microbiomes for healthy aging. Annu Rev Food Sci Technol 15(1):283–305

Tang Z, Wang Q, Zhao Z, Shen N, Qin Y, Lin W, Huang L (2023) Evaluation of fermentation properties antioxidant capacity in vitro and in vivo and metabolic profile of a fermented beverage made from apple and cantaloupe. LWT 179:114661

Tiwari R, Latheef SK, Ahmed I, Iqbal HM, Bule MH, Dhama K, Farag MR (2018) Herbal immunomodulators-a remedial panacea for designing and developing effective drugs and medicines: current scenario and future prospects. Curr Drug Metab 19(3):264–301

Walker AW, Lawley TD (2013) Therapeutic modulation of intestinal dysbiosis. Pharmacol Res 69(1):75–86

Wang Y, Xiang Y, Xin VW, Wang XW, Peng XC, Liu XQ, Xin HW (2020) Dendritic cell biology and its role in tumor immunotherapy. J Hematol Oncol 13:1–18

Wang Y, Tuccillo F, Lampi AM, Knaapila A, Pulkkinen M, Kariluoto S, Katina K (2022) Flavor challenges in extruded plant-based meat alternatives: a review. Compr Rev Food Sci Food Saf 21(3):2898–2929

Wang S, Shan Y, Zhang S, Zhang L, Jiao Y, Xue D, Yi H (2023) *Lactobacillus paracasei* subsp *paracasei* X12 strain induces apoptosis in HT-29 cells through activation of the mitochondrial pathway. Nutrients 15(9):2123

Wasiak J, Głowacka P, Pudlarz A, Pieczonka AM, Dzitko K, Szemraj J, Witusik-Perkowska M (2024) Lactic acid bacteria-derived Postbiotics as adjunctive agents in breast cancer treatment to boost the antineoplastic effect of a conventional therapeutic comprising tamoxifen and a new drug candidate: an Aziridine–hydrazide hydrazone derivative. Molecules 29(10):2292

Wong CC, Yu J (2023) Gut microbiota in colorectal cancer development and therapy. Nat Rev Clin Oncol 20(7):429–452

Yao Y, Cai X, Fei W, Ye Y, Zhao M, Zheng C (2022) The role of short-chain fatty acids in immunity inflammation and metabolism. Crit Rev Food Sci Nutr 62(1):1–12

Yoo JY, Groer M, Dutra SVO, Sarkar A, McSkimming DI (2020) Gut microbiota and immune system interactions. Microorganisms 8(10):1587

Zhan ZS, Zheng ZS, Shi J, Chen J, Wu SY, Zhang SY (2024) Unraveling colorectal cancer prevention: the vitamin D-gut flora-immune system nexus. World J Gastroenterol 16(6):2394

Zhang J, Xiao Y, Wang H, Zhang H, Chen W, Lu W (2023) Lactic acid bacteria-derived exopolysaccharide: formation, immunomodulatory ability, health effects, and structure-function relationship. Microbiol Res 274:127432

Zhao J, Liao Y, Wei C, Ma Y, Wang F, Chen Y, Tang D (2023a) Potential ability of probiotics in the prevention and treatment of colorectal cancer. Clin Med Insights Oncol 17:11795549231188225

Zhao LY, Mei JX, Yu G, Lei L, Zhang WH, Liu K, Hu JK (2023b) Role of the gut microbiota in anticancer therapy: from molecular mechanisms to clinical applications. Sig Transduct Target Ther 8(1):201

Zhong L, Zhang X, Covasa M (2014) Emerging roles of lactic acid bacteria in protection against colorectal cancer. World J Gastroenterol 20(24):7878

Felicia Murunwa Nemaguvhuni,
Afsatou Ndama Traore, Mapula Razwinani,
and Amidou Samie

Abstract

Gold nanoparticles (AuNPs) were synthesized using extracts from Spirostachys Africana. The AuNPs were characterised using Ultraviolet-Visible spectrophotometry (UV-Vis), X-ray diffraction (XRD), Dynamic light scattering (DLS), Transmission electron microscopy (TEM), and Fourier-transform infrared spectroscopy (FTIR). Their antimicrobial activity was tested against World Health Organization (WHO) bacterial priority pathogens. We also assessed their cytotoxicity, anti-inflammatory, and antioxidant activities. The UV-Vis analysis indicated AuNPs formation with surface plasmon resonance (SPR) peaks at 541. The XRD showed face-centred cubic structures with crystalline sizes ranging from 9 to 10 nm. DLS measurements displayed a monodisperse distribution while TEM showed spheroidal AuNPs with sizes of 6–32 nm. The FTIR identified the hydroxyl, carboxyl, and amine groups, involved in reducing and capping of the Au + ions. Antimicrobial assays demonstrated significant antibacterial effects against S. aureus and P. aeruginosa, with MICs of 0.64 and 1.2 mg/mL respectively. Ethanol extracts-conjugated AuNPs exhibited selective cytotoxicity against MCF-7 cancer cells, compared normal cells with an IC_{50} of 2.67 mg/mL. These AuNPs significantly decreased NO production, indicating anti-inflammatory potential. Methanol-conjugated AuNPs exhibited good antioxidant activity with an IC50 of 38 µg/mL. AuNPs synthesised from S. africana extracts exhibit promising antibacterial activity against AMR pathogens, as well as significant cytotoxic, anti-inflammatory, and antioxidant activities.

Keywords

Antimicrobial resistance · *Spirostachys africana* · Gold nanoparticles · Anti-cancer activity · Antimicrobial activity · Biological activity · Anti-inflammatory activity · Antioxidant activity

10.1 Introduction

Antimicrobial resistance (AMR) poses a significant threat to global health, contributing to approximately 700,000 deaths annually. A figure that could escalate to 10 million by 2050 according to the World Health Organization (WHO)

F. M. Nemaguvhuni · A. N. Traore · M. Razwinani · A. Samie (✉)
Department of Biochemistry and Microbiology, Faculty of Science, Engineering and Agriculture, University of Venda, Thohoyandou, South Africa
e-mail: samie.amidou@univen.ac.za

© The Author(s) 2026
A. Shonhai et al. (eds.), *Advances in Biochemistry and Molecular Biology to meet Africa's Needs*, Advances in Experimental Medicine and Biology 1507, https://doi.org/10.1007/978-3-032-24254-9_10

(Mancuso et al. 2021). In 2019 alone, over 1.2 million deaths were attributed to antibiotic-resistant (ABR) bacterial infections (Tang et al. 2023), underscoring the urgent need for effective interventions. AMR occurs when microorganisms (bacteria, viruses, fungi, and parasites) adapt to survive despite the presence of antimicrobial agents designed to eliminate them, primarily due to the misuse of antibiotics (Salam et al. 2023; Yang et al. 2020), making infections increasingly difficult to treat and resulting in severe illnesses or death.

Of particular concern are the WHO priority ESKAPE pathogens (*Enterococcus faecium, Staphylococcus aureus, Klebsiella pneumoniae, Acinetobacter baumannii, Pseudomonas aeruginosa,* and *Enterobacter species*), which demonstrate multidrug resistance and cause life-threatening infections from hospital-acquired pneumonia to bloodstream infections (Idris and Nadzir 2023). The urgent need for innovative solutions to combat AMR has become evident.

One promising approach is the development of Metallic nanoparticles (MNPs) derived from medicinal plants. They are highly versatile and are used in many fields, including medicine, electronics and consumer products (Zuhrotun et al. 2023). Gold nanoparticles (AuNPs) are particularly valuable due to their biocompatibility and adjustable optical properties, making them suitable for drug delivery and disease diagnosis (Botteon et al. 2021; Wang et al. 2016). Recent research highlights the antibacterial properties of AuNPs, which disrupt bacterial cell walls and inhibit DNA replication, making them effective against ABR pathogens (Arafa et al. 2018). Furthermore, AuNPs excel at delivering therapeutic agents directly to targeted sites while minimizing damage to healthy cells (Yayehrad et al. 2022).

Traditional synthesis methods for AuNPs are often toxic or expensive. In contrast, green synthesis using medicinal plants offers a cost-effective and environmentally friendly alternative (Wintachai et al. 2019). This method utilizes phytochemical compounds in plants as reducing and stabilizing agents for nanoparticle synthesis

(Dada et al. 2019). Medicinal plants are traditionally used to treat infections and contain beneficial phytochemicals like coumarins, terpenoids, glycosides, flavonoids, phenols, polyphenols, vitamins, and steroids (Al-Snafi 2015; Shaikh et al. 2022).

The current study focuses on *S. africana*, a medicinal plant used to treat various illnesses such as diarrhoea, stomach ulcers and diabetes, just to cite a few (Mabogo 1990; Lennox and Bamford 2015). The plant extracts were used to synthesize and characterize AuNPs for combating AMR. Despite the success of green synthesis with other plants, *S. africana* has not been explored for this purpose, making this study a pioneering effort in addressing the AMR crisis.

10.2 Methodology

10.2.1 Plant Material Preparation

The stem bark and leaves of *S. africana* were harvested in Ha-Matsa village in the Nzhelele region of Limpopo province, South Africa. The plant materials were washed, cut into smaller pieces, and left to dry for 2 weeks at room temperature. They were then ground into a fine powder using an electric grinder model FZ102 (Tianjin Taisite Instruments, Tianjin, China) and stored in a sealed glass jar in the dark at ambient temperature. The plant materials were extracted using various solvents, such as methanol, ethanol, and acetone (Rochelle Chemical, Gauteng, South Africa). About 50 g of the ground bark and leaves were soaked in 500 mL of each solvent for 72 h, with frequent shaking. The crude extracts were filtered using Munktell filter paper 3HW (Lasec, Gauteng, South Africa) and concentrated by evaporating the solvents using a rotary evaporator (Buchi Rotavapor R-300, Cole-Parmer, Vernon Hills, IL, USA) under pressure at 40 °C (Agyepong et al. 2014). For water extracts, the filtrate was first frozen at −80 °C and then placed in a freeze dryer (EPIC Freeze Dryer, Millrock Technology, Kinston, NY, USA) for 24 h. All the plant extracts were stored at 4 °C until use.

10.2.2 Green Synthesis of AuNPs Mediated by *S. africana* Extracts

Gold nanoparticles (SA-AuNPs) were synthesised using a microwave irradiation method (Ayinde et al. 2019). Initially, 2 mL of *S. africana* extracts were added to 8 mL of 1 mM $HAuCl_4$ solution in a 250 mL beaker at a ratio of 1:4 (v/v), and the mixture was thoroughly mixed. Various plant extract concentrations (10, 50, and 100 mg/mL) were utilised in the synthesis process to optimise the nucleation intensity of the NPs. Subsequently, the beaker with the solution mixture was placed on the turntable of a domestic microwave oven (Russell Hobbs 20 L (RHEM 21 L), operating at 700 W and a frequency of 2450 MHz. The exposure duration ranged from 30 to 90 s to achieve complete bio-reduction, with a noticeable colour change from brown to deep purple or red wine, indicating SA-AuNPs formation. The produced SA-AuNPs samples were labelled accordingly and stored at 4 °C for subsequent characterisation and assessment of biological activities.

10.2.3 Characterisation of *S. africana* Crude Extracts and Its Green Synthesised AuNPs

10.2.3.1 Ultraviolet-Visible Spectrophotometry Analysis

The formation of SA-AuNPs was validated by measuring the absorbance of the reaction mixture using UV-Vis spectrophotometry (Baharara et al. 2018; Jalab et al. 2021). The SA-AuNPs samples were diluted to a 1:24 ratio with sterile distilled water. Dilutions were performed in 2 mL centrifuge tubes, and then 200 µL of the diluent was transferred into a 96-well plate in triplicate. The plate containing the NP samples underwent scanning in a wavelength range from 200 to 800 nm using a SpectraMax M3 spectrophotometer (Molecular Devices, California, USA) at a resolution of 2 nm. However, this procedure was done for all the samples except for the acetone extracts-based SA-AuNPs, where 2 mL of the diluted

solution was dispersed in a glass cuvette to read for absorbance, as the acetone solvent was not suitable for the plastic 96-well plate used. Absorption peaks in the UV-Vis spectra confirming the production of nanoparticles were observed, and their values were recorded in Microsoft Excel. Subsequently, graphs were plotted using OriginPro 2023b software.

10.2.3.2 X-Ray Diffraction Unit (XRD) Analysis

The synthesised SA-AuNPs were first filtered, and the pellets were analysed by an XRD (Pan Analytical, X-pert pro, Netherlands). The XRD measurements were conducted using a Cu-Kα radiation source ($\lambda = 1.5406$ Å) in the scattering range m (2θ) of 20–80 on an instrument operating at a voltage of 45 kV and a current of 40 mA. This was done to ascertain the synthesised NPs' crystalline nature, phase variety, and grain size. The Debye-Scherrer formula was applied to calculate the average crystalline size:

$$D = \frac{k\lambda}{\beta \cos\theta}$$

Where: D represents the average crystallite size (measured in nm), k stands for the Scherrer constant, which has a fixed value of 0.9, λ denotes the wavelength of the X-ray source, θ signifies the peak position (measured in radians), and β represents the full width at half maximum (FWHM) (Dubey et al. 2010).

10.2.3.3 Transmission Electron Microscopy (TEM) Analysis

TEM was employed to study SA-AuNPs, offering high resolution and magnification for direct visualisation of the particles and providing detailed information about their size, shape, and distribution (Cascio et al. 2015). The TEM analysis was conducted using a FEI Technai-12 electron microscope (FEI, Eindhoven, Netherlands) operated at an acceleration voltage of 120 kV. The samples were prepared by immersing a copper grid in the nanoparticle solution and quickly removing it to enable drying, forming a thin film. Obtained images from TEM were analysed to determine particle sizes using ImageJ software by measuring the diameter (nm) of each particle. Following this, the acquired values were plotted using OriginPro 2023 software. Notably, in this analysis, 100 NPs from distinct regions were examined for each sample.

10.2.3.4 Fourier-Transform Infrared Spectroscopy (FTIR) Analysis

The functional groups that were responsible for biological activities present in *S. africana* extracts and SA-AuNPs were identified by FTIR using the Bruker ALPHA FT-IR Spectrophotometer (Bruker Scientific Private Limited, Mumbai, India). The analysis was done by placing 500 µL of the plant extracts or NP samples on top of the FTIR spectrophotometer chamber. Percentage transmittance measurements were recorded within the wavelength range of 400–4000 cm^{-1}, as outlined by Patel et al. (2020).

10.2.4 Preparation of Microorganisms

Four bacterial pathogens listed by the WHO as critical priority pathogens of ABR were used in the present study. *Acinetobacter baumannii* (ATCC BAA-1605), *Pseudomonas aeruginosa* (ATCC 27853), *Escherichia coli* (ATCC BAA-2469), and *Staphylococcus aureus* (NCTC 12493). All these bacterial strains were purchased from Anatech Instruments (Gauteng, South Africa). They were initially plated on their selective respective media and incubated at 37 °C for 24 h. The resulting colonies were preserved in the Muller-Hinton (MH) broth (Separations, Gauteng, South Africa) with 15% glycerol in Eppendorf tubes and stored at −20 °C. Before conducting the bioassays, microbial suspensions were inoculated in 5 mL of MH broth and incubated for 24 h at 37 °C. The bacterial suspensions were then adjusted to 0.5 McFarland standard (1.5 × 10^8 CFU/mL) before antimicrobial testing.

10.2.5 Antimicrobial Activity of *S. africana* Extracts and Nanoparticles

10.2.5.1 Agar Well-Diffusion Method

The antibacterial activity of SA-AuNPs was assessed using the agar well-diffusion method following the protocol established by Ghavam (2021). Initially, MH agar plates were prepared to culture bacterial strains. A 200 µL volume of the test organisms with turbidity adjusted to 0.5 McFarland standard was inoculated over the surface of MH agar using a sterile spreader and allowed to dry for 30 minutes. Six evenly spaced wells with a diameter of 8 mm were aseptically created by gently punching the agar with a sterile plastic pipette tip. Each well was filled with 50 µL of each SA-AgNPs (10.14 mg/mL). The plates were left for 1 h to allow diffusion of the treatment samples into the inoculated agar. Finally, the agar plates were incubated at 37 °C for 24 h. The antibiotic lomefloxacin hydrochloride (10 mg/mL) (Sigma-Aldrich, Darmstadt, Germany) was used as a positive control, while 10% dimethyl sulfoxide (DMSO) (Rochelle Chemical, Gauteng, South Africa) served as the negative control. To assess the efficacy of the experiment, the diameter of the clear zone, indicating the inhibition of bacterial growth, was meticulously measured in millimetres using a ruler. For the precision and reliability of the results, the experiment was repeated three times for each test organism. The growth inhibition zones were subsequently recorded and presented in graphs.

10.2.5.2 Microdilution Assay

The minimum inhibitory concentrations (MICs) of SA-AuNPs against the priority pathogens (*A. baumannii*, *P. aeruginosa*, *E. coli*, and *S. aureus*) were determined using the microdilution assay (Samie et al. 2005). The procedure was as follows: In a standard 96-well plate, each well was filled with 100 µl of MH broth as the growth medium. The plate's first row received negative control (10% DMSO), positive control (lomefloxacin hydrochloride at 10 mg/mL), and SA-AuNPs to the rest of the remaining wells in the same row. A two-fold serial dilution was performed by mixing contents of the first row and transferring 100 μL to the next well within the same column until reaching the last well within the same column, with the final 100 μL discarded. The resulting SA-AuNP concentrations ranged from 10.14 to 0.08 mg/mL after 50% serial dilution. Subsequently, 100 μL of the bacterial culture was added to the wells. The plates were then incubated for 24 h at 37 °C. The next day, 40 μL of a 0.2% Iodo Nitro Tetrazolium (INT) solution (Inqaba Biotech, Gauteng, South Africa) was added to each well and incubated at 37 °C for 30 min and observed for colour change.

10.2.5.3 Determination of Minimum Bactericidal Concentration (MBC)

For minimum bactericidal concentration (MBC), approximately 5 μL of each sample was taken from the microdilution wells (described in the previous section) with no observable bacterial growth during the MIC tests and sub-cultured onto MH agar plates, subsequently incubated at 37 °C for 24 h and checked the next day for growth. The MBC is defined as the minimum concentration at which no bacterial growth is observed on the agar (Quelemes et al. 2013).

10.2.6 Cytotoxicity and Anti-inflammatory Assays

10.2.6.1 Cell Lines

The SA-AuNPs samples exhibiting higher antimicrobial activity were tested for cytotoxicity against human oestrogen-positive breast cancer cells (MCF7), normal human mammary epithelial cells (MCF10), and the RAW 264.7 murine macrophage cells. The RAW 264.7 cells were additionally used to assess anti-inflammatory activity. All cell lines were provided by the Department of Biochemistry, Genetics, and Microbiology at the University of Pretoria (Pretoria, South Africa).

10.2.6.2 Cell Culture

Dulbecco Modified Eagles Medium F12 (DMEM/F12) and Alpha Minimum Essential Medium (α-MEM) media (Thermo Fisher Scientific, Waltham, MA, USA) were used for cell culture. The media were prepared following the protocol provided by ScienCell[MT] Research Laboratories and supplemented with 1% Penicillin-Streptomycin, Amphotericin B/Fungizone, and 10% Fetal bovine serum (FBS) (Thermo Fisher Scientific, Waltham, MA, USA). The cell lines used included: Raw 264.7 (catalogue No ATCC TIB-71), MCF7 (Catalogue No ATCC HTB-22) and MFC10 (Catalogue No ATCC CRL-10318) and were originally stored in liquid nitrogen. The cells were gently transferred into a T25 flask containing 10 mL supplemented DMEM/α-MEM media and incubated at 37 °C in a humidified incubator with 5% CO_2. The MCF7 cells were cultured in α-MEM, while MCF10 and RAW 264.7 cells were cultured in DMEM/F12. The cultured cells were monitored daily until they reached 80% confluence after 7 days.

10.2.6.3 Cytotoxicity Assay

The cytotoxicity of SA-AuNPs was determined on the MCF-7, MCF-10, and RAW 264.7 cells using the 3-(4,5-dimethylthiazol-2-yl)-2,5-diphenyltetrazolium bromide (MTT) method (Vijayarathna and Sasidharan 2012). Initially, 50 μL of media was added to all wells of a 96-well plate. Subsequently, 50 μL of the SA-AuNPs samples were added to the first row of each column of the plate, while three columns were reserved for the negative control. Serial dilution was performed by mixing the contents within each column across the 96-well plate, discarding 50% of the solution from the last well. The resulting sample concentrations ranged from 10.14 to 0.08 mg/mL, respectively. Following this, 50 μL of cell suspension (5000 cells) was seeded into all wells of the 96-well plate, bringing the final volume to 100 μL in each well. The negative control consisted of untreated cells to establish baseline cell viability. The plate was then incubated at 37 °C in a humidified 5% CO_2 incubator for 48 h. After incubation, 15 μL of MTT solution (Promega, USA) (5 mg/mL in phosphate-buffered saline [PBS]) was added to each well and incubated for 4 h. Subsequently, 100 μL of DMSO was added, and the plate was incubated overnight in a sealed container with a humidified atmosphere at room temperature to solubilise the formazan crystals completely. Absorbance was measured at 595 nm using a SpectraMax M3 multi-mode microplate reader (Molecular Devices, California, USA). The nanoparticle background interference was corrected by subtracting the absorbance of nanoparticles alone from that of nanoparticles combined with cells and the percentage cell viability was calculated using the following formula:

$$\%\text{Cell viability} = \frac{\text{Average absorbance of drug wells}}{\text{Average absorbance of control wells}} \times 100$$

The results were expressed as the mean ± standard error and further determined the half-maximal inhibitory concentration (IC_{50}), defined as the concentration of the examined SA-AuNP that resulted in a 50% inhibition of cell growth compared to the control. The IC_{50} was calculated using a non-linear regression curve in GraphPad Prism 8 software (14).

10.2.6.4 Anti-inflammatory Activity

The anti-inflammatory activity of SA-AuNPs was evaluated using RAW 264.7 macrophage cells. The cells were initially plated at a density of 5000 cells per well in a 96-well plate and left to incubate overnight. Subsequently, the cells were treated with the samples at varying concentrations ranging from 10.14 to 0.08 mg/mL, following 50% serial dilution across the plate. The cells with the media alone served as a negative control. After 4 h of incubation, inflammation was induced by adding 1 μg/mL of lipopolysaccharide (LPS) (Promega, USA), and the cells were incubated for an additional 24 h. The nitric oxide (NO) levels were determined using the Griess reaction method (Promega, USA). Approximately 50 μL of the cell supernatant was collected and mixed with 50 μL of sulfanilamide solution (1% sulfanilamide in 5% phosphoric acid) for a 10-min reaction. Subsequently, 50 μL

of NED solution (0.1% N-1-naphthylenediamine) (Promega, USA) was introduced for another 10-minute reaction period, and the absorbance was measured at 595 nm using a SpectraMax M3 spectrophotometer. The concentration of nitrate released into the culture medium was then calculated from the slope of the standard curve (Bahrami et al. 2021; Kang et al. 2022).

10.2.7 Antioxidant Activity

The free radical scavenging activity of SA-AuNPs was assessed by the 2,2-diphenyl-1-picrylhydrazyl (DPPH) method (Nemudzivhadi and Masoko 2015). Briefly, 100 μL of deionised water was introduced into each well of a 96-well plate followed by the addition of 100 μL of SA-AuNPs (2.5–0.02 mg/mL) to the first well of each column of the 96-well plate in triplicate. Subsequently, 50% serial dilutions were performed. Vitamin C (1 mg/mL) was used as an antioxidant standard control, and the solvents used to prepare the extracts for synthesised SA-AuNPs were employed as the blanks. An additional 100 μL of a 0.1 mM DPPH methanolic solution (Inqaba Biotech, Gauteng, South Africa) was introduced into all the wells of the 96-microplate. The plate was then kept in the dark for 30 min. Following this incubation period, the absorbance was measured at 517 nm using VersaMax Microplate Reader (Molecular Devices, Sunnyvale, USA). The percentage scavenging activity (DPPH reduced) was calculated as follows:

$$\%\text{Scavenging activity} = \frac{A_{\text{control}} - A_{\text{sample}}}{A_{\text{control}}} \times 100$$

Where A_{control} is the absorption of the blank solution, and A_{sample} is the absorption of the extract solution. The IC_{50} values (the extract concentrations at which 50% radical scavenging activity occurs) were determined from the graph using linear regression analysis and thereafter generated a table of DPPH scavenging activity using those values.

10.2.8 Statistical Analysis

Data were recorded in Microsoft Excel. All biological experiments were performed in triplicate ($n = 3$), and results are presented as mean ± standard error. Statistical significance was determined using one-way and two-way analysis of variance (ANOVA) with a significance level of $p < 0.05$. When significant differences were found, means were compared using Tukey's post hoc test via GraphPad Prism 8 software. All statistical analyses were conducted using GraphPad Prism 8 software. Graphs were plotted using Excel and OriginPro 2023b software.

10.3 Results

10.3.1 Green Synthesis of AuNPs Mediated by *S. africana* Extracts

The synthesis of SA-AuNPs was confirmed by a visible change in colour within the reaction mixture, transitioning from yellowish-brown to a deep purple within a few seconds (Fig. 10.1). The consistent nature of this colour change indicates the successful reduction of Au^{3+} ions into NPs. Stable SA-AuNPs with enhanced activity were found after increasing the concentration of plant extracts and employing microwave-assisted synthesis for 30–90 s. However, the 90-s microwave synthesis yielded superior results as compared to reactions lasting 30 or 60 s.

10.3.2 Characterisation of SA-AuNPs

10.3.2.1 Ultraviolet-Visible Spectrophotometry Analysis

The UV-Vis analysis of SA-AuNPs, as depicted in Fig. 10.2, included samples exhibiting optimal synthesis SA-AuNPs, synthesised by 1 mM $HAuCl_4$ and 10 mg/mL extracts after 90 s microwave-assisted nanoparticle synthesis. Notably, methanol extracts-conjugated SA-AuNPs (SABM-AU) exhibited an SPR peak at 533 nm, while ethanol extracts-conjugated SA-AuNPs (SABE-AU) displayed peaks at 541 nm, and acetone extracts-conjugated SA-AuNPs (SABA-AU) at 574 nm. These results confirmed that the synthesis of SA-AuNPs was successful.

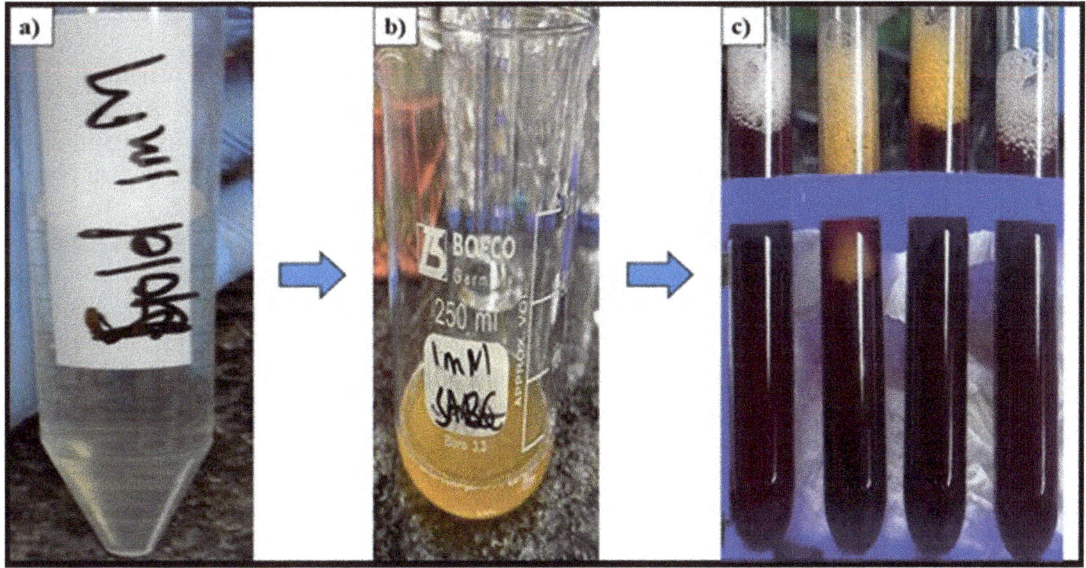

Fig. 10.1 Synthesis of the gold nanoparticle from *S. africana* extracts showing the colour change indicating the production of AuNPs: (**a**) Colourless 1 mM HauCL₄ solution prior to the addition of the *S. africana* methanol extracts, (**b**) brown colour of *S. africana* plant extracts mixed with HAuCL₄ solution, (**c**) complete transformation of the reaction mixture into deep purple shades for the 90 s incubation period

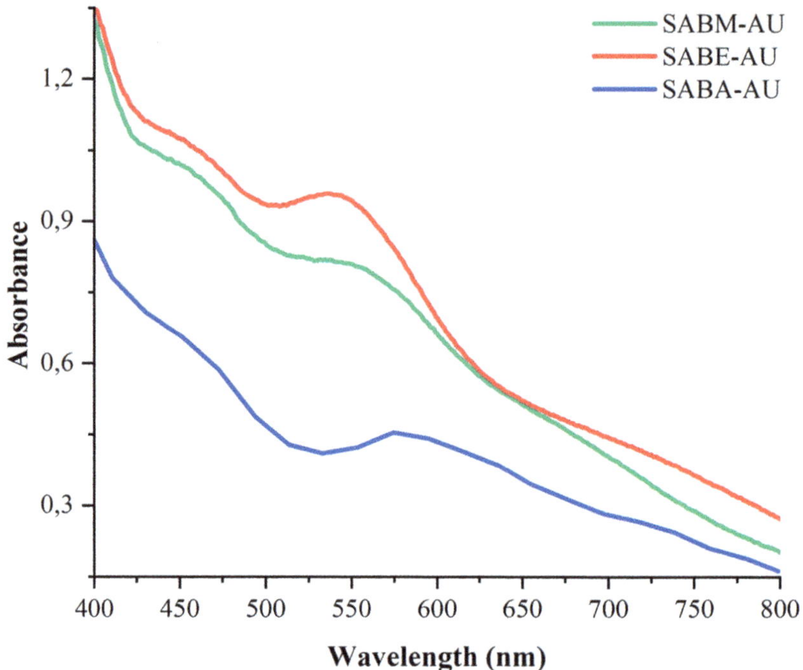

Fig. 10.2 Ultraviolet-visible spectra of SA-AuNPs; SABM-AU (bark methanol SA-AuNPs), SABE-AU (bark ethanol SA-AuNPs), SABA-AU (bark acetone SA-AuNPs) for 90 s

The impact of the plant extract concentration and reaction time on the synthesis of SA-AuNPs can be seen in supplementary.

10.3.2.2 X-Ray Diffraction Unit (XRD) Analysis of SA-AuNPs

XRD spectroscopy showed four diffraction peaks with 2θ values which were observed at 37°, 43°, 63°, and 76° with planes at (111), (200), (220), and (311), respectively (Fig. 10.3) for the SA-AuNPs samples, as determined using Bragg's law. Samples SABE30S100, SABE60S100, and SABA60S100 lacked a peak at an angle of 43°. These diffraction indices align with the cubic face-centred (FCC) structure of AuNPs, as specified by the Joint Committee on Powder Diffraction Standards (JCPDS) (JCPDS 01–1167). The average crystalline size of SA-AuNPs samples ranged from 8 to 10 nm (Table 10.1).

10.3.2.3 Transmission Electron Microscopy (TEM) Analysis of SA-AuNPs

TEM was used to investigate the shape and size of SA-AuNPs. Figure 10.4 displays TEM micrograph images of SA-AuNPs at different magnifications (200, 100, and 20 nm), along with their corresponding particle size distribution portrayed in histogram graphs. The SA-AuNPs have demonstrated various shapes such as spheroidal and star-shaped, with a core size range of 6–32 nm, and an average particle size distribution of 14.1 ± 6.4 nm.

10.3.2.4 Fourier-Transform Infrared Spectroscopy (FTIR) Analysis

The FTIR spectral analysis results confirmed the functional groups and chemical compounds present in *S. africana* extracts (Fig. 10.5). The extracts had a broad peak at 3310 cm⁻¹, indicating the presence of O-H (hydroxy) groups. A narrow peak at 1632 suggests C=C stretching (alkenes) or N-H (amide) groups. Smaller peak at 1011 cm⁻¹, which indicates C-O stretching (alcohols). Moreover, an absorption peak at 950 cm⁻¹ was also seen and is associated with C-H (aromatic) in-plane bending (Devi and Sathishkumar 2017; Nandiyanto et al. 2019). The FTIR spectral analysis for SA-AuNPs displayed similar results to those of the extracts, albeit with minor shifts in absorption bands, indicating that these functional groups are also present in the SA-AuNPs and are involved in reducing and capping Au⁺ ions.

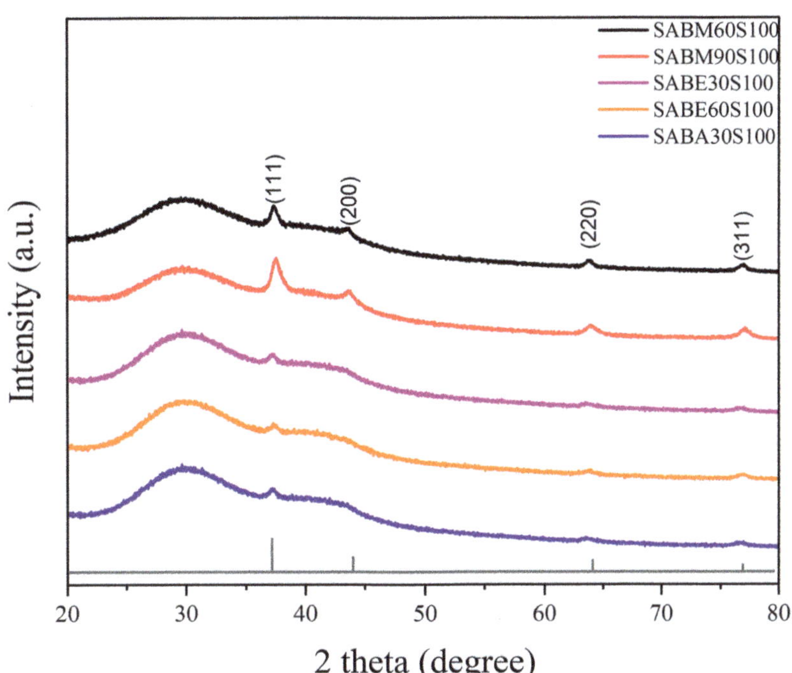

Fig. 10.3 XRD spectrum of SA-AuNPs synthesised at different reaction times. Bark methanol-AuNPs at 60 s (SABM60S100), bark methanol-AuNPs at 90 s (SABM90S100), bark ethanol-AuNPs at 30 s (SABE30S100), bark ethanol-AuNPs at 60 s (SABE60S100), and bark acetone-AuNPs at 30 s (SABA30S100)

Table 10.1 Average crystalline size of gold nanoparticles (SA-AgNPs)

Sample	Average crystalline size (nm)
SABM60S100	9 ± 0.7
SABM90S100	8 ± 0.6
SABE30S100	9 ± 0.8
SABE60S100	10 ± 0.7
SABE90S100	17 ± 0.7
SABA30S100	9 ± 0.6

Key: *SA S. africana*, *B* bark, *M* methanol, *E* ethanol, *A* acetone, *30S* 30 seconds, *60S* 60 seconds, *90S* 90 seconds

10.3.3 Antimicrobial Activity of SA-AuNPs

10.3.3.1 Agar Well-Diffusion Assay

Antimicrobial activity results by agar-well diffusion revealed that SA-AuNPs exhibited higher efficacy against *S. aureus*, followed by *P. aeruginosa*, with the lowest activity observed against *A. baumannii*, and *E. coli* showed resistance, as there was no inhibition observed. The most significant activity was observed in samples SABM90S100 and SABE90S100, both showing a zone inhibition diameter of 25 mm against *S. aureus*. SABE90S100 is also seen to have activity against *P. aeruginosa* (16 mm). The difference in antimicrobial activity between SA-AuNPs and the positive control was statistically significant

($p = 0.0011$) for SABE90S100 against *S. aureus*, suggesting the potential of SA-AuNPs as antimicrobial agents. Conversely, lower activity was observed against *A. baumannii*, and none of the samples were effective against *E. coli*. The overall results are illustrated in Fig. 10.6, showcasing the zone of inhibition in the form of a bar graph.

10.3.3.2 Microdilution Assay

The MIC results for SA-AuNPs revealed varying levels of activity against the tested bacterial strains, with *S. aureus* emerging as the most sensitive among the strains examined (Fig. 10.7). Notably, samples SABM60S100 and SABE90S100 exhibited the highest activity against *S. aureus*, with the lowest MIC value recorded at 0.64 mg/mL. Similarly, SABE90S100

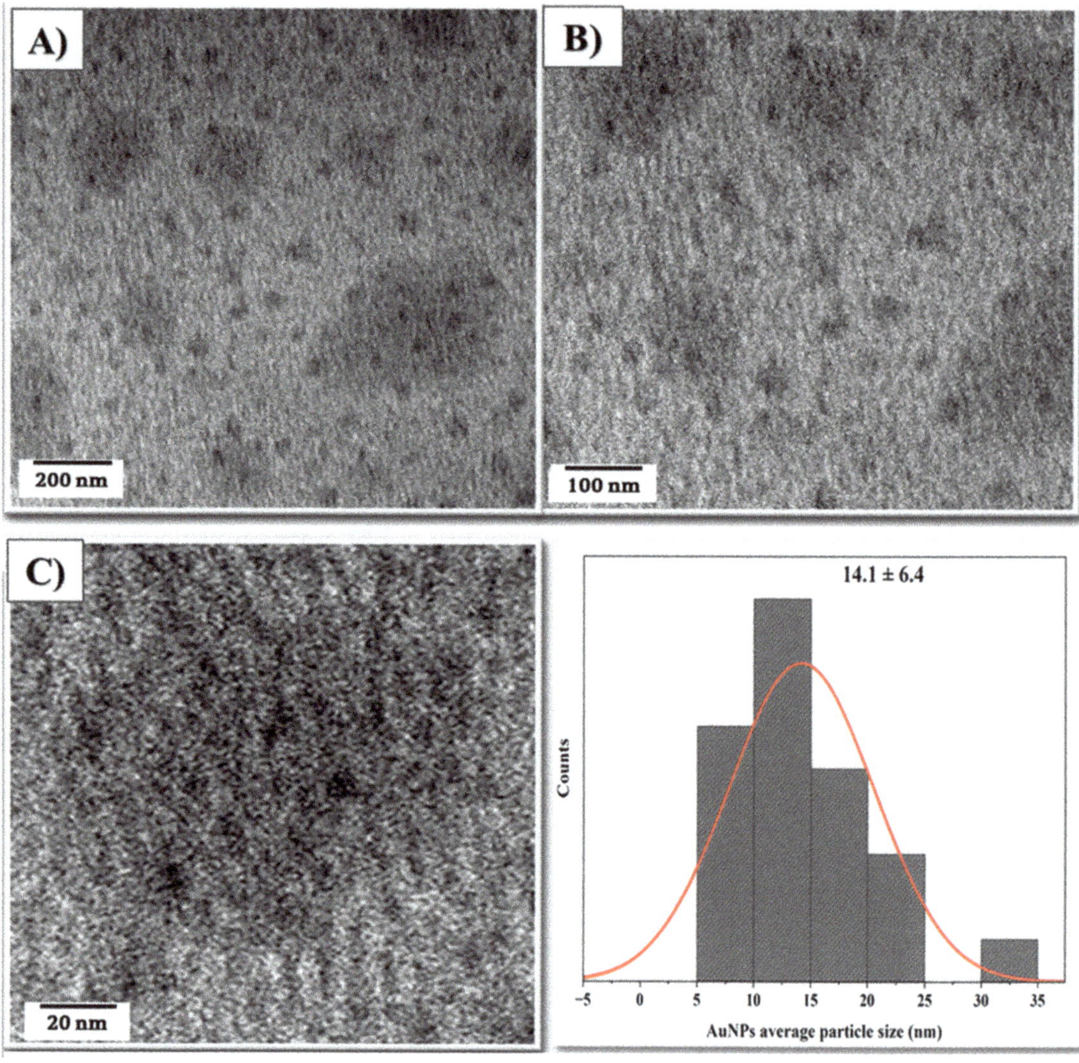

Fig. 10.4 TEM micrograph images of SA-AuNPs at various magnifications (**a** = 200 nm, **b** = 100 nm, **c** = 20 nm) along with their corresponding average particle size distributions

and SABA90S100 demonstrated notable efficacy against *E. coli*, both displaying a MIC value of 1.27 mg/mL. Additionally, SABA90S100 exhibited activity against *P. aeruginosa*, with a MIC value of 1.27 mg/mL. None of the tested SA-AuNPs samples showed activity against *A. baumannii.*

10.3.3.3 Minimum Bactericidal Concentration (MBC)

The MBC results for SA-AuNPs showed that sample SABE90S100 exhibited the lowest MBC

of 0.63 mg/mL against *S. aureus*. Sample SABM90S00 showed activity against both *S. aureus* and *E. coli*, with MBC values of 2.5 mg/mL for both (Fig. 10.8). No killing activity was observed against *A. baumannii* or *P. aeruginosa.*

10.3.4 Cytotoxicity Assay

The cytotoxicity of the SA-AuNP sample SABE90S100 (SABE-AU) was evaluated against MCF-7, MCF-10, and RAW 264.7 macrophage

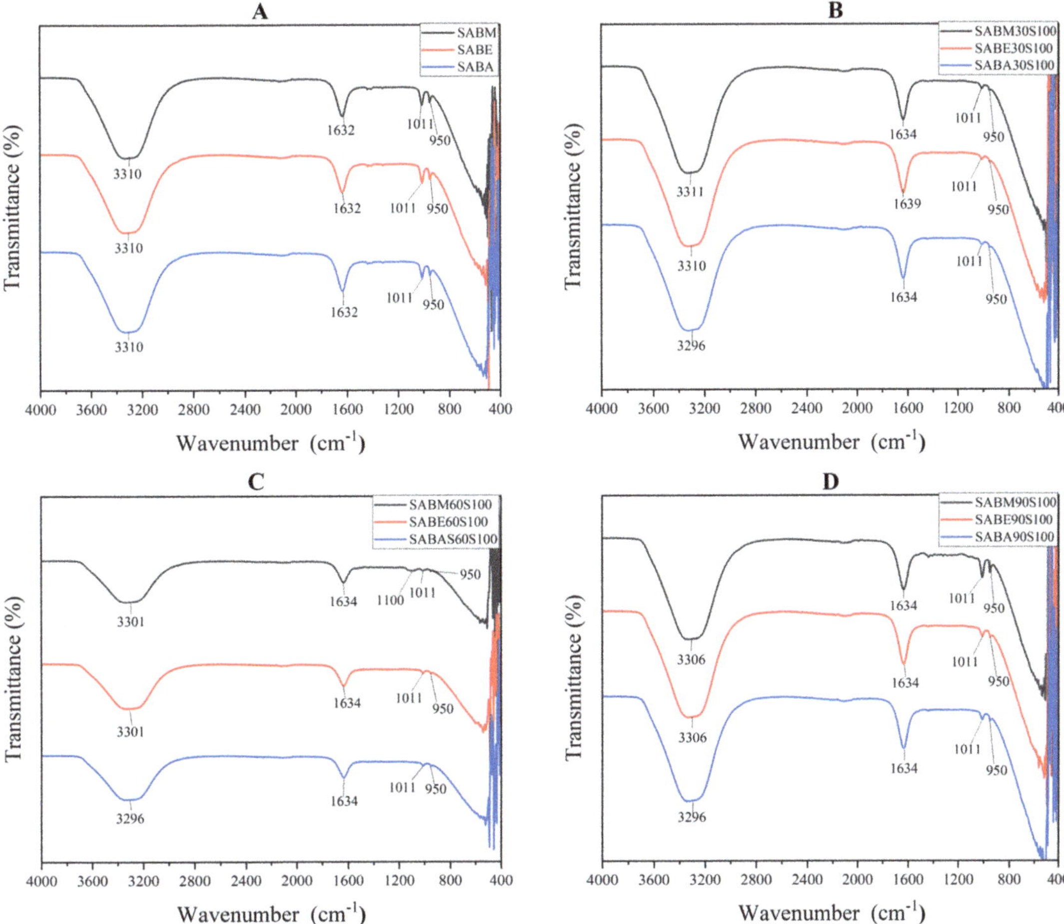

Fig. 10.5 Fourier-transform infrared (FTIR) for *S. africana* bark extracts (**a**) and synthesised SA-AuNPs by microwave radiation at 30 s (**b**), 60 s (**c**), and 90 s (**d**)

cells using the MTT assay. SABE-AU was selected because it showed the strongest antimicrobial activity, with the lowest MIC and MBC values among all formulations. After 24 h of incubation at 37 °C, treated cells exhibited characteristic cytotoxic morphological changes, including rounding and shrinkage, as observed under a light microscope (Fig. 10.9).

The cytotoxic effects of SABE-AU on MCF-7 breast cancer cells, MCF-10 normal breast epithelial cells, and RAW 264.7 macrophages were expressed as percentage viability (Fig. 10.10). SABE-AU induced a significant, concentration-dependent decrease in cell viability over 0.08–10.14 mg/mL in all three cell lines com-

pared to untreated controls ($p < 0.0001$), with greater cytotoxicity toward MCF-7 cells than MCF-10 and RAW 264.7 cells at all concentrations. At 10.14 mg/mL, cell viability was 33% (MCF-7), 36% (MCF-10) and 37% (RAW 264.7), while at 1.27–2.53 mg/mL it ranged from 34–41%, 59–77% and 62–82% for MCF-7, MCF-10 and RAW 264.7 cells, respectively; at 0.08–0.16 mg/mL, viability remained ≥71% in all cell lines. The IC_{50} values were 2.67 mg/mL for MCF-7, 6.40 mg/mL for MCF-10, and 6.58 mg/mL for RAW 264.7 cells, indicating higher sensitivity of MCF-7 cancer cells to SABE-AU than normal breast epithelial and macrophage cells.

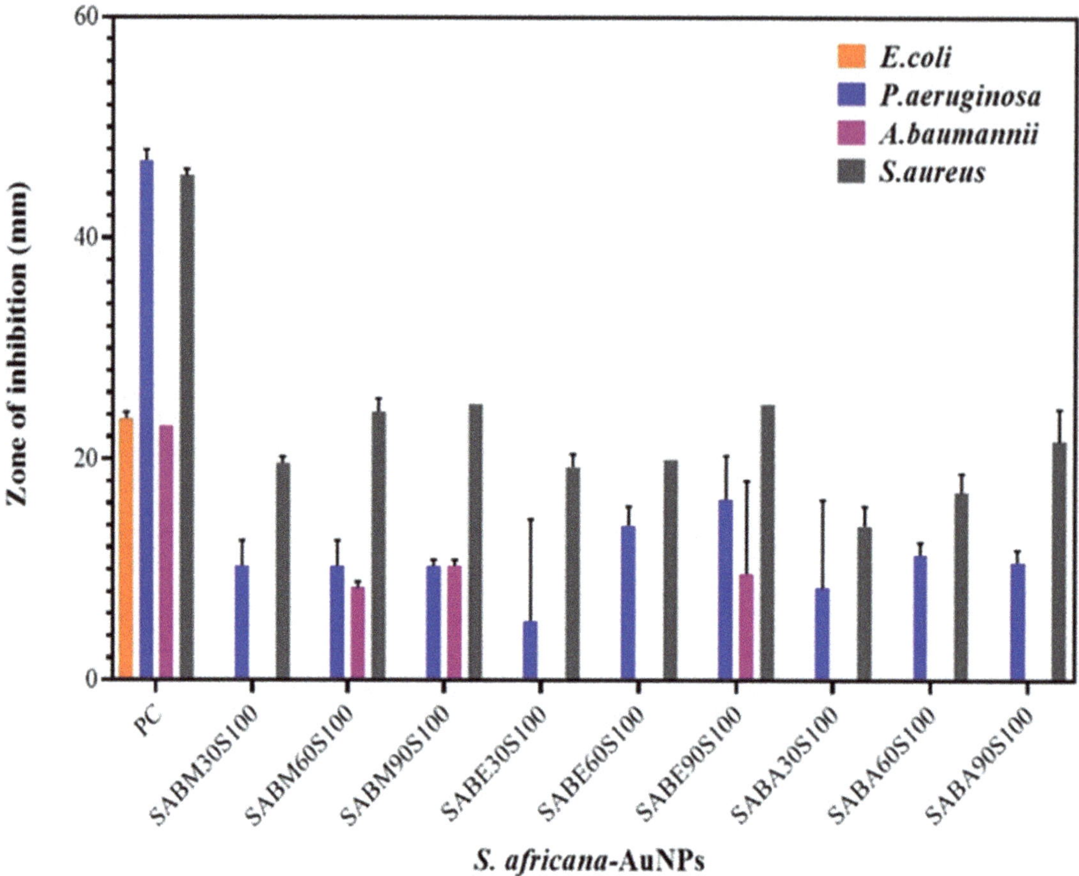

Fig. 10.6 Antimicrobial activity of SA-AuNPs (10.14 mg/mL) against *E. coli*, *P. aeruginosa*, *A. baumannii*, and *S. aureus*, shown by the zone of inhibition (mm). SA-AuNPs synthesised by bark methanol-AuNPs at 30 seconds (SABM30S100), 60 seconds (SABM60S100), 90 seconds (SABM90S100), ethanol-AuNPs at 30 seconds (SABE30S100), 60 seconds (SABE60S100), 90 seconds (SABE90S100), and acetone-AuNPs at 30 seconds (SABA30S100), 60 seconds (SABA60S100) and 90 seconds (SABA90S100). $p = 0.0113$ for the SABE90S100 activity against *S. aureus,* which showed the highest activity. PC is the positive control (Lomefloxacin hydrochloride) ($n = 3$, One-way ANOVA, Tukey's post hoc test)

10.3.5 Anti-inflammatory Assay

The RAW 264.7 macrophage cells are known to play a critical role in regulating inflammatory diseases. In the current study, we aimed to investigate the potential effects SA-AuNPs on the inflammatory response of macrophages when stimulated with lipopolysaccharides (LPS). The results of this study demonstrated that the treatment with the tested samples efficiently inhibited the production and secretion of nitric oxide NO induced by LPS (Fig. 10.11). Sample SABE-AU exhibited a significant decrease in NO production across all concentrations (10.14–0.08 mg/mL) levels tested compared to the control (LPS treated solely) ($p < 0.05$). These findings suggest that SABE-AU possesses potential anti-inflammatory effects across all tested concentrations.

10.3.6 Antioxidant Activity

The antioxidant activities of SA-AuNPs were evaluated using the DPPH assay, and the calculated IC_{50} values (µg/mL) of DPPH free radical

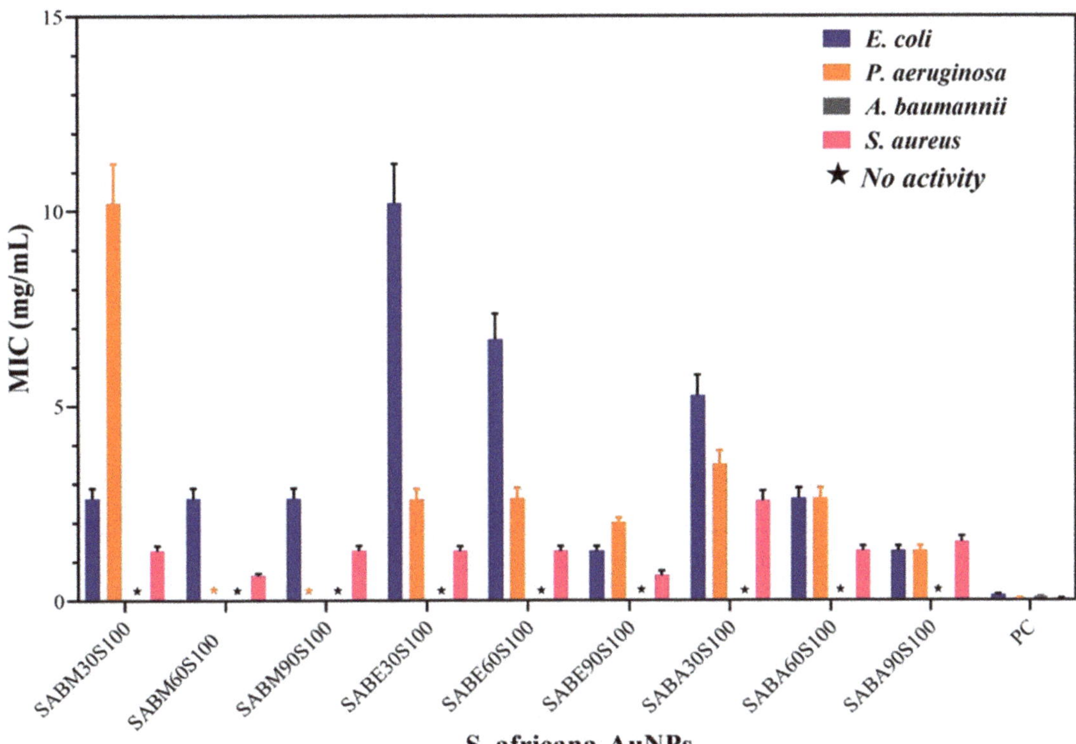

Fig. 10.7 Minimum Inhibitory Concentration (MIC) in mg/mL of SA-AuNPs synthesised with bark methanol extract at 30 seconds (SABM30S100), 60 seconds (SABM60S100), 90 seconds (SABM90S100), ethanol-AuNPs at 30 seconds (SABE30S100), 60 seconds (SABE60S100), 90 seconds (SABE90S100), and acetone-AuNPs at 30 seconds (SABA30S100), 60 seconds (SABA60S100) and 90 seconds (SABA90S100). PC positive control (Lomefloxacin hydrochloride) (One-way ANOVA, Tukey's post hoc test, $p > 0.05$)

scavenging activity are presented in Table 10.2. The antioxidant activity of the tested samples was primarily assessed by observing the colour change from purple (after the addition of the DPPH solution) to yellow (after a few minutes). It is well-established that a lower IC_{50} value indicates higher antioxidant activity (Albanese et al. 2019). Among all the tested SA-AuNPs, sample methanol conjugated AuNPs (SABM-AU) exhibited good antioxidant activity with an IC_{50} value of 38 µg/mL, compared to the positive control Ascorbic acid (0.1 µg/mL). Meanwhile, ethanol conjugated AuNPs (SABE-AU) did not have good antioxidant activity, indicated by a higher IC_{50} value of 247 µg/mL and this is in line with the cytotoxicity data.

10.4 Discussion

Medicinal plants have a long history of use to treat various diseases (Kavitha and Nadu 2021). Their potential to inhibit microbial growth depends on the presence of various secondary metabolites (Asong et al. 2023). The emergence of AMR in bacteria has led to a global crisis, and medicinal plants have gained attention as a promising source of new antibacterial drugs (Asong et al. 2023). Plant-based AuNPs have been recognized for their diverse applications and potential as eco-friendly solutions to combat the challenge of AMR (Anand et al. 2022). The present study synthesised and characterised SA-AuNPs using

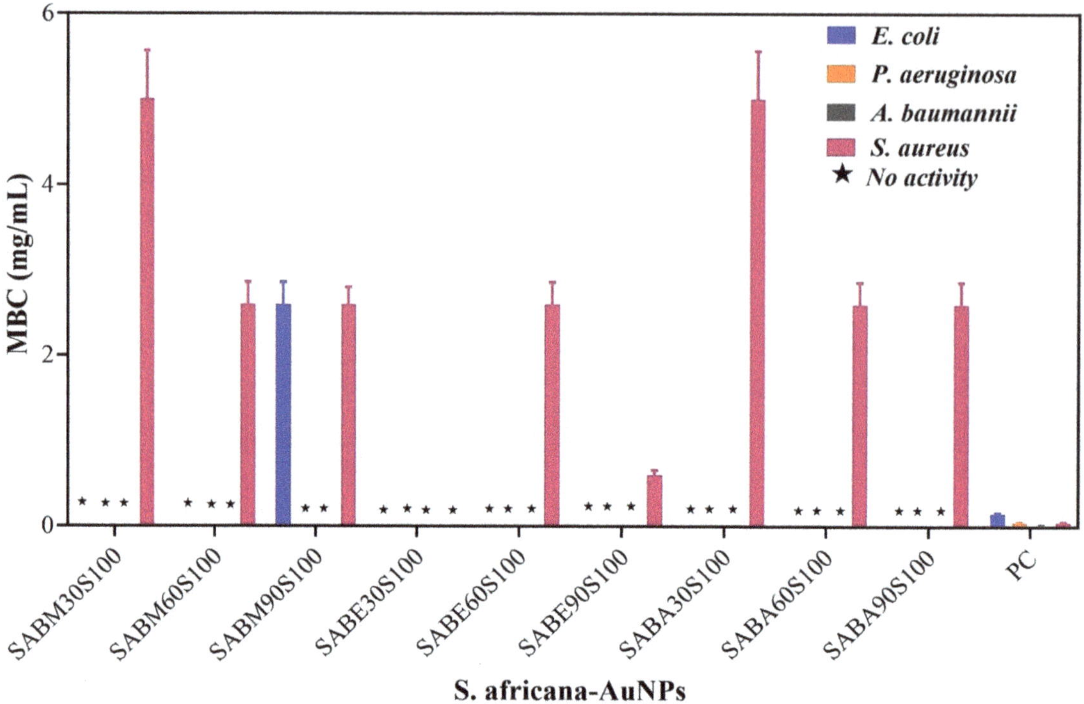

Fig. 10.8 Minimum Bactericidal Concentration (MBC) in mg/mL of SA-AuNPs synthesised by bark methanol-AuNPs for 30 seconds (SABM30S100), 60 seconds (SABM60S100), 90 seconds (SABM90S100), ethanol-AuNPs for 30 seconds (SABE30S100), 60 seconds (SABE60S100), 90 seconds (SABE90S100), and acetone-AuNPs for 30 seconds (SABA30S100), 60 seconds (SABA60S100) and 90 seconds (SABA90S100). PC positive control (Lomefloxacin hydrochloride). $n = 3$. One-way ANOVA, Tukey's post hoc test, $p > 0.05$, $n = 3$

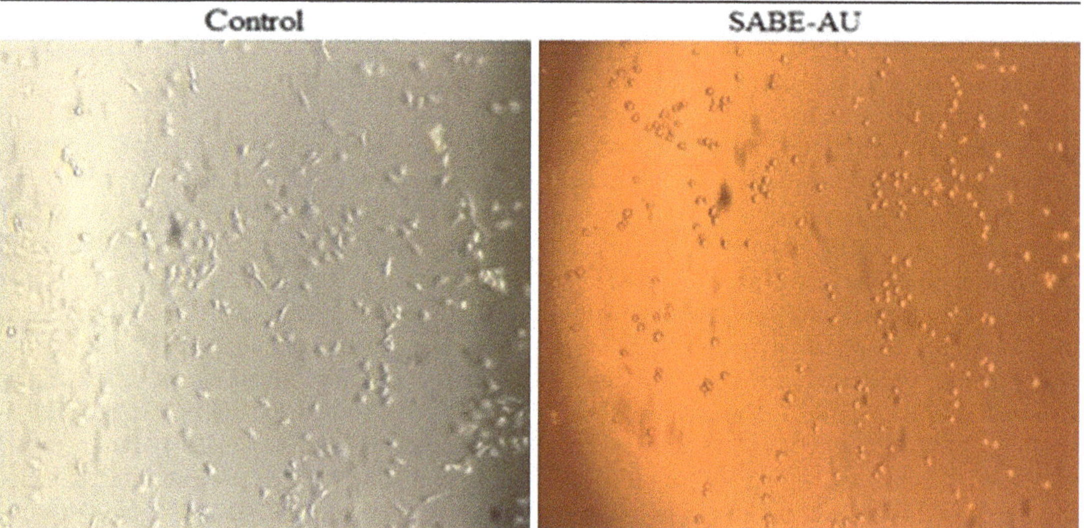

Fig. 10.9 Changes in cell morphology caused by treatment with ethanol-conjugated SA-AuNPs (SABE-AuNPs at the concentration of 1 mg/mL) on the cells after 24 h and the untreated cells as a control

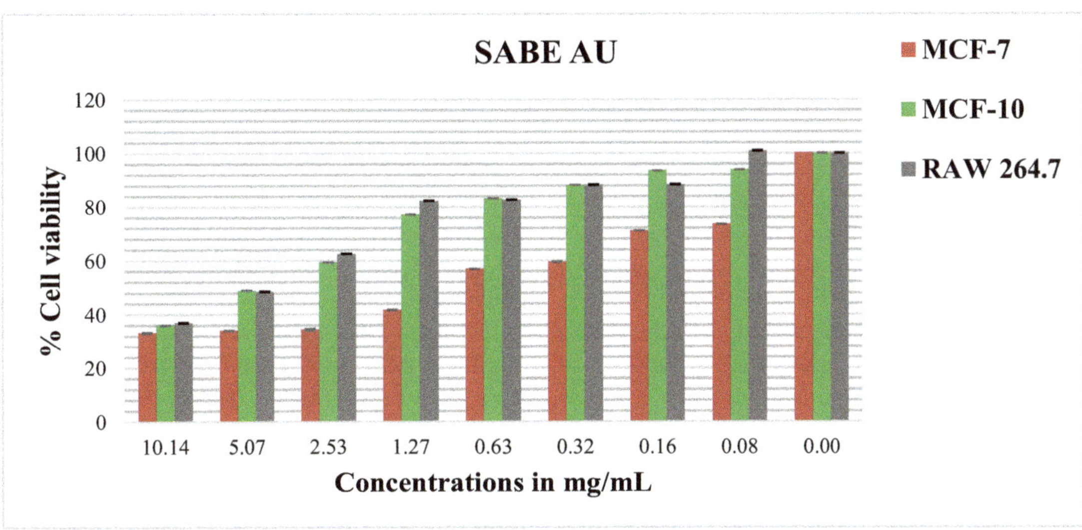

Fig. 10.10 Cytotoxicity effects of SA-AuNPs (SABE-AU) on MCF-7 breast cancer cells, MCF-10 normal cells and RAW 264.7 macrophage cells by MTT assay (% cell viability). Data represent mean ± SD ($n = 3$) (two-way ANOVA, Tukey's post hoc test, $p < 0.05$)

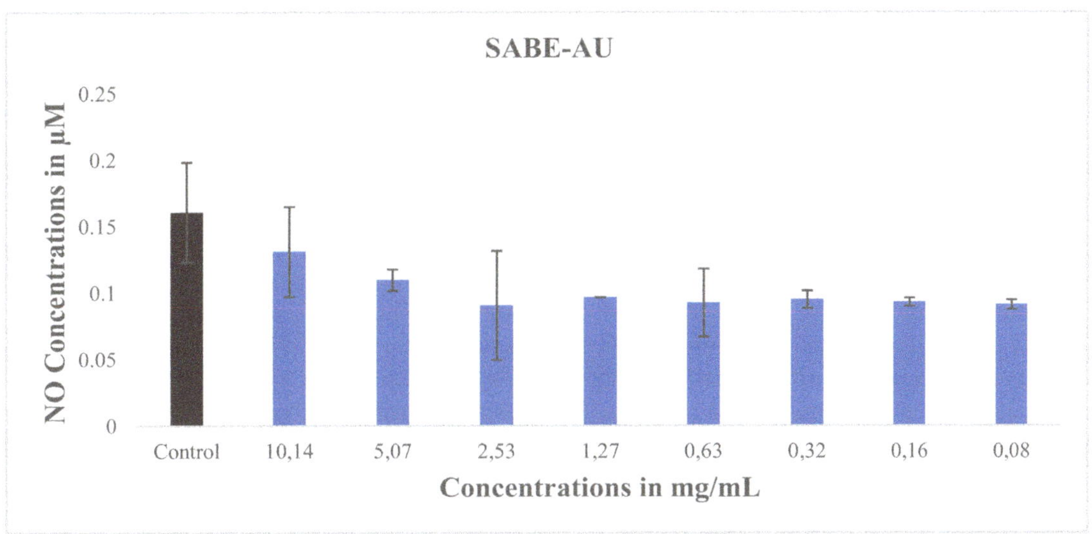

Fig. 10.11 The inhibitory effects of different concentrations of gold (SABE-AU) nanoparticles on lipopolysaccharide (LPS) induced inflammatory response on nitric oxide (NO) production in RAW 264.7 macrophage cells. Two independent experiments portray results as means ± standard error of the means (SEMs), $n = 3$

Table 10.2 Antioxidant activities of *S. africana* extracts-based gold nanoparticles (SA-AuNPs) presented as DPPH scavenging activity IC_{50} values (μg/mL)

Treatment	DPPH scavenging activity IC_{50} (μg/mL)
Gold nanoparticles (SA-AuNPs)	
SABM-AU	38 ± 3.5
SABA-AU	88 ± 4.1
SABE-AU	247 ± 0.8
Ascorbic acid	10 ± 4.5

Notes: SABM-AU = bark-methanol conjugated AuNP, SABA-AU = bark-acetone conjugated AuNP. $n = 3$

extracts from *S. africana*, evaluating their antimicrobial activity against ABR bacteria, which are listed as priority pathogens by the WHO. Additionally, the study assessed the cytotoxicity, anti-inflammatory, and antioxidant properties of the SA-AuNPs and extracts.

The study successfully utilised methanol, ethanol, and acetone extracts of the *S. africana* plant to synthesise AuNPs. Optimization of various parameters was conducted to achieve excellent activities. For SA-AuNPs, the colour changed to deep purple within a few seconds when synthesized by microwave radiation, indicating a successful production of AuNPs (Singh and Mijakovic 2022). The colour change is due to the surface plasmon vibrations and reduction of Au^{3+} ions by the phytoconstituents from the extracts (Liaqat et al. 2022). Our study found that the optimal method for the preparation of SA-AuNPs with good biological activity is to use increased extracts' concentration and 1 mM $HAuCl_4$ through microwave synthesis for 90 s. The combination of phytochemicals and microwave heating facilitates the rapid synthesis of AuNPs, offering a novel method for the production of medically relevant nanoparticles (Yasmin et al. 2014). Saifuddin et al. (2009) also demonstrated that using plant extracts-stabilized microwave radiation can increase the speed of the formation of the AuNPs, which can then be formed in as little as 1 min. Additionally, microwave radiation ensures uniform heating of aqueous solutions, preventing particle aggregation during synthesis.

In the present study, the SPR bands at wavelengths of 533 nm, 541 nm, and 574 nm observed in SA-AuNPs samples confirmed the production of AuNPs (Perveen et al. 2021). This is consistent with the typical absorption spectrum of Au colloids, which exhibits a distinct absorption band in the visible region between 500 and 600 nm (Yadi et al. 2022). A study on *Hibiscus rosa-sinensis* stabilised AuNPs showed that changing plant extract concentration, synthesising with microwaves, and adjusting exposure time influenced absorption spectra. It was noted that increased plant extract concentration led to the formation of ideal AuNPs, while a 90-s exposure produced

small NPs with a distinct peak around 520 nm (Yasmin et al. 2014). These results underscore the crucial role of plant extract concentration in controlling NP size and shape.

Additional peaks were also observed in the UV-Vis spectra of samples SABM-AU and SABE-AU before the 500 nm wavelength, possibly due to the presence of phytochemical compounds involved in Au^{3+} ion reduction. A study by Niranjan Dhanasekar et al. (2015) suggested that these peaks may arise from aromatic amino acids such as tryptophan, tyrosine, or phenylalanine, as well as peptide linkages, which help stabilise the NPs. A broad SPR band at wavelength 574 nm in SABA-AU represents the polydisperse nature of AuNPs (Perveen et al. 2021). These findings agree with earlier reports identifying a broad peak in the 550–600 nm range (Guadie Assefa et al. 2017). This is further validated because toward a higher wavelength (red-shift) where we find larger and aggregated NPs. Consistent with earlier reports that found the SPR peaks at 527, 541, and 543 nm, which suggested that AuNP size increases as the SPR shifts to longer wavelengths (red-shift) (Nur and Nasir 2008). Moreover, these reports suggested that NPs with wavelengths around 500 to 550 nm exhibit a hexagonal shape. Overall, these findings emphasise the importance of optimising reaction parameters like extract type and concentration, metal salt concentration, reaction time, and temperature during the synthesis to achieve the desired properties of AuNPs.

The X-ray diffraction (XRD) patterns observed in our study confirm the crystalline nature of SA-AuNPs. Our results are in parallel with the calculated phases from the reference library JCPDS No. 01–1167, indicating a face-centred cubic (FCC) structure, similar to previous studies (Arshi et al. 2011; Singh et al. 2012). However, sample SABA60S100 lacked a peak at 43°, indicating possible variations in particle characteristics; this was consistent with results obtained by Salayová et al. (Salayová et al. 2021). The Bragg reflections propose that the X-ray scattering FCC nature of NPs could act as capping agents, influencing NP structures (Dilbar et al. 2023).

Another interesting observation in all samples is that the lattice planes corresponding to (200), (220), and (311) were noted to exhibit weak intensity and broadening peaks compared to the (111) Bragg's reflection. This suggests that the nanocrystals synthesised biologically display a high level of anisotropy, with the NPs predominantly oriented along the (111) direction (Singh et al. 2012). This correlates with the UV-Vis results obtained. The average crystalline size of the SA-AuNPs samples ranged from 8 to 10 nm, indicating relatively uniform particle sizes across different synthesis conditions (Zahra et al. 2023).

Analysis by TEM confirmed the presence of SA-AuNPs with various particle core sizes, distribution, and morphology. The SA-AuNP (SABE-AU) were smaller (6–32 nm) with spheroidal and star-shaped morphology, suggesting that these NPs have a relatively small size and a narrow size distribution, which is desirable for many applications, including drug delivery, due to their increased surface area to volume ratio, which enhances their interaction with biological systems (Fouda et al. 2022). The results are comparable to AuNP by active compounds in ginger aqueous extract, resulting in spherical NPs ranging from 5 to 53 nm, with an average size of 15.11 ± 8.5 nm (Hatipoğlu 2022). The particle size distribution range observed in TEM validated the monodispersed nature in SA-AuNPs and supported the conclusions drawn based on the UV-Vis, and XRD analysis.

The FTIR analysis identified potential functional groups responsible for the reduction of Au^+ ions. The crude extracts of *S. Africana* bark and leaves revealed the presence of O-H (hydroxy) groups, C=C stretching (alkenes) or N-H (amide) groups, C-O stretching (alcohols), and C-H (aromatic) in-plane bending. These functional groups were also seen in SA-AuNPs samples but with minor shifts in absorption bands. The observed frequency shift suggests that hydroxyl, alkenes, or amide groups, alcohols, and hydrocarbons played crucial roles in reducing and stabilizing metal ions in the synthesised SA-AuNPs, thereby indicating successful synthesis (Keskin et al. 2023). Corresponding studies with other plant extracts like *D. kaki* and *E. senticosus* have also

shown the involvement of functional groups such as hydroxyl, carboxyl, and amine groups in the reduction and capping of Au^+ ions into NPs (Abbai et al. 2016; Ahn et al. 2019). In addition, proteins and amino acids present in plant extracts also play a significant role in the capping, coating, and protection of NPs (Sher et al. 2022).

For antimicrobial activity, SA-AuNPs specifically bark-methanol conjugated AuNPs (SABM90S100) and bark-ethanol conjugated AuNPs (SABE90S100) displayed the most significant inhibition activity against *S. aureus* and *P. aeruginosa*. Lower activity was observed against *A. baumannii*, and resistance was noted in *E. coli*. These findings suggested that these SA-AuNPs may have selective antimicrobial effects against specific bacterial strains, particularly *S. aureus* and *P. aeruginosa* while showing limited efficacy against others like *A. baumannii* and *E. coli*. Studies have shown that AuNPs synthesised via green approaches exhibit strong antimicrobial activities against both gram-positive (*S. aureus*) and gram-negative (*E. coli*) bacteria, including resistant strains (Rani et al. 2023; Dove et al. 2023). Researchers have discovered that plant-synthesised AuNPs generally possess superior antimicrobial activity due to factors such as morphology, surface area, particle size and shape, and surface polarity. The observed susceptibility of *P. aeruginosa* to AuNPs underscores their promising antimicrobial potential, particularly against antibiotic-resistant strains (Rani et al. 2023).

The microdilution assay was used to determine the MIC of SA-AuNPs. We observed varying activity levels against the tested bacterial strains, with *S. aureus* emerging as the most sensitive, followed by *E. coli*, and lastly *P. aeruginosa*. Intriguingly, none of the tested samples showed activity against *A. baumannii*. Bactericidal activity of SA-AuNPs showed that sample SABE90S100 killed *S. aureus* with the MBC of 0.63 mg/mL. No bactericidal effect was observed against *P. aeruginosa*. The differential effectiveness of SA-AuNPs against various bacterial strains can be attributed to several factors, including the size, shape, and surface charge of the NPs. Other metal ions in the samples may also influence their antimicrobial activity (Yakoup et al. 2024). The lack of activity against *A. baumannii* may be due to the unique characteristics of this bacterium, such as its ability to produce biofilms and its resistance to various antibiotics (Elemike et al. 2017).

The antimicrobial activity of gold nanoparticles is often attributed to the release of Au ions by the AuNPs, which disrupt cell walls and secondary metabolites adsorbed onto the AuNPs (Sylvester 2011). The combination of the phytochemicals, which also have antimicrobial activities, may create a synergistic effect. Additionally, bioactive compounds within the organic matrix might contribute to the heightened antibacterial activity of AuNPs by potentially enhancing their interaction with the cell wall (Salayová et al. 2021). Smaller NPs have a greater bactericidal effect compared to larger ones because of their larger surface area, facilitating interaction and absorption more easily. These findings collectively underscore the diverse antimicrobial effects of *S. africana* extract-conjugated AuNPs, positioning them as promising candidates against AMR bacteria.

The overall cytotoxicity data showed that the crude *S. africana* extracts reduced the viability of both MCF-7 cancer cells and normal MCF-10 and RAW 264.7 cells, indicating limited intrinsic selectivity at the extract level. In contrast, the corresponding SA-AuNPs (SABE-AU) displayed enhanced and more selective toxicity, with a lower IC_{50} in MCF-7 and higher IC_{50} values in MCF-10 and RAW 264.7, demonstrating greater susceptibility of breast cancer cells than normal epithelial and macrophage cells. This shift suggests that green synthesis on a gold core can refine the biological profile of the plant material, improving anticancer selectivity relative to the parent extracts.

Direko and colleagues (Direko et al. 2019) reported that *S. africana* extracts possess anti-inflammatory and anticancer properties, showing that concentrations below 10 µg/mL inhibited MCF-7 growth while stimulating MCF-10 cells and NO production. Polyphenolic compounds, particularly ellagic acid, demonstrated safety towards healthy cells but toxicity towards cancerous cells, suggesting their pivotal role in observed toxicity (Lee et al. 2019). In the present study, conjugation of these phytochemicals to AuNPs appears to preserve their anticancer potential but superimposes nanoparticle-specific effects, such as enhanced cellular internalisation, localised ROS generation, and mitochondrial disruption, that collectively favour killing of MCF-7 over normal cells, without completely eliminating toxicity in the latter. The relatively small but real separation in IC_{50} values suggests that SABE-AU could be useful where localised or controlled delivery is feasible, while systemic application would require careful dose optimisation to avoid off-target damage. Similar observations by Barabadi and colleagues (Grewal et al. 2022) support the view that there could be more to the mechanisms of action of nanoparticle cytotoxicity that remain poorly understood and further research is needed in order to clarify the differential cytotoxicity of green synthesised nanoparticles.

The observed cytotoxicity effect of SA-AuNPs can be explained by the complex interplay between the surface properties of the NPs and their concentration-dependent behaviours. When NPs are not modified to prevent spontaneous agglomeration, higher concentrations can promote their aggregation, leading to the formation of nanoaggregates where individual particles are closely packed together. Recent studies have shown that AuNPs can be effective in inducing cytotoxic effects on cancer cells by interfering with the cellular electron transfer chain in mitochondria, leading to increased levels of reactive oxygen species (ROS) and oxidative stress, ultimately causing cellular toxicity (Grewal et al. 2022). Additional studies reported that NPs create an unfavourable environment for cancer cells by increasing ROS and malondialdehyde levels while decreasing catalase and glutathione peroxidase enzyme activity, resulting in cell death (Selvi et al. 2016). Liu and others discovered another pathway for inducing apoptosis using NPs, where the primary cause of cell death is the upregulation of endoplasmic reticulum stress-related messenger ribonucleic acid and protein expression (Liu et al. 2021). Anti-inflammatory effects highlighted the impact of various samples on the NO production in RAW 264.7 microphage cells stimulated with LPS, a typical inducer of inflammation. Nitric oxide (NO) is a substance that the body naturally produces in response to infection or inflammation. It is a well-known inflammatory mediator released from activated macrophages during these immune responses (Liu et al. 2021). Sample SABE-AU demonstrated a non-significant decrease in NO production across all concentration levels, suggesting potential weak anti-inflammatory effects. These findings underscore the complex interplay between different compounds and inflammation regulation (Liu et al. 2021). Further research is essential to elucidate the underlying mechanisms and determine these samples' therapeutic potential and safety profiles in managing inflammatory conditions.

The observed anti-inflammatory effects are attributed to various compounds, including O-glycosides, quercetin, flavonoid-C-glycosides, tannins, and variations of cinnamic acid. These compounds have been shown to inhibit the production of inflammatory factors and interfere with LPS binding to Toll-like receptors, thereby activating the antioxidative nuclear factor erythroid 2-related factor 2 (Nrf2) and increasing the expression of antioxidant proteins (Rod-In et al. 2021).

Bacterial infection can cause damage to various tissues in the body, particularly affecting epithelial cells, through microbial activity and the host's inflammatory response. Macrophages release pro-inflammatory chemicals, exacerbating tissue damage if the injury persists. Effective tissue repair requires inflammation reduction. Research suggests plant extracts and their bioactive compounds possess significant anti-inflammatory potential, offering promise in treating various inflammatory diseases, including RTIs (Nascimento et al. 2017). For instance, sorghum extracts inhibit nitric oxide (NO), interleukin-6 (IL-6), and reactive oxygen species (ROS) production in activated macrophages, with ethanolic extracts showing superior anti-inflammatory activity due to their tannin content (Matotoka et al. 2023). Thus, reducing NO production with

plant extracts may offer therapeutic benefits in diseases driven by excessive NO levels, including bacterial infections.

The antioxidant activities of SA-AuNPs by the DPPH assay showed that bark-methanol conjugated AuNPs (SABM-AU) exhibited strong antioxidant activity with an IC_{50} value of 38 µg/mL while the ethanol extract nanoparticles showed even stronger activity with IC_{50} of 2.4 µg/mL According to Setha et al. (Hong et al. 2020), antioxidants can be classified as very strong ($IC_{50} < 50$ µg/mL), strong (IC_{50}: 50–100 µg/mL), moderate (IC_{50}: 101–150 µg/mL), and weak (IC_{50}: 250–500 µg/mL). Concerning the antioxidant activity, it is known that oxidative stress may occur due to an imbalance between ROS production and the body's ability to neutralize them, potentially leading to cell harm. Antioxidative enzymes such as superoxide dismutase, glutathione peroxidase, and antioxidants like glutathione play a vital role in maintaining this balance and protecting cellular health. Free radicals, highly reactive species, can lead to oxidative stress when not adequately regulated by antioxidants. Oxidative stress can damage lipids, proteins, and DNA, contributing to various human diseases and disorders (Setha et al. 2013). To counteract oxidative stress, the body must balance free radicals and antioxidants. However, plant compounds with antioxidative properties, like flavonoids and polyphenols, play a crucial role in scavenging free radicals and preventing cellular damage. Plant phytochemical compounds offer significant benefits in combating oxidative stress and promoting overall health and well-being, prompting investigation into their antioxidant potential.

Phytochemical compounds of *S. africana* reported to have anti-inflammatory and antioxidant properties include saponins, flavonoids, and glycosides (Rod-In et al. 2021; Direko et al. 2019). Flavonoids were notably abundant in the extracts of *S. africana* in the current study, potentially contributing to the observed antioxidant activity, consistent with findings by Huo et al. (More and Makola 2020). Additionally, terpenoids, specifically triterpenoids, were identified as key contributors to the antioxidant activities in the tested crude extracts (Direko et al. 2019).

Total soluble phenolics in plant extracts may contribute to their antioxidant properties (Huo et al. 2018). Similarly, the dichloromethane fraction of the brown seaweed *Sargassum siliquastrum* exhibited strong antioxidant activity, with concentrations ranging between 0.4 and 50 μg/mL, potentially due to the presence of phenolic compounds (Kamatou et al. 2010). The crude extract and ethyl acetate-soluble fraction of the red algae *Polysiphonia urceolata* also demonstrated high antioxidant activity with concentrations lower than 50 μg/mL, correlated with their total phenolic content and reducing power (Lim et al. 2002).

The difference in antioxidant activity between AuNPs was attributed to their biosynthesised nature, confirmed by FT-IR spectral results (Duan et al. 2006). This was evident in the current research, where SA-AuNPs exhibited more peaks, indicating the presence of higher concentrations of phytochemicals, enhancing their antioxidant efficacy against DPPH free radicals. These findings collectively suggest that the antioxidant properties of the SA-AuNPs were likely due to their flavonoids, terpenoids, glycosides, phenolic content, and reducing power.

10.5 Conclusions

The study demonstrated the successful synthesis of gold nanoparticles (SA-AuNPs) from *S. africana* extracts, highlighting their potential in combating antimicrobial resistance (AMR) as well as displaying good antioxidant activity. Optimizing reaction parameters, including extract concentration, metal salt ratio, temperature, and reaction time, resulted in smaller, monodispersed nanoparticles with spheroidal and star-shaped morphologies that exhibited strong antimicrobial activity against both Gram-negative and Gram-positive bacteria. Additionally, SA-AuNPs showed selective toxicity toward MCF-7 cancer cells at lower concentrations compared to normal cells, indicating their dose-dependent potential for targeted cancer treatment. The nanoparticles also displayed antioxidant, anti-inflammatory, and cytotoxic properties, making them promising candidates for multi-functional therapeutic applications. Future research should focus on elucidating the mechanisms of nanoparticle formation, isolating bioactive compounds from *S. africana* and optimizing other biological activities such as anti-cancer activities. Furthermore, in vivo studies are recommended to validate safety and scalability for clinical applications.

Conflict of Interest Author MFN declares that she has no conflict of interest. Author ANT declares that she has no conflict of interest. Author MR declares that she has no conflict of interest. Author AS declares that he has no conflict of interest.

Ethical Approval This chapter does not contain any studies with human, or animal participants performed by any of the authors.

Funding The authors would like to acknowledge the support of the following sponsors towards the 28th SASBMB congress: the University of Venda (UniVen), the International Union of Biochemistry and Molecular Biology (IUBMB), Diplomics (https://www.diplomics.org.za), the National Research Foundation (NRF) of South Africa (Grant No: KIC24043211994), and the South African Medical Research Council (SAMRC).

References

Abbai R, Mathiyalagan R, Markus J, Kim YJ, Wang C, Singh P, Ahn S, Farh MEA, Yang DC (2016) Green synthesis of multifunctional silver and gold nanoparticles from the oriental herbal adaptogen: Siberian ginseng. Int J Nanomed 11:3131–3143

Agyepong N, Agyare C, Adarkwa-Yiadom M, Gbedema SY (2014) Phytochemical investigation and antimicrobial activity of *Clausena anisata* (Willd), Hook. Afr J Tradit Complement Altern Med 11(3):200–209

Ahn EY, Jin H, Park Y (2019) Green synthesis and biological activities of silver nanoparticles prepared by *Carpesium cernuum* extract. Arch Pharm Res 42:926–934

Albanese L, Bonetti A, D'Acqui LP, Meneguzzo F, Zabini F (2019) Affordable production of antioxidant aqueous solutions by hydrodynamic cavitation processing of silver fir (Abies alba Mill.) needles. Foods 8(2):65–74

Al-Snafi AE (2015) Therapeutic properties of medicinal plants: a review of plants with hypolipidemic, hemostatic, fibrinolytic and anticoagulant effects (part 1). Asian J Pharma Sci Technol 5(4):271–284

Anand U, Carpena M, Kowalska-Góralska M, Garcia-Perez P, Sunita K, Bontempi E, Dey A, Prieto MA, Pročków J, Simal-Gandara J (2022) Safer plant-based nanoparticles for combating antibiotic resistance in bacteria: a comprehensive review on its potential applications, recent advances, and future perspective. Sci Total Environ 821:153472–153478

Arafa MG, El-Kased RF, Elmazar MM (2018) Thermoresponsive gels containing gold nanoparticles as smart antibacterial and wound healing agents. Sci Rep 8(1):13674

Arshi N, Ahmed F, Anwar MS, Kumar S, Koo BH, Lu J, Lee CG (2011) Novel and cost-effective synthesis of silver nanocrystals: a green synthesis. Nano Brief Rep Rev 6(04):295–300

Asong JA, Frimpong EK, Seepe HA, Katata-Seru L, Amoo SO, Aremu AO (2023) Green synthesis of characterized silver nanoparticle using *Cullen tomentosum* and assessment of its antibacterial activity. Antibiotics 12(2):203–211

Ayinde WB, Gitari WM, Samie A (2019) Optimisation of microwave-assisted synthesis of silver nanoparticle by *Citrus paradisi* peel and its application against pathogenic water strain. Green Chem Lett Rev 12(3):225–234

Baharara J, Ramezani T, Hosseini N, Mousavi M (2018) Silver nanoparticles synthesised coating with *Zataria multiflora* leaves extract induced apoptosis in hela cells through p53 activation. Iranian J Pharma Res 17(2):627

Bahrami M, Ghazavi A, Ganji A, Mosayebi G (2021) Anti-inflammatory activity of *S. Marianum* and *N. Sativa* extracts on macrophages. Rep Biochem Mol Biol 10(2):288–301

Botteon CEA, Silva LB, Ccana-Ccapatinta GV, Silva TS, Ambrosio SR, Veneziani RCS, Bastos JK, Marcato PD (2021) Biosynthesis and characterization of gold nanoparticles using Brazilian red propolis and evaluation of its antimicrobial and anticancer activities. Sci Rep 11(1):1974

Cascio C, Geiss O, Franchini F, Ojea-Jimenez I, Rossi F, Gilliland D, Calzolai L (2015) Detection, quantification and derivation of number size distribution of silver nanoparticles in antimicrobial consumer products. J Anal At Spectrom 30(6):1255–1265

Dada AO, Adekola FA, Dada FE, Adelani-Akande AT, Bello MO, Okonkwo CR, Inyinbor AA, Oluyori AP, Olayanju A, Ajanaku KO, Adetunji CO (2019) Silver nanoparticle synthesis by *Acalypha wilkesiana* extract: phytochemical screening, characterization, influence of operational parameters, and preliminary antibacterial testing. Heliyon 5(10):e02517. https://doi.org/10.1016/j.heliyon.2019.e02517

Devi GK, Sathishkumar K (2017) Synthesis of gold and silver nanoparticles using *Mukia maderaspatna* plant extract and its anticancer activity. IET Nanobiotechnol 11(2):143–151

Dilbar S, Sher H, Ali H, Ullah R, Ali A, Ullah Z (2023) Antibacterial efficacy of green synthesized silver nanoparticles using *Salvia nubicola* extract against *Ralstonia solanacearum*, the causal agent of vascular wilt of tomato. ACS Omega 8(34):31155–31167

Direko P, Mfengwana H, Mashele S, Sekhoacha M (2019) Investigating the angiogenic modulating properties of *Spirostachys africana* in MCF-7 breast cancer cell line. Int J Pharmacol 15(8):970–977

Dove AS, Dzurny DI, Dees WR, Qin N, Nunez Rodriguez CC, Alt LA, Ellward GL, Best JA, Rudawski NG, Fujii K, Czyż DM (2023) Silver nanoparticles enhance the efficacy of aminoglycosides against antibiotic-resistant bacteria. Front Microbiol 13:1064095

Duan XJ, Zhang WW, Li XM, Wang BG (2006) Evaluation of antioxidant property of extract and fractions obtained from a red alga, Polysiphonia urceolata. Food Chem 95(1):37–43

Dubey SP, Lahtinen M, Sillanpää M (2010) Green synthesis and characterizations of silver and gold nanoparticles using leaf extract of *Rosa rugosa*. Colloids Surf A Physicochem Eng Asp 364(1–3):34

Elemike EE, Onwudiwe DC, Fayemi OE, Ekennia AC, Ebenso EE, Tiedt LR (2017) Biosynthesis, electrochemical, antimicrobial and antioxidant studies of silver nanoparticles mediated by *Talinum triangulare* aqueous leaf extract. J Clust Sci 28:309–330

Fouda A, Eid AM, Guibal E, Hamza MF, Hassan SED, Alkhalifah DHM, El-Hossary D (2022) Green synthesis of gold nanoparticles by aqueous extract of Zingiber officinale: characterization and insight into antimicrobial, antioxidant, and in vitro cytotoxic activities. Appl Sci 12(24):12879–12887

Ghavam M (2021) Relationships of irrigation water and soil physical and chemical characteristics with yield, chemical composition and antimicrobial activity of Damask rose essential oil. PLoS One 16(4):e0249363

Grewal J, Kumar V, Rawat H, Gandhi Y, Singh R, Singh A, Babu G, Srikanth N, Mishra SK (2022) Cytotoxic effect of plant extract-based nanoparticles on cancerous cells: a review. Environ Chem Lett 20(4):2487–2507

Guadie Assefa A, Adugna Mesfin A, Legesse Akele M, Kokeb Alemu A, Gangapuram BR, Guttena V, Alle M (2017) Microwave-assisted green synthesis of gold nanoparticles using Olibanum gum (*Boswellia serrate*) and its catalytic reduction of 4-nitrophenol and hexacyanoferrate (III) by sodium borohydride. J Clust Sci 28:917–935

Hatipoğlu A (2022) Green synthesis of silver nanoparticles and their antimicrobial effects on some food pathogens. Süleyman Demirel Univ J Sci Technol 26(1):106–114

Hong S, Pangloli P, Perumal R, Cox S, Noronha LE, Dia VP, Smolensky D (2020) A comparative study on phenolic content, antioxidant activity and anti-inflammatory capacity of aqueous and ethanolic extracts of sorghum in lipopolysaccharide-induced RAW 264.7 macrophages. Antioxidants 9(12):1297–1312

Huo Y, Singh P, Kim YJ, Soshnikova V, Kang J, Markus J, Ahn S, Castro-Aceituno V, Mathiyalagan R, Chokkalingam M, Bae KS (2018) Biological synthesis of gold and silver chloride nanoparticles by

Glycyrrhiza uralensis and in vitro applications. Artif Cells Nanomed Biotechnol 46(2):303–312

Idris FN, Nadzir MM (2023) Multi-drug resistant ESKAPE pathogens and the uses of plants as their antimicrobial agents. Arch Microbiol 205(4):115

Jalab J, Abdelwahed W, Kitaz A, Al-Kayali R (2021) Green synthesis of silver nanoparticles using aqueous extract of *Acacia cyanophylla* and its antibacterial activity. Heliyon 7(9):e08033

Kamatou GP, Viljoen AM, Steenkamp P (2010) Antioxidant, antiinflammatory activities and HPLC analysis of South African *Salvia* species. Food Chem 119(2):684–688

Kang SG, Lee GB, Vinayagam R, Do GS, Oh SY, Yang SJ, Kwon JB, Singh M (2022) Anti-inflammatory, antioxidative, and nitric oxide-scavenging activities of a quercetin nanosuspension with polyethylene glycol in LPS-induced RAW 264.7 macrophages. Molecules 27(21):7432–7438

Kavitha R, Nadu T (2021) Phytochemical screening and GC-MS analysis of bioactive compounds present in ethanolic extracts of leaf and fruit of *Trichosanthesis dioica* roxb. Int J Pharm Sci Res 12(5):2755–2764

Keskin C, Ölçekçi A, Baran A, Baran MF, Eftekhari A, Omarova S, Khalilov R, Aliyev E, Sufianov A, Beilerli A, Gareev I (2023) Green synthesis of silver nanoparticles mediated *Diospyros kaki* L. (Persimmon): determination of chemical composition and evaluation of their antimicrobials and anticancer activities. Front Chem 11:1187808

Lee YJ, Ahn EY, Park Y (2019) Shape-dependent cytotoxicity and cellular uptake of gold nanoparticles synthesized using green tea extract. Nanoscale Res Lett 14:1–14

Lennox SJ, Bamford M (2015) Use of wood anatomy to identify poisonous plants: charcoal of *Spirostachys africana*. S Afr J Sci 111(3–4):1–9

Liaqat N, Jahan N, Anwar T, Qureshi H (2022) Green synthesised silver nanoparticles: optimisation, characterisation, antimicrobial activity, and cytotoxicity study by hemolysis assay. Front Chem 10:952006

Lim SN, Cheung PCK, Ooi VEC, Ang PO (2002) Evaluation of antioxidative activity of extracts from a brown seaweed, *Sargassum siliquastrum*. J Agric Food Chem 50(13):3862–3866

Liu H, Lai W, Liu X, Yang H, Fang Y, Tian L, Li K, Nie H, Zhang W, Shi Y, Bian L (2021) Exposure to copper oxide nanoparticles triggers oxidative stress and endoplasmic reticulum (ER)-stress induced toxicology and apoptosis in male rat liver and BRL-3A cell. J Hazard Mater 401:123349

Mabogo DEN (1990) The ethnobotany of the VhaVenda. MSc Dissertation. University of Pretoria, Pretoria

Mancuso G, Midiri A, Gerace E, Biondo C (2021) Bacterial antibiotic resistance: the most critical pathogens. Pathogens 10(10):1310

Matotoka MM, Mashabela GT, Masoko P (2023) Phytochemical content, antibacterial activity, and antioxidant, anti-inflammatory, and cytotoxic effects of traditional medicinal plants against respiratory tract bacterial pathogens. Evidence-Based Complement Altern Med 31:1243438. https://doi.org/10.1155/2023/1243438

More GK, Makola RT (2020) In-vitro analysis of free radical scavenging activities and suppression of LPS-induced ROS production in macrophage cells by *Solanum sisymbriifolium* extracts. Sci Rep 10(1):6493–6508

Nandiyanto ABD, Oktiani R, Ragadhita R (2019) How to read and interpret FTIR spectroscope of organic material. Indonesian J Sci Technol 4(1):97–118

Nascimento AM, Maria-Ferreira D, Dal Lin FT, Kimura A, de Santana-Filho AP, Werner MFDP, Iacomini M, Sassaki GL, Cipriani TR, de Souza LM (2017) Phytochemical analysis and anti-inflammatory evaluation of compounds from an aqueous extract of *Croton cajucara* Benth. J Pharm Biomed Anal 145:821–830

Nemudzivhadi V, Masoko P (2015) Antioxidant and antibacterial properties of *Ziziphus mucronata* and *Ricinus communis* leaves extracts. Afr J Tradit Complement Altern Med 12(1):81–89

Niranjan Dhanasekar N, Ravindran Rahul G, Badri Narayanan K, Raman G, Sakthivel N (2015) Green chemistry approach for the synthesis of gold nanoparticles using the fungus *Alternaria* sp. J Microbiol Biotechnol 25(7):1129–1135

Nur H, Nasir SM (2008) Gold nanoparticles embedded on the surface of polyvinyl alcohol layer. Malays J Fundam Appl Sci 4(1):245–252

Patel N, Kasumbwe K, Mohanlall V (2020) Antibacterial screening of *Gunnera perpensa*-mediated silver nanoparticles. J Nanotechnol 2020(1):4508543

Perveen K, Husain FM, Qais FA, Khan A, Razak S, Afsar T, Alam P, Almajwal AM, Abulmeaty MM (2021) Microwave-assisted rapid green synthesis of gold nanoparticles using seed extract of *Trachyspermum ammi*: ROS mediated biofilm inhibition and anticancer activity. Biomolecules 11(2):197–202

Quelemes PV, Araruna FB, De Faria BE, Kuckelhaus SA, Da Silva DA, Mendonça RZ, Eiras C, dos S. Soares MJ, Leite JRS (2013) Development and antibacterial activity of cashew gum-based silver nanoparticles. Int J Mol Sci 14(3):4969–4981

Rani P, Varma RS, Singh K, Acevedo R, Singh J (2023) Catalytic and antimicrobial potential of green synthesized Au and Au@Ag core-shell nanoparticles. Chemosphere 317:137841

Rod-In W, Talapphet N, Monmai C, Jang AY, You S, Park WJ (2021) Immune enhancement effects of Korean ginseng berry polysaccharides on RAW264. 7 macrophages through MAPK and NF-κB signalling pathways. Food Agric Immunol 32(1):298–309

Saifuddin, N., Wong, C.W., And Nur Yasumira, A.A. 2009, Rapid biosynthesis of silver nanoparticles using culture supernatant of bacteria with microwave irradiation. E-J Chem 6(1), p. 61-70. http://wwwe-journalsnet

Salam MA, Al-Amin MY, Salam MT, Pawar JS, Akhter N, Rabaan AA, Alqumber MAA (2023) Antimicrobial resistance: a growing serious threat for global public health. Healthcare 11:1946. [online]

Salayová A, Bedlovičová Z, Daneu N, Baláž M, Lukáčová Bujňáková Z, Balážová Ľ, Tkáčiková Ľ (2021) Green synthesis of silver nanoparticles with antibacterial activity using various medicinal plant extracts: morphology and antibacterial efficacy. Nano 11(4):1005–1011

Samie A, Obi CL, Bessong PO, Namrita L (2005) Activity profiles of fourteen selected medicinal plants from Rural Venda communities in South Africa against fifteen clinical bacterial species. Afr J Biotechnol 4(12):1443–1451

Selvi BCG, Madhavan J, Santhanam A (2016) Cytotoxic effect of silver nanoparticles synthesized from *Padina tetrastromatica* on breast cancer cell line. Adv Nat Sci Nanosci Nanotechnol 7(3):035015

Setha B, Gaspersz FF, Idris APS, Rahman S, Mailoa MN (2013) Potential of seaweed *Padina* spp. as a source of antioxidant. Int J Sci Technol Res 2(6):221–224

Shaikh IA, Alshabi AM, Alkahtani SA, Orabi MA, Abdel-Wahab BA, Walbi IA, Habeeb MS, Khateeb MM, Shettar AK, Hoskeri JH (2022) Apoptotic cell death via activation of DNA degradation, caspase-3 activity, and suppression of Bcl-2 activity: an evidence-based *Citrullus colocynthis* cytotoxicity mechanism toward MCF-7 and A549 cancer cell lines. Separations 9(12):411

Sher N, Alkhalifah DHM, Ahmed M, Mushtaq N, Shah F, Fozia F, Khan RA, Hozzein WN, Aboul-Soud MA (2022) Comparative study of antimicrobial activity of silver, gold, and silver/gold bimetallic nanoparticles synthesized by green approach. Molecules 27(22):7895–7904

Singh P, Mijakovic I (2022) Green synthesis and antibacterial applications of gold and silver nanoparticles from *Ligustrum vulgare* berries. Sci Rep 12(1):7902–7908

Singh C, Baboota RK, Naik PK, Singh H (2012) Biocompatible synthesis of silver and gold nanoparticles using leaf extract of *Dalbergia sissoo*. Adv Mater Lett 3(4):279–285

Sylvester PW (2011) Optimization of the tetrazolium dye (MTT) colorimetric assay for cellular growth and viability. In: Drug Design Discov Methods Protoc, pp 157–168

Tang KWK, Millar BC, Moore JE (2023) Antimicrobial resistance (AMR). Br J Biomed Sci 80:11387

Vijayarathna S, Sasidharan S (2012) Cytotoxicity of methanol extracts of *Elaeis guineensis* on MCF-7 and Vero cell lines. Asian Pac J Trop Biomed 2(10):826–829

Wang C, Singh P, Kim YJ, Mathiyalagan R, Myagmarjav D, Wang D, Jin CG, Yang DC (2016) Characterization and antimicrobial application of biosynthesized gold and silver nanoparticles by using *Microbacterium resistens*. Artif Cells Nanomed Biotechnol 44(7):1714–1721

Wintachai P, Paosen S, Yupanqui CT, Voravuthikunchai SP (2019) Silver nanoparticles synthesized with *Eucalyptus critriodora* ethanol leaf extract stimulate antibacterial activity against clinically multidrug-resistant *Acinetobacter baumannii* isolated from pneumonia patients. Microb Pathog 126:245–257

Yadi M, Azizi M, Dianat-Moghadam H, Akbarzadeh A, Abyadeh M, Milani M (2022) Antibacterial activity of green gold and silver nanoparticles using ginger root extract. Bioprocess Biosyst Eng 45(12):1905–1917

Yakoup AY, Kamel AG, Elbermawy Y, Abdelsattar AS, El-Shibiny A (2024) Characterization, antibacterial, and cytotoxic activities of silver nanoparticles using the whole biofilm layer as a macromolecule in biosynthesis. Sci Rep 14(1):364–378

Yang W, Chen X, Li Y, Guo S, Wang Z, Yu X (2020) Advances in pharmacological activities of terpenoids. Nat Prod Commun 15(3):1934578X20903555

Yasmin A, Ramesh K, Rajeshkumar S (2014) Optimization and stabilization of gold nanoparticles by using herbal plant extract with microwave heating. Nano Converg 1(1):12–18

Yayehrad AT, Wondie GB, Marew T (2022) Different nanotechnology approaches for ciprofloxacin delivery against multidrug-resistant microbes. Infect Drug Resist 5(15):413–426

Zahra ST, Rasheed S, Sajjad S, Ali MA, Sultan B (2023) Theranostics applications of plant-based nanoparticles. Mater Chem Mech 1(2):64–82

Zuhrotun A, Oktaviani DJ, Hasanah AN (2023) Biosynthesis of gold and silver nanoparticles using phytochemical compounds. Molecules 28(7):3240